ADVANCES IN CHEMICAL PHYSICS

VOLUME LXXXIX

EDITORIAL BOARD

Advances in CHEMICAL PHYSICS

Edited by

I. PRIGOGINE

University of Brussels
Brussels, Belgium
and
University of Texas
Austin, Texas

and

STUART A. RICE

Department of Chemistry
and
The James Franck Institute
The University of Chicago
Chicago, Illinois

VOLUME LXXXIX

AN INTERSCIENCE® PUBLICATION
JOHN WILEY & SONS, INC.
NEW YORK • CHICHESTER • BRISBANE • TORONTO • SINGAPORE

This text is printed on acid-free paper.

An Interscience® Publication

Library of Congress Catalog Number: 58-9935

ISBN 0-471-05157-8

10 9 8 7 6 5 4 3 2 1

CONTRIBUTORS TO VOLUME LXXXIX

MARTIN GRANT, Centre for the Physics of Materials and Physics Department, McGill University, Montréal, Québec, Canada

HONG GUO, Centre for the Physics of Materials and Physics Department, McGill University, Montréal, Québec, Canada

PETER HÄNGGI, Department of Physics, University of Augsburg, Augsburg, Germany

PETER JUNG, Department of Physics, University of Augsburg, Augsburg, Germany

MOHAMED LARADJI, Centre for the Physics of Materials and Physics Department, McGill University, Montréal, Québec, Canada

IAIN R. MCNAB, Department of Physics, University of Newcastle upon Tyne, United Kingdom

UDAYAN MOHANTY, Department of Chemistry, Boston College, Chestnut Hill, Massachusetts

JOHN ROSS, Department of Chemistry, Stanford University, Stanford, California

IGOR SCHREIBER, Department of Chemical Engineering, Prague Institute of Chemical Technology, Prague, Czech Republic

JANET D. STEMWEDEL, Department of Chemistry, Stanford University, Stanford, California

MARTIN J. ZUCKERMAN, Centre for the Physics of Materials and Physics Department, McGill University, Montréal, Québec, Canada

INTRODUCTION

Few of us can any longer keep up with the flood of scientific literature, even in specialized subfields. Any attempt to do more and be broadly educated with respect to a large domain of science has the appearance of tilting at windmills. Yet the synthesis of ideas drawn from different subjects into new, powerful, general concepts is as valuable as ever, and the desire to remain educated persists in all scientists. This series, *Advances in Chemical Physics*, is devoted to helping the reader obtain general information about a wide variety of topics in chemical physics, a field which we interpret very broadly. Our intent is to have experts present comprehensive analyses of subjects of interest and to encourage the expression of individual points of view. We hope that this approach to the presentation of an overview of a subject will both stimulate new research and serve as a personalized learning text for beginners in a field.

I. PRIGOGINE
STUART A. RICE

CONTENTS

THE SPECTROSCOPY OF H_3^+

IAIN R. McNAB

Department of Physics, The University of Newcastle upon Tyne, Newcastle upon Tyne, NE1 7RU, UK

CONTENTS

Advances in Chemical Physics, Volume LXXXIX, Edited by I. Prigogine and Stuart A. Rice.
ISBN 0-471-05157-8

I. INTRODUCTION

This chapter summarizes the progress that has been made in obtaining and interpreting spectra of H_3^+, and its deuterium isotopomers, together with their astrophysical observation. Many reviews on the theory [1–5], spectroscopy [6–9], and astrophysical detection [5] of H_3^+ have appeared, but there has been no previous review in which the two areas of spectroscopy of H_3^+, at low and high energies, are examined from a single viewpoint.

As the simplest polyatomic molecular system H_3^+ plays the same role in the development of calculations of polyatomic molecule properties that is enjoyed by the hydrogen atom for atomic structure, and the hydrogen molecular ion for diatomic molecules [10]. It is frequently used as a test of new calculational procedures, where results can be contrasted with the most accurate calculations that can be achieved for a polyatomic molecule.

The H_3^+ ion is usually formed by the reaction $H_2^+ + H_2 \rightarrow H_3^+ + H$, which is exothermic by about 1.7 eV (14,700 cm^{-1}); it is an extremely stable species (in the absence of collisions with other atoms or molecules). Moreover, in hydrogen plasmas it is the dominant species at high pressures. The H_3^+ ion is therefore of interest in any environment where hydrogen plasmas occur, whether in the laboratory, in the interstellar medium (ISM), or in planetary atmospheres. In particular, the chemistry of the ISM is predicted to be almost entirely driven by ion–molecule reactions and H_3^+ is thought to play an essential part in these reactions because it is an extremely powerful protonating agent.

J.J. Thompson [11] first identified H_3^+ in 1912. Thompson eliminated the possibilities that his observations could be due to, for example, carbon with four charges using chemical arguments. Thompson pointed out that "The existence of this substance is interesting from a chemical point of view, as it is not possible to reconcile its existence with the ordinary conceptions about valency, if hydrogen is regarded as always monovalent." The bonding in H_3^+ became an outstanding problem of theoretical chemistry. Twenty years after its discovery it was still not known whether H_3^+ had a linear or triangular geometry, and no convincing arguments had been put forward for either. In an attempt to settle this pressing question, Hirschfelder and co-workers (see II.A.1 for references) investigated the structure of H_3^+ in a series of five papers. They reached the conclusion that in its ground electronic state the equilibrium configuration of molecular H_3^+ was bent, and had a geometry between that of a right triangle and an equilateral triangle. The final paper concludes with the remark that "The vibrational frequencies of H_3^+ are estimated but their exact value as well as the exact configuration of the stable state are somewhat in doubt. Two of these frequencies should be infrared active and susceptible to direct experimental measurement."

Subsequent *ab initio* calculations of the H_3^+ structure were refined enough to conclude that the ground-state configuration was that of an equilateral triangle [12], but the first experimental confirmation of this predicted geometry was due to Gaillard et al. [13] in 1978 using Coulomb explosion measurements. Photographs of protons from electron-stripped H_3^+ (and other measurements) were clearly consistent with the geometry of an equilateral triangle.

The infrared (IR) spectrum of H_3^+ predicted by Hirschfelder in 1938 was not obtained until 1980, despite many exhaustive searches by Herzberg [14]. A side benefit of these "unsuccessful" searches was the discovery of the Rydberg spectrum of H_3 [15, 16]. The IR spectrum of

H_3^+ was finally discovered by Oka [6] in 1980 after a $4\frac{1}{2}$-year search! His experiment located the spectrum by means of a tunable IR laser in absorption in a liquid nitrogen cooled glow discharge. Oka detected 15 lines in H_3^+, which were assigned overnight (!) by Watson using a program that he had written to analyze the H_3 data of Herzberg [17]. The H_3^+ ion was shown to be an equilateral triangle in its ground state.

At the same time that the spectrum of H_3^+ was located the first spectrum of the deuterium analogue D_3^+ was found by Wing and co-workers [18], who used an ion beam–laser technique. Eight transitions were observed and four were assigned using high quality *ab initio* constants that were calculated by Carney and Porter [19] and reported in the following article.

In 1985 the first possible astrophysical detection of H_3^+ was reported; a rotational transition of H_2D^+. Subsequent observations suggest that this was not a transition of H_2D^+, but recently another plausible detection of H_2D^+ has been made. Since 1985 H_3^+ has been found in the auroras of Jupiter, Uranus, and Saturn, and possibly also in the remnants of supernova 1987a and in the interstellar medium.

In 1982, Carrington et al. [20] discovered an IR spectrum of H_3^+, involving levels within $1100\,cm^{-1}$ of the dissociation limit. Transitions were detected by monitoring fragmentation of $H_3^+ \rightarrow H^+ + H_2$, which increased at resonance (when the frequency of the radiation was equal to a transition frequency of the molecule). Further investigations by Carrington and Kennedy [21] showed that the spectrum was due to transitions resulting from predissociating states that could lie up to $4000\,cm^{-1}$ above the dissociation limit, and had lifetimes of microseconds. Over 26,500 lines were recorded in the regions $874–1094\,cm^{-1}$, and a pseudo-low-resolution spectrum revealed four intense "clumps" separated by $52\,cm^{-1}$. Experiments on the isotopomers H_2D^+ and D_2H^+ showed that different spectra were seen according to whether resonances were monitored using protons or deuterons. At the time of the discovery of these spectra, no theoretical explanations of their origin or properties were available.

The main features of the predissociation spectrum are now thought to be understood, mainly due to a series of audacious classical calculations by Pollak, Schlier and their co-workers [2]. The spectrum arises in an energy region where classical dynamics are chaotic and the low-resolution "clumping" of the data is associated with remnants of a stable periodic orbit (the horseshoe orbit), which corresponds to a high-amplitude bending motion of quasilinear H_3^+. The long lifetime of the quasibound states is due to trapping behind centrifugal barriers. The difference between spectra of H_2D^+ and D_2H^+ taken by monitoring protons and deuterons can be understood in terms of the properties of these centrifugal barriers as a function of total angular momentum.

The spectroscopy of H_3^+ forms a bridge between different types of spectroscopic analysis. The spectra of the low-vibrational energy levels of H_3^+, where it is close to its equilibrium geometry, are wonderful examples of the utility of assigning molecular spectra using least-squares fitting to effective vibration–rotation Hamiltonians. At the high-energy limit, the molecule exhibits classical chaos, and the spectrum of these levels is probably the second molecular spectrum to be obtained in a region of energy where classical chaos pertains (see Section V.E). In this region of energy, the traditional spectroscopic analysis approach of refining effective Hamiltonians is now believed to be possible although nobody has yet attempted such an analysis for the H_3^+ spectrum. However, the behavior of the system and its characteristic dynamics have been almost entirely elucidated using classical trajectory calculations. Although an exact assignment of the individual lines (over 26,500) has not yet been possible, the fundamental dynamics of the molecule and its behavior in this region are now believed to be understood. This analysis has been substantially confirmed by the further experiments of Carrington et al. [22].

The spectra in both energy regions may be understood by *ab initio* variational calculations of energy levels using the full Born–Oppenheimer (clamped nuclei) Hamiltonian for nuclear motion. This work has been pioneered by Tennyson and co-workers [4],who have now solved the bound state problem for H_3^+. In their quantum calculations they find the same "horseshoe" motions that are thought to drive the dynamics of H_3^+ at dissociation, but they have not yet extended their calculations into the regions above the first dissociation limit; they have not solved the model in the continuum. With that extension it may be possible to assign the most intense lines in the predissociation spectrum.

Finally, I note that in IR spectroscopy energies (strictly energies/hc) and transition wavenumbers (the reciprocal of the transition wavelength) are traditionally quoted in reciprocal centimeters (cm^{-1}) ($1\,cm^{-1} \triangleq 30\,GHz \triangleq 0.124\,meV \triangleq 4.56\,\mu E_h \triangleq 1.44\,K$, where $\triangleq$ means "is equivalent to").

II. THEORY

Both the low-energy spectra and the predissociation spectrum of H_3^+ can now be understood within a unified theoretical framework, and the predictions made within this framework may enable experiments to unify the two energy regimes. All of the theory that is described below rests upon the Born–Oppenheimer approximation, but in Section II.C we see that theory now needs to go beyond this limitation in order to achieve agreement with the available experimental data.

A compilation of significant energy levels for understanding the structure of H_3^+ is shown in Fig. 2.1; Tables I and II list some of the most significant calculations of H_3^+ properties.

A. Born–Oppenheimer Approximation

The Born-Oppenheimer approximation asserts that it is reasonable to solve separately for the electronic motion and the nuclear motion of a molecule. The usual justification for this is that electrons are so much lighter than nuclei that they will adjust their position on a time scale infinitesimal compared with the movement of the nuclei [23], although this will be less true for H containing molecules than any others. While the Born–Oppenheimer approximation has been put on a rigorous footing for diatomic molecules [24], it is only recently that the full consequences have been explored for polyatomic molecules. This work has been reviewed by Sutcliffe [25].

The separation of electronic and nuclear motion leads naturally to the concept of nuclei moving across a potential energy surface, which may be employed in classical, semiclassical, and quantal calculations. However, these surfaces must first be calculated. That is the electronic wave function must be found for many different nuclear geometries, and fitted to some functional form. Provided electronic states are well separated, the nuclear motion is usually insufficient to couple them together, and it is a good approximation to solve for separate electronic states. For H_3^+ there is an avoided crossing that brings the two lowest energy $^1A_1$ electronic states very close together and results in the need for diabatic calculations and/or surface fittings in this region of the potential.

1. *Calculations of Equilibrium Geometry*

The geometry of H_3^+ was unknown from the time of its discovery until late in this century. This led Henry Eyring to describe the H_3^+ problem as [26] "the scandal of modern chemistry." The theoretical problem was computational; the three center integrals required in the electronic structure calculation cannot be solved analytically, and in the 1930s the technology did not exist with which they could be readily determined.

In a series of five papers, Hirschfelder and co-workers [27–31], attempted to find the geometry of the equilibrium ground state of H_3^+ by using low-order molecular orbital (MO) wave functions for H_3^+ and examining the energy of H_3^+ at various nuclear configurations. This

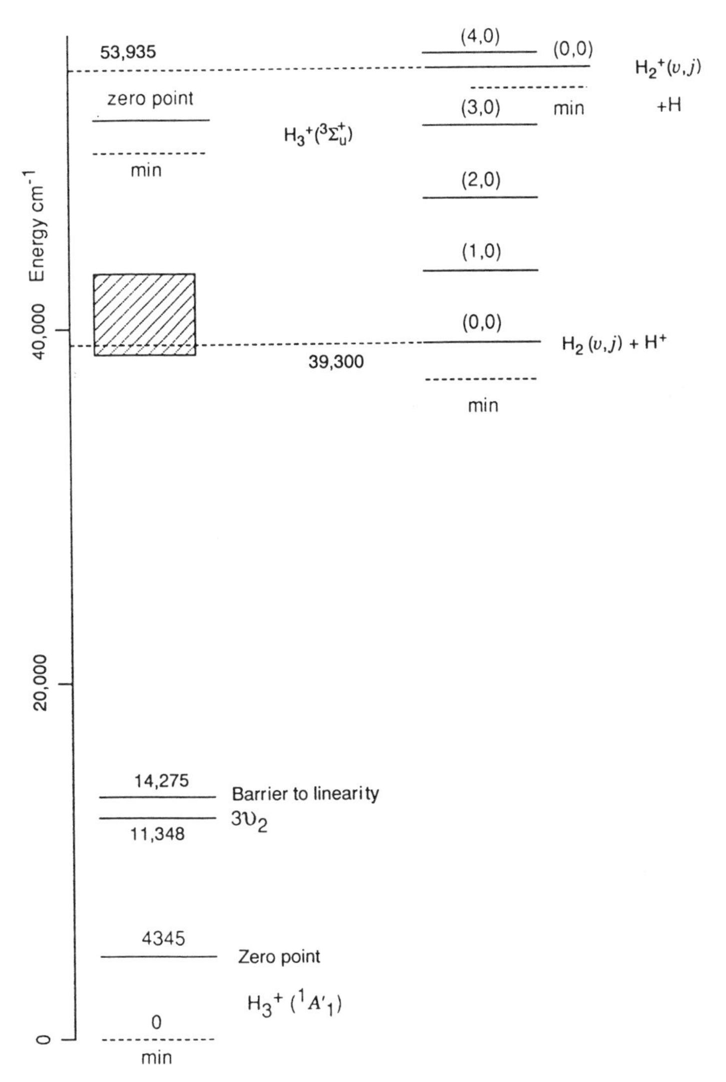

Figure 2.1. Significant energies of H_3^+ (after Kennedy [32a]). The highest transition so far observed in the laboratory at low energies is $3\nu_2 \leftarrow 0$. The shaded box represents the energy range believed to dominate in the predissociation spectrum; it is not known through which dissociation channel the predissociation occurs.

TABLE I
Landmark Calculations of H_3^+ Properties Near Equilibrium

Year and Reference	Dissociation Energy	ν_1 (cm^{-1})[a]	ν_2 (cm^{-1})[b]	Equilibrium Distance (r_e) (bohr)
1938 Hirschfelder [31]	$-1.293\ E_h$ $-283{,}780$ cm^{-1}	1550	1100	1.79
1964 Christoffersen [12]	$-1.3326\ E_h$ $-292{,}471$ cm^{-1}	3354	2790	1.6575
1974 Carney and Porter [43]	$-1.3441\ E_h$ $-294{,}995$ cm^{-1}	3471	2814	1.65
1976 Carney and Porter [45]	$-1.33519\ E_h$ $-293{,}040$ cm^{-1}	3185	2516	1.6585
1980 Carney [58]		3221	2546	
1986 Meyer et al. [48]		3178.2 3178.4 after adjustment	2518.9 2521.3 after adjustment	1.6501
1991 Anderson [36]	$-1.343835\ E_h$ $-293{,}937.7$ cm^{-1}			1.6500

[a] The wavenumber ν_1 was measured by Ketterle et al. [39] to be 3178 cm^{-1}.
[b] The wavenumber ν_2 was measured by Oka [17] to be 2521 cm^{-1}.

TABLE II
Landmark Calculations of H_3^+ Away from the Equilibrium Geometry of the Ground Electronic State

Year and Reference	What Was Calculated	Results	
1974 Schaad and Hicks [55]	Bound excited state of H_3^+ ($^3\Sigma_u^+$)		
1977 Ahlrichs et al. [56]	Equilibrium properties of $^3\Sigma_u^+$ state	$D_e = 0.01343\ E_h$ (2947 cm^{-1}) $r_{12} = r_{23} = 2.4568a_0$	Symmetric stretch: (σ_g^+) = 1233 cm^{-1} Asymmetric stretch: (σ_u^+) = 826 cm^{-1} Degenerate bend: (π_u) = 715 cm^{-1}
1988 Miller and Tennyson [202]	High angular momentum dissociation limit of electronic ground state	Found $J_{max} = 46$ (on MBB pe surface)	
1989 Wormer and de Groot [57]	Full pe surface for triplet state ($^3\Sigma_u^+$)		
1990 Henderson and Tennyson [241]	All $J = 0$ bound states of H_3^+ supported by electronic ground state	881 ± 10 bound states (on MBB pe surface)	

geometry was found by a heroic evaluation of integrals, both analytically, and also (for the final study, that of equilateral H_3 and H_3^+) with a differential analyzer of the Moore school of electrical engineering [32]. To read these papers is to appreciate our fortune in being able to calculate integrals to arbitrary degrees of accuracy (in most cases). In some of this work they were guided by the earlier work of Coulson [33], who in the absence of firm calculations, had argued that

> In H_3^+, we are only concerned with two electrons, and thus, in the ground state, only one type of orbital is filled: the two electrons will have the same space wave function but opposed spins. We are to choose that configuration of the nuclei which gives this orbital greatest bonding energy. If a, b, c is the linear molecule, then the lowest orbital will only represent resonance between a–b and b–c; the resonance a–c is too small to make any appreciable contribution to the bonding. But in the triangular model, all three resonances contribute, and we may there-fore expect greater bonding.

This argument ignored the increased nuclear repulsion that arises in the triangular geometry.

Despite computational difficulties Hirschfelder and co-workers [27–32] succeeded in a remarkable accomplishment. In the fifth paper of the series [31], the conclusion was reached that "The energy of the ground state of H_3^+ is lower for the right triangle (-180 kcal) than for the equilateral triangle (-173 kcal), which indicates that the stable configuration for the H_3^+ lies between the right and the equilateral triangular configurations." Other important conclusions reached in these papers were that the formation energy of H_3^+ from two hydrogen atoms and a proton was 184 kcal, and therefore that the reaction $H_2^+ + H_2 \rightarrow H_3^+ + H$ is very exothermic, and that H_3^+ should possess two IR active modes.

2. *Calculations of Potential Energy Surfaces*

Techniques for calculating potential energy (pe) surfaces of small molecules are well known, and have been reviewed elsewhere [34]. For an introduction to the concept of the pe surface (and that of H_3^+) the reader may consult the book by Murrell et al. [35]. Christoffersen [12] was the first person to conduct a high level *ab initio* study of the electronic energy and electron density of H_3^+. He showed that H_3^+ had a genuine three-center two-electron bond with an electron density that was shared equally by the three protons, leading to an equilateral triangle geometry. The most accurate calculation of the energy of equilibrium geometry H_3^+ to-date is that of Anderson [36], using quantum Monte Carlo calculations (see Table I); his paper contains a very full list of theoretical energies obtained for H_3^+ since 1938. Although quantum Monte Carlo calculations give very accurate energies, they are not well adapted to calculating pe surfaces, and we shall not consider them further.

The electronic pe surfaces of H_3^+ are complicated by the presence of two dissociation channels: H_2 $(^1\Sigma_g^+) + H^+$ and $H_2^+(^2\Sigma_u^+) + H$. This leads to an avoided crossing between the two lowest electronic states of the $^1A_1$ symmetry, which occurs at $r \approx 2.5a_0$ [see Fig. 2.2(a) for coordinates]. This avoided crossing was first studied by Preston and Tully [37] using the diatomics in molecules (DIM) pe surface [38, 41] and was the first avoided crossing between polyatomic pe surfaces to be considered with detailed *ab initio* calculations [40]; it is important in calculations of accurate pe surfaces for the ground state.

That there must be an avoided crossing may be seen by considering the nature of the potential curves corresponding to the two dissociation limits $H_2 + H^+$ and $H_2^+ + H$. The presence of a H atom at infinity lowers the total energy of the H_2^+ potential by $0.5E_h$ ($\approx 10^5$ cm^{-1}) and the curves therefore cross at $r \approx 2.5a_0$, as shown in Fig. 2.2. If the third nucleus is brought closer, the state correlating with $H_2^+ + H$ splits into two surfaces that have $^1\Sigma^+$ and $^3\Sigma^+$ symmetries in linear configurations. Since the surface that correlates with $H_2 + H^+$ is also $^1\Sigma^+$ in the linear configuration, an avoided crossing between the two singlet surfaces occurs. Each surface therefore has a region in which the electronic configuration is a superposition of $H_2^+ + H$ *and* $H_2 + H^+$.

Although many *ab initio* surfaces have been calculated for H_3^+, so far no surface has been published that is both of spectroscopic accuracy and well behaved at dissociation. Numerous model calculations of the behavior of H_3^+ have therefore been performed using the DIM [38, 41] neglecting overlap pe surface as implemented by Preston and Tully [37] in their investigations of the surface-crossing problem. The DIM surface is well behaved at all geometries and is often used for qualitative calculations, but is not accurate. The DIM approximation expresses the electronic structure of H_3^+ in terms of the electronic structure of the fragments $H_2 + H^+$, and $H_2^+ + H$, and allows the nonadiabatic coupling between the first two $^1A_1$ states to be calculated in a straightforward manner. The two energy levels of the lowest $^1A_1$ states are given by the lowest two eigenvalues of a symmetric 3×3 Hamiltonian matrix with elements given by [41]:

$$\mathcal{H}_{11} = E_1 + \tfrac{1}{2}(^gE_2 + {}^uE_2 + {}^gE_3 + {}^uE_3) - 2E_H$$

$$\mathcal{H}_{12} = \tfrac{1}{2}(^gE_3 - {}^uE_3)$$

$$\mathcal{H}_{13} = \tfrac{1}{2}(^gE_3 - {}^uE_2)$$

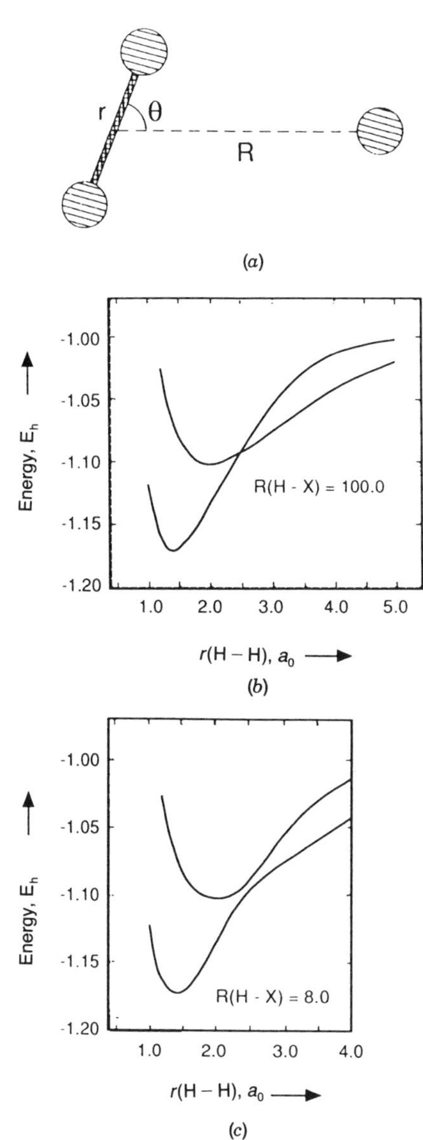

Figure 2.2. Origin of the avoided intersection of the lowest ${}^1A_1$ H_3^+ electronic surfaces (after Bauschlicher et al. [40]). (*a*) The scattering (or Jacobi) coordinates shown in (*b*) and (*c*). (*b*) $H_2 + H^+$ at $R = 100$ and $H_2^+ + H$ at $R = 100\ a_0$; the presence of a H atom at infinity lowers the H_2^+ curve by 0.5 E_h, and the curve therefore crosses the H_2 curve, which has the same symmetry. (*c*) At $R = 8.0\ a_0$ the avoided intersection can be seen at $r = 2.5a_0$.

$$\mathcal{H}_{22} = E_2 + \tfrac{1}{2}({}^gE_1 + {}^uE_1 + {}^gE_3 + {}^uE_3) - 2E_H$$

$$\mathcal{H}_{23} = \tfrac{1}{2}({}^gE_1 - {}^uE_1)$$

$$\mathcal{H}_{33} = E_3 + \tfrac{1}{2}({}^gE_1 + {}^uE_1 + {}^gE_2 + {}^uE_2) - 2E_H$$

where E_i are the electronic energies of the ${}^1\Sigma_g^+$ ground state of H_2, which are evaluated at internuclear distances R_i. The parameters ${}^gE_i$, and ${}^uE_i$ are the energies of the ${}^2\Sigma_g^+$ ground state and the ${}^2\Sigma_u^+$ first excited state of H_2^+ evaluated at R_i and E_H is the energy of the 1*s* state of H ($-0.5E_h$). The three internuclear distances (R_1, R_2, R_3) are implied by the values of diatomic fragment energies used, and the matrix elements can be considered as being generated by three protons with electrons fixed to each two in turn; that is, $\mathscr{H}_{11}$ is generated from the energy of the hydrogen bond at distance R_1 and the two associated H_2^+ energies of the ground and first excited states at R_2 and R_3 with the energy of the two hydrogen atoms subtracted. The parameters $\mathscr{H}_{22}$ and $\mathscr{H}_{33}$ are generated by cyclic permutation.

Where it is possible to calculate energies on a grid of points completely spanning a pe surface, it is then necessary to fit these energies to some functional form that will yield a value of the potential at any geometry [42]. This procedure can be a major difficulty in using a set of calculated points, and is more of an art than a science. The perfectly fit surface should vary smoothly from its equilibrium geometry (where it should display near harmonic behavior) to the region of dissociation, reflect any permutation symmetries that may be present, and have saddle points as appropriate. The ground-state surface of H_3^+, for example, has a saddle point at the energy region where the linear geometry becomes energetically accessible [14,275 cm^{-1} above the pe minimum (see Fig. 2.1)].

It is not yet known (other than heuristically) how to construct a pe surface in order to achieve a given level of accuracy for a particular problem: how many points on the surface must be calculated, where in configuration space they should be placed, or to what functional form they should be fitted are all problems under active investigation. In principle all these problems could be avoided if the pe were to be calculated directly at each configuration for which it was required in other calculations. This would be very expensive and is unlikely to be feasible in the near future.

The *ab initio* calculations of Carney and Porter [43] were instrumental in the location of the H_3^+ spectrum. They calculated high quality pe, electric dipole, and quadrupole surfaces at 74 grid points close to the equilibrium geometry and used a Simons–Parr–Finlan (SPF) [44] fit to their calculated points. They calculated band positions by generating force fields, and used their dipole and quadrupole moment functions to evaluate transitions strengths and Einstein *A* coefficients [45].

The first *full* analytical pe surface for the ground state of H_3^+ was a fit by Giese and Gentry [46] to a functional form of a sum of three diatomic H_2 potentials, using three sets of previously calculated *ab initio* points (including those of Carney and Porter). They treated the avoided intersection by arguing that for large R [scattering coordinates (see Fig. 2.2(a)], the intersection is sharp, but the probability of an electronic transition between the two surfaces is high, and therefore diabatic behavior is expected, so that the appropriate form for the potential is one that continues smoothly through the crossing. They therefore fit the lower sheet of the adiabatic surfaces for small separations of H^+ and H_2 (small R), but fitted the diabatic surface for $R > 4a_0$. This distinction is only meaningful for $r > 2.5a_0$ when the avoided crossing can be accessed.

Schinke, Dupuis and Lester, [47] calculated a complete *ab initio* surface for H_3^+ (the SDL surface), to a *common* accuracy that was sufficient for scattering calculations, but did not seek improvements in energy over previous calculations. They followed a similar fitting procedure to Giese and Gentry, considering the system to be diabatic for $R > 4a_0$. Murrell et al. [35] comment that, given the difficulty in providing a good fit to the partly adiabatic, partly diabatic lower sheet of the H_3^+ surface, that it is simpler to use a many-valued representation, and that such a representation will almost certainly be necessary to describe the rotation–vibration states near the dissociation limit. They give one such approximate representation, but this procedure does not appear to have been generally followed.

Meyer et al. [48] calculated a pe surface using full configuration interaction (CI) for H_3^+ in order to find spectroscopic transition frequencies near the equilibrium geometry. The surface was both more accurate and more extensive than that of Carney and Porter [43]. As they were uninterested in the regions of the surface where the avoided crossing becomes a problem, they did not calculate points in this region, and did not attempt to fit it. Their surface did, however, contain the barrier to linear H_3^+. This was the first H_3^+ surface to be calculated to a spectroscopic accuracy, and has been used by many other workers to calculate vibration–rotation energy levels and spectroscopic constants [49]. The paper contains a number of surfaces, including the one usually called the MBB potential, which was calculated with full configuration interaction. In the same paper, Meyer et al. [48] also calculated an *ab initio* dipole surface, which has been of great utility in calculating the expected intensities in various spectroscopic transitions. The pe surface was fitted to the following functional form:

$$V = \sum_{n,m,k} V_{n,m,k} S_a^n S_e^{2m+3k} \cos(3k\varphi) \qquad n + 2m + 3k \leq N$$

where $V_{n,m,k}$ are the coefficients determined by fitting and the value of N determines the order of the fit. Coordinates (S_a, S_e, φ) are related to atom–atom distances (R_{12}, R_{23}, R_{31}) by two transformations:

$$R' = \frac{1 - e^{-\beta[(R-R_e)/R_e]}}{\beta}$$

which changes from atom–atom distances to Morse coordinates, and a second:

$$S_a = \frac{(R'_{12} + R'_{23} + R'_{31})}{\sqrt{3}}$$

$$S_x = \frac{(2R'_{12} - R'_{23} - R'_{31})}{\sqrt{6}} = S_e \cos(\varphi)$$

$$S_y = \frac{(R'_{23} - R'_{31})}{\sqrt{2}} = S_e \sin(\varphi)$$

which changes to symmetry-adapted deformation coordinates. The MBB potential has one term of the fit adjusted ($S_e^2 \rightarrow S_e^2 \times 1.0019$) in order to reproduce the observed H_3^+ bending fundamental. This parameterization of the H_3^+ surface has also been used to Lie and Frye [50] to fit the calculated *ab initio* points of Frye et al. [51], which also gives a surface of spectroscopic accuracy. Some confusion has arisen in the literature due to an incorrect implementation of the MBB surface, which came into general use [52] and was responsible for several discrepancies in vibration–rotation energies that were ostensibly calculated with the same surface.

Potential energy surfaces for excited electronic states of H_3^+ have been calculated by Kawaoka and Borkman [53] (6 excited states calculated in equilateral geometries only), Talbi and Saxon [54] (pe surfaces and transition moments for 13 excited singlet states of H_3^+), and Schaad and Hicks [55] (20 excited electronic states of H_3^+).

The calculations of Schaad and Hicks [55] were the first to locate an excited electronic state of H_3^+ that was bound with respect to vibration, the linear triplet state ${}^3\Sigma_u^+$, which dissociates to $H_2^+({}^2\Sigma_g^+) + H({}^2S)$. In order to determine whether the minimum of the triplet was deep enough to contain a vibrational level Schaad and Hicks used a harmonic oscillator approximation to the pe surface near the minimum, and found that it did support vibrational levels. The most accurate calculation of the equilibrium properties of the ${}^3\Sigma_u^+$ state is by Ahlrichs et al. [56] and their results are given in Table II. They also used a harmonic approximation to determine the vibrational properties and found that the energy of formation was

$$H_2^+(^2\Sigma_g^+) + H(^2S) \rightarrow H_3^+(^3\Sigma_g^+)$$

$$\Delta E = -0.01343E_h \qquad (-8.43\ \text{kcal mol}^{-1})$$

A full pe surface for the triplet state $^3\Sigma_u^+$ has been provided by Wormer and de Groot [57]. They calculated more than 400 points and gave an analytic fit based on 240 points, but do not claim spectroscopic accuracy. They found that the surface was very flat, so the molecule is floppy, and has a high probability of tunneling between three symmetry-related minima. No calculations of vibration–rotation energies were attempted, and they comment that the harmonic approximation used by previous researchers is inappropriate given the anharmonicity of the surface. No accurate calculations of vibration–rotation levels supported by this surface have been published, and no spectroscopic observations have been reported, which involve the $^3\Sigma_u^+$ state. Such calculations and observations would be extremely interesting.

B. Vibration–Rotation Levels in Ground State H_3^+

The state of a system is fully described by its wave function, and in many systems the state can be described using both exact quantum numbers (true constants of the motion), and approximate quantum numbers. For molecules an example of an exact quantum number is the total angular momentum. All vibrational quantum numbers are approximate quantum numbers; how "good" a vibrational quantum number is depends on the degree of coupling with other states. Such couplings may arise, for example, from vibration–rotation interaction (it is not possible to separate vibration and rotation without making approximations) and anharmonicity.

As we shall see later, the distinction between exact quantum numbers and approximate quantum numbers is important in discussing the predissociation spectrum of H_3^+.

1. *Normal Modes of* H_3^+

All high level *ab initio* calculations on H_3^+ (D_3^+) find that the molecule is an equilateral triangle (with $r_0 \approx 1.6a_0$), and therefore of D_{3h} symmetry. A normal mode analysis of this configuration results in the two possible vibrational modes (see Fig. 2.3), the totally symmetric (breathing) A_1' mode (ν_1), which is Raman active, and the doubly degenerate E' mode (ν_2), which is IR active. Both H_2D^+ and D_2H^+ have three normal modes; ν_1 is the symmetric breathing mode, ν_2 is the symmetric bending

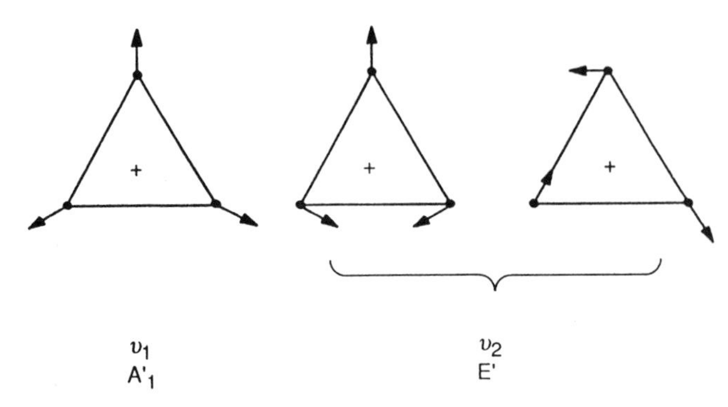

Figure 2.3. The normal modes of H_3^+ at equilibrium; ν_1 is Raman active, and ν_2 is IR active. Transitions in the $\nu_1 \leftarrow 0$ band have been detected by IR spectroscopy [99]. This illustrates the limitations of using transition rules that are derived by using point groups that treat molecules as near-rigid bodies.

mode, and ν_3 is the antisymmetric bending mode (more like antisymmetric stretching in H_2D^+), all of which are IR active. The importance of normal modes in vibrational analysis is that normal-mode coordinates diagonalize the kinetic term and the harmonic limit of the pe term of the classical Hamiltonian simultaneously. The Schrödinger equation for nuclear motion can therefore be separated into a set of one-dimensional (1-D) equations.

Frequencies of the vibrational modes of H_3^+ have been estimated in many *ab initio* calculations, but the first calculations to take account of anharmonicity were those of Carney and Porter [45] who used an SPF [44] expansion. Their initial estimate of ν_2 was $2814\,cm^{-1}$, which was subsequently refined [45] to $2516\,cm^{-1}$, and finally [58] to $2546\,cm^{-1}$. They found that the vibrational anharmonicity was large. Carney and Porter [59–61] also gave vibrational energies for the deuterium and tritium isotopomers of H_3^+.

2. *Quantum Numbers, Selection Rules, and All That*

Although many readers will be familiar with the quantum numbers used for triatomic molecules [62, 63] they will be reproduced here for clarity, together with some quantum numbers introduced specifically for X_3 molecules. The quantum numbers and approximate selection rules used for molecules close to their equilibrium geometry rest upon the assumption that at low energies vibrational motion is well described by small (harmonic) displacements from this equilibrium geometry, and that a separation can be made (at least in a first approximation) between vibrational and rotational motions within a given electronic state. If a potential were electrically and mechanically harmonic there would be no intensity in overtone and combination bands; the very intense overtones and combinations that have been observed in H_3^+ provide experimental proof that such anharmonicity is present, even near the equilibrium geometry.

a. Vibration–Rotation Quantum Numbers. The quantum numbers commonly used in describing H_3^+ (and other X_3 molecules) are the number of quanta (v_1, v_2) in each of the two normal modes ν_1 $(\Gamma_v = A_1')$ and ν_2 $(\Gamma_v = E')$. The doubly degenerate mode (ν_2) has unit vibrational angular momentum $(\zeta = -1)$ [63] when singly excited [64], and the vibrational angular momentum of mode ν_2 is designated by quantum number ℓ_2; ℓ_2 steps in two's and when ν_2 is excited with v_2 quanta, $|\ell_2|$ can assume the values v_2, $v_2 - 2 \cdots 1$ or 0 [65]. Particular vibrational levels of H_3^+ are therefore designated as $v_n \nu_n(\ell_n)$, for example, $3\nu_2(\ell_2 = 1)$, and these can form combinations and differences in the usual manner.

Since H_3^+ (D_3^+) is a symmetric top, rotational energy levels are specified by the rotational angular momentum J, its projection on the molecular axis K, and the vibrational angular momentum ℓ_2. For large values of J and high excitation, K is not likely to be a good quantum number [66], and K is not a good quantum number in the ν_2 state even for low J [67].

The molecular ions H_2D^+ and D_2H^+ are the lightest asymmetric top molecules known. The number of quanta in each of the three vibrational modes is specified analogously to H_3^+, as is rotational angular momentum. In common with all asymmetric rigid rotors, the rotational levels are specified by J_τ or $J_{K_A K_C}$, where $\tau = K_A - K_C$ and K_A and K_C are the values of $|K|$ for the limiting prolate and oblate symmetric rotors with which the energy level correlates [63]; τ is not strictly speaking a quantum number, but it does provide a 1:1 correspondence with useful labeling for the pattern of energy levels for a given J value.

b. Good Quantum Numbers. For H_3^+ quantum numbers that are constants of the motion are the total angular momentum (F) and parity $(\Pi = \pm)$. These quantum numbers are good in all states because the Hamiltonian is invariant with respect to rotation in space and space inversion (space is isotropic). Because the Hamiltonian is invariant to permutation of the three identical nuclei, symmetry labels for vibration–rotation wave functions are also good quantum numbers. The vibration–rotation wave functions may be assigned symmetries of A_1, A_2, or E (using Mulliken's notation [68]) according to their behavior under the complete nuclear permutation group S_3 which is isomorphous with the point group D_{3h} [69]. States form $\Gamma_v = A_1, A_2$ pairs, when $\ell_2 = 3n$ (i.e., $\ell_2 = 3, 6, 9, \ldots$) and is nondegenerate. States of ℓ_2 not equal to $3n$ (i.e., $\ell_2 = 1, 2, 4, 5, 7, 8, \ldots$) have $\Gamma_v = E$ symmetry. Vibrational states of $\ell_2 = 0$ are singly degenerate and have symmetry $\Gamma_v = A_1$.

c. Selection Rules. Symmetry arguments allow selection rules to be obtained for vibration–rotation transitions, but these rules are only obeyed when the quantum numbers involved remain good. For one photon electric dipole transitions the rigorous selection rules on any transitions are [69]

$$\Delta F = 0, \pm 1 \ (F'' = 0 \not\leftrightarrow F' = 0)$$

$$\Delta \Pi \neq 0 \ (\text{i.e.}, + \leftrightarrow -)$$

where F is the *total* angular momentum of the molecule ($F = J + I$) and Π is the parity. The electronic motion (J) and the nuclear motion (I) are coupled together by the hyperfine interaction. Strictly speaking, neither J nor I are good quantum numbers, but the hyperfine interaction is very small near equilibrium. However, for highly excited states, where the D_{3h} symmetry may be broken dynamically, nuclear spin symmetry violation is expected to be more prominent [66]. As J and I are almost completely uncoupled,

$$\Delta J = 0, \ \pm 1 \ (J'' = 0 \not\leftrightarrow J' = 0) \qquad \text{and} \qquad \Delta I = 0$$

are also good selection rules. There are two species of H_3^+ according to nuclear spin, ortho ($I = \frac{3}{2}$) and para ($I = \frac{1}{2}$), and the selection rule $\Delta I = 0$ therefore forbids ortho–para transitions.

As mentioned above, the symmetry of the wave function with respect to nuclear permutation provides good quantum numbers and application of group theoretical arguments show that rovibrational transitions of H_3^+ obey the selection rules $E - E$ and $A_{1,2} - E$ only, where $A_{1,2}$ and E are the symmetry species (Γ_{vr}) of the rovibrational wave functions.

The use of approximate wave functions leads to approximate selection rules. The only way of accurately determining intensities in vibration–rotation transitions is by *ab initio* calculations using full vibration–rotation wave functions and an accurate electric dipole surface. The **approximate** wave functions used to describe low-energy states of H_3^+ are $|v_1, v_2, \ell_2\rangle |J, K\rangle$ and under the molecular symmetry group operations (123), E^*, and (23) these have the properties [70]:

Property A $(123)|v_1, v_2, \ell_2\rangle |J, K\rangle = \exp[2\pi i(K - \ell_2)/3]|v_1, v_2, \ell_2\rangle |J, K\rangle$
Property B $E^*|v_1, v_2, \ell_2\rangle |J, K\rangle = (-1)^K |v_1, v_2, \ell_2\rangle |J, K\rangle$
Property C $(23)|v_1, v_2, \ell_2\rangle |J, K\rangle = (-1)^J |v_1, v_2, -\ell_2\rangle |J, -K\rangle$

The behavior under E^* determines the parity of the wave function; (123) and (23) denote permutations of identical nuclei. The dipole moment μ is unaffected by interchanging like nuclei and so for electric dipole transitions, using Property A we find:

$$(\text{let } |v_1, v_2, \ell_2\rangle|J, K\rangle = |\gamma, g\rangle \quad \text{where } g = K - \ell_2)$$

$$\langle\gamma, g|\mu|\gamma', g'\rangle = \langle\gamma, g|(123)^{-1}(123)\mu(123)^{-1}(123)|\gamma', g'\rangle$$

$$= \langle\gamma, g|\exp(-2\pi i g/3)\mu \exp(2\pi i g'/3)|\gamma', g'\rangle$$

$$\therefore \qquad 1 = \exp(2\pi i[g' - g]/3)$$

$$\therefore \qquad g' - g = \Delta g = \Delta(K - \ell_2) = 3n \text{ (where } n \text{ is an integer)}$$

The dipole moment μ changes sign under E^* and using a similar argument and property B yields the selection rule $\Delta K = \text{odd}$ [70], which is a restatement of the selection rule on parity, with parity given by $(-1)^K$. In general the strongest transitions are those with the smallest value of ΔK and n [71].

Transforming the above functions to a Wang-type basis [72] (except for $\ell_2 = K = 0$) yields

$$|v_1, v_2, \ell_2, J, K, s\rangle = \{|v_1, v_2, \ell_2\rangle|J, K\rangle + s|v_1, v_2, -\ell_2\rangle|J, -K\rangle\}/\sqrt{2}$$
$$s = \pm 1$$

where s is the Wang index. These have the more convenient property

Property D $\quad (23)|v_1, v_2, \ell_2, J, K, s\rangle = (-1)^J s|v_1, v_2, \ell_2, J, K, s\rangle$

for $K \geq 0$, $s = \pm 1$, and matrix elements connect functions of the same s only. For $\ell_2 = K = 0$ the function $|v_1, v_2, \ell_2 = 0, J, K = 0, s\rangle$ satisfies property D with $s = 1$. The use of these basis functions allows the energy matrix to be partitioned according to the symmetry species of the point group D_{3h}.

3. *The Structure of Vibration–Rotation Energy Levels*

An energy level formula for the rotational levels of H_3^+ near equilibrium is given by [63]:

$$F_r(J, K, \ell_2) = BJ(J+1) + (C - B)K^2 - 2\zeta CK\ell_2 + E_c(J, K, \ell_2)$$

where F is the rotational term value (energy of the state in reciprocal centimeters), B and C are the rotational constants for the vibrational levels, ζ is the Coriolis coupling constant (hence the inclusion of the vibrational angular momentum ℓ_2 in the expression for rotational energy), and $E_c(J, K, \ell_2)$ denotes the higher order correction(s) to the rotational energy (which may be represented in many ways). For H_3^+, Oka [17] determined that:

$$B(\nu_2) = 43.568\ \text{cm}^{-1}$$

$$C(\nu_2) = 20.708\ \text{cm}^{-1}$$

$$\langle \zeta C \rangle(\nu_2) = -18.527\ \text{cm}^{-1}$$

Watson [70] derived an effective Hamiltonian for states near equilibrium using contact transformations (a perturbation theory approach [73]) of his Eckart-type Hamiltonian [74], and found that the largest term coupling vibration and rotation was

$$\mathcal{H}_\ell = q[q_+^2 J_-^2 + q_-^2 J_+^2]/4$$

where q is the ℓ-doubling constant and $q_\pm$ and $J_\mp$ are the vibrational and rotational ladder operators for ℓ_2 and K. This term, $\mathcal{H}_\ell$, is responsible for mixing between the states $|J, K \pm 1, \ell_2 = \pm 1\rangle$ and $|J, K \mp 1, \ell_2 = \mp 1\rangle$. When the levels are degenerate it causes ℓ-type doubling, otherwise it causes ℓ-type resonance. The $\mathcal{H}_\ell$ term dominates the observed patterns of energy levels; it is a severe perturbation. It is large for two reasons: due to the small mass and the large anharmonicity of H_3^+, q is very large ($\approx -5.4\ \text{cm}^{-1}$), and because of the equal masses and equilateral triangle geometry, $B \approx 2C$ and also $C\zeta = -C$, which causes the two interacting levels to be always nearly degenerate. The strong mixing of these two levels causes K and ℓ_2 to no longer be good quantum numbers, but $K - \ell_2$ is a nearly good quantum number; the effective Hamiltonian has, *to a good approximation*, energies that are diagonal in the quantum number $G = |K - \ell_2|$, note that when $\ell_2 = 0$, $G = |K|$.

The K spacing of the bands is determined by the parameter

$$\delta\nu_K = 2[C(1 - \zeta) - B]$$

and is small because $\zeta = -1$ and due to planarity $B \approx 2C$ as discussed above. The same parameter, $\delta\nu_K$, also determines the separation between levels that are connected by the matrix elements of the ℓ-type doubling term:

$$F(\ell_2 = +1, K = G + 1) - F(\ell_2 = -1, K = G - 1) \approx 2G\delta\nu_K$$

ℓ-type resonances are large for all K values, and are approximately independent of K.

In general, levels can be described by the label U and the quantum numbers J and G [70]. The parameter U is not a quantum number but has values $U = +1$ for the upper level of the $\ell = 1$ dyad and $U = -1$ for the lower level. Watson's notation of U has been extended by Miller and Tennyson [75] by taking $|U| = |\ell|$ for the $2\nu_2$, $\ell = 2$ and $3\nu_2$, $\ell = 3$ levels so that $U = -2$, $+2$ and $U = -3$, $+3$ label states of lower and higher energy. For $G = 3$ an additional index is necessary to describe the split components A_1, A_2 and this is the Wang index, $s = \pm 1$.

Transitions in X_3 molecules are either designated by giving the quantum numbers for both levels involved (J, G, U or J, K with a prime indicating the upper level and a double prime indicating the lower level), or by using standard spectroscopic notation; $\Delta J = -1, 0, +1$ corresponding to P, Q, R, with numbers in parentheses indicating the J value of the lower level and K. For example,

$$R(2,2) = |J' = 3, G' = 2, U' = 1\rangle \leftarrow |J'' = 2, K'' = 2\rangle$$

4. *Influence of Nuclear Spin*

One of the identifying characteristics of the $\nu_2 \leftarrow 0$ H_3^+ spectrum is the absence of a state with zero angular momentum, $J = K = 0$, which is a feature of any system composed of three identical nuclei that are fermions of spin $\frac{1}{2}$. For the usual case of a totally symmetric electronic wave function the Pauli principle demands that the wave function for a system must be antisymmetric with respect to the interchange of fermions. This is usually achieved by the combination of a symmetric space function and an antisymmetric spin function, or vice versa. However, when the space function for three fermions is symmetric under both the permutation of two, or all-three nuclei, it cannot satisfy the Pauli principle for the two operations simultaneously for any spin function.

The direct result of these considerations is that in its ground vibrational level, H_3^+ contains a zero-point vibrational energy of $4345.3\ cm^{-1}$, and in addition to this a rotational energy of $63.9\ cm^{-1}$. The minimum possible energy of an H_3^+ molecule is therefore $4409\ cm^{-1}$. This corresponds to a rotational temperature of 92 K, which will be possessed even in interstellar environments with a temperature of 5–10 K.

The symmetry of the vibrations determines the nuclear-spin statistical weights. These result in alternating intensities which can be useful when assigning spectral lines. For the cases of H_3^+ and D_3^+ the statistical weights are [70], for nuclear spin i_x;

$\Gamma_{e,v,r}$	Bosons, General Case	$i_x = 1$ (D_3^+)	Fermions, General Case	$i_x = \frac{1}{2}$ (H_3^+)
A_1', A_1''	N_1	10	N_2	0
A_2', A_2''	N_2	1	N_1	4
E', E''	N_3	8	N_3	2

where

$$N_1 = (2i_x + 1)(2i_x + 2)(2i_x + 3)/6$$

$$N_2 = (2i_x + 1)(2i_x)(2i_x - 1)/6$$

$$N_3 = (2i_x)(2i_x + 1)(2i_x + 2)/3$$

We see that the A_1' and A_1'' levels of H_3^+ have zero statistical weight as expected from the above discussion. However, these missing levels are obtained in conventional variational calculations that neglect nuclear spin and although they are often listed in order to facilitate comparisons between different calculations they must be excluded from spectral analysis.

C. Approaches to the Assignment of Spectroscopic Data

Molecular spectroscopists aim to obtain information about molecular structure, but are hindered by the inversion problem: It is not possible to obtain from measured transition frequencies an analytical expression for the pe surface of a molecule. In spectroscopy the transition frequencies correspond to differences between the energies of particular levels of a molecule which, within the Born–Oppenheimer approximation, are determined from solution of the vibration–rotation Hamiltonian of the molecule on a given pe surface. In the absence of a direct analytic method for obtaining the desired pe surface one solves the problem iteratively, that is, one assumes a reasonable pe and dipole surface, solves for the vibration rotation levels, and calculates transition frequencies and intensities. These calculations are often of great use in assigning spectra. The second part of the problem is to compare the calculated frequencies and intensities with those determined experimentally. If a 1:1 correspondence can be found between some, or all, of the experimental and theoretically produced data, then a process of refinement of the pe surface can take place (see below). The process of calculating molecular spectra for small molecules has recently been reviewed by Tennyson [76], and methods of calculating vibration–rotation states for "floppy molecules" have been reviewed by Bačić and Light [77].

Another procedure that is often adopted to fit experimental data is to use an effective Hamiltonian, and for almost all studies of H_3^+ near equilibrium (and many other molecules) the Hamiltonian used is that formulated by Watson (see below). Strictly speaking the effective Hamiltonian (in this case) is a solution for vibration–rotation only, which is usually based upon a particular model, for example, an anharmonic oscillator, or set of coupled anharmonic oscillators, to allow for interactions such as the Coriolis coupling.

In order to calculate molecular properties it is desirable to use a coordinate system that is embedded within the molecule; if the problem could be completely solved, then any coordinate system that travels with the molecule would suffice. However, some coordinate systems are more useful than others. To transform the complete laboratory-fixed Hamiltonian to one that moves with the molecule and has coordinates that are internal to the molecule is difficult. This problem was first solved by Eckart [78] in 1935. His solution has been reviewed by Sutcliffe [79]. It relies upon the fact that in a deep potential well the nuclei will have a fairly rigid equilibrium geometry, whose shape is determined by the potential minimum. Following from this Eckart embedded his coordinate system in the molecule in such a way that the rotation spectrum could be understood in terms of the rotation of this rigid body. However, the picture is appropriate only when the internal motions of the nuclei correspond to small displacements from equilibrium. Eckart showed that in this model vibrational motion could be described in terms of products of harmonic oscillator functions. The variables of these functions are the normal coordinates, which are linear in combinations of displacements. For an introduction to normal coordinate analysis the reader may consult Herzberg Vol. II [63].

In 1968 Watson [74, 80] derived a quantal Hamiltonian incorporating the Eckart conditions. The Eckart conditions [78] had previously been used as the basis of a quantum mechanical Hamiltonian by Wilson and Howard [81], which was made Hermitian by Darling and Dennison [82] in 1940, but Watson's derivation led to a form of vibration–rotation Hamiltonian that was amenable to calculation. Watson's Hamiltonian has been used both for variational calculations of vibration–rotation energies, and has also been subjected to a series of contact transformations [70] leading to a form suitable for use in least-squares fitting of spectroscopic data. Developments in computer technology made numerical calculations using such a Hamiltonian feasible, and the first such calculations were attempted in the work of Carney and Porter [43] in 1974 and the work of

Whitehead and Handy [83] in 1975. The philosophy adopted by Carney and Porter was to solve for nuclear motion in an effective potential that depended on the 5th degree SPF *ab initio* force field (V_{SPF}) and also upon the Watson term (U_{W}), which has its source in the kinetic energy (ke) operator and involves both mass- and geometry-dependent terms:

$$U_{\mathrm{W}} = -\frac{\hbar^2}{8} \sum_{\alpha} \mu_{\alpha\alpha}$$

where the $\mu_{\alpha\alpha}$ depend on the instantaneous moments ($I_{\alpha\alpha}$) and products ($I_{\alpha\beta}$) of inertia of the molecule. The Watson term is quite small for molecules that are not nearly linear. Specifically, the Hamiltonian used by Carney and Porter was

$$\begin{aligned}
\mathcal{H} = {} & \frac{1}{2m}(\mathbf{p}_1^2 + \mathbf{p}_2^2 + \mathbf{p}_3^2) + \mathbf{V}(q_1, q_2, q_3) + \frac{1}{2}\frac{(q_2\mathbf{p}_3 - q_3\mathbf{p}_2)^2}{I_{zz} - m(q_2^2 + q_3^2)} \\
& - \frac{\hbar^2}{8}\left[\frac{I_{zz}}{I_{xx}I_{yy} - I_{xy}^2} + \frac{1}{I_{zz} - m(q_2^2 + q_3^2)}\right] + \frac{1}{2}\frac{I_{yy}\mathbf{J}_x^2}{I_{xx}I_{yy} - I_{xy}^2} \\
& + \frac{1}{2}\frac{I_{xx}\mathbf{J}_y^2}{I_{xx}I_{yy} - I_{xy}^2} + \frac{1}{2}\frac{\mathbf{J}_z^2}{I_{zz} - m(q_2^2 + q_3^2)} \\
& - \frac{1}{2}\frac{I_{xy}\mathbf{J}_x\mathbf{J}_y}{I_{xx}I_{yy} - I_{xy}^2} - \frac{(q_2\mathbf{p}_3 - q_3\mathbf{p}_2)\mathbf{J}_z}{I_{zz} - m(q_2^2 + q_3^2)}
\end{aligned}$$

where the fourth term corresponds to U_{W}. The coordinate system is defined by placing z along the C_3 symmetry axis, and the positive y direction is parallel to one interatomic vector, q_i are "rectilinear internal symmetry coordinates" [45].

The problem of constructing Hamiltonians for internal motions lies in the transformation from the laboratory frame to a molecule-fixed frame (the embedding problem), to which Eckart gave one solution that is valid at energies near equilibrium. The Jacobian of such transformations is necessarily nonlinear, and coordinate dependent. It will vanish for particular values of the coordinates, and in these regions the molecule fixed coordinates are undefined. Any coordinate system that succeeds in separating rotational motion from other motions causes the Jacobian to have a region in which it becomes zero [84]. In the case of the Eckart approach, and hence the Watson Hamiltonian, this zero in the Jacobian is found when vibrations have a sufficiently large magnitude to cause one or more of the generalized moments of inertia to vanish. For H_3^+ these regions occur when the molecule can become linear and thus there is a natural limit to the utility of the Watson Hamiltonian in assigning the spectrum; the Hamiltonian must fail when the barrier to linearity is reached. The limitations of this approach have been subject to attack by Tennyson and Sutcliffe who attempted to frame a theory in a coordinate system (for triatomic molecules), which can be defined in such a way as to cause placings of the singular regions wherever desired. Although it is not possible to choose a system without singular regions this approach enables them to be placed in a region that is not damaging to the problem under consideration. The derivation of the Hamiltonian for nuclear motion for any set of internal coordinates has been reviewed by Sutcliffe [85].

In the 1990 paper by Oka and co-workers [86] the comment is made that: "Probably the best strategy for this molecule is to provide the *ab initio* theorists with our experimentally determined energy levels, so that they can further adjust their potentials and accurately predict new rovibrational transitions." This is proposed in contrast to the usual strategy of fitting the observed transitions to a spectroscopic perturbation Hamiltonian and using the parameters in the Hamiltonian to predict further transitions. The reason for this change in philosophy is that to fit the data for H_3^+, which is abnormally anharmonic, requires almost as many parameters in the fitting Hamiltonian as there are experimental data points. This may completely determine the parameters, but it renders them useless for predictions of unobserved levels of the system at higher energies. Where this approach is, and has been, of enormous utility is in taking a limited set of spectral data, assigning it, refining the constants in the effective Hamiltonian, and by a process of iteration going on to assign more spectral data in the same energy region. In this manner many of the spectra of H_3^+ and its isotopomers have been fitted by Watson. Such perturbation Hamiltonians also give valuable insights into the relative importance of various physical effects in determining the energy level structure. For H_3^+ the approach has been found to be less useful as the barrier to linearity is reached. Other Hamiltonians have been developed by Jensen and co-workers [87–89] and used for calculating and assigning spectra of H_3^+ and its isotopomers.

The direct variational solution of the nuclear motion problem has been used by both Jensen and co-workers [87–89], and Tennyson and Sutcliffe [90] to calculate and assign transition frequencies in H_3^+ and its isotopomers. For H_3^+ this approach has been found of greater utility than the solution of perturbation Hamiltonians, particularly for the higher vibrational levels.

Such variational calculations of vibration–rotation energies, transition frequencies, and intensities have been stimulated by further refinements in computational procedures that have occurred simultaneously with the work on the embedding problem: two-step variational procedures (an approach that has solved the problem of dealing with rotationally excited states of triatomic molecules) and the use of finite-element methods. These techniques were recently reviewed by Tennyson [76].

The two-step variational procedure of Tennyson and Sutcliffe [91] first solves a series of problems that neglect coupling terms in the Hamiltonian between vibration and rotation. Wigner rotation matrices [92] are used to carry the rotational motion and for a particular value of the rotational quantum number J, $2J+1$ functions are required ($-J \leq k \leq J$). It has been found that it is extremely advantageous to solve first a vibrational problem for each value of $K = |k|$. The solutions for the K part of the problem are then ordered according to energy, and the lowest of them used as a new basis for expanding the complete problem, including couplings. The solution is rapidly convergent because only a small part of the problem has been left out in the first step, and therefore the functions are well tailored for the problem being solved. This two-step variational procedure has enabled the calculation of vibration–rotation levels up to dissociation for H_3^+ [93, 94]. What has not yet been solved is the calculation of the quasibound levels of the molecule, which are germane to the high-energy H_3^+ spectra described below (Section V). This problem is qualitatively different, being a scattering problem rather than a bound-state problem.

The use of finite-element methods has also speeded up calculations significantly [76]. The finite elements are based on coordinates rather than functions and led to the "discrete variable representation" (DVR), which has given an order of magnitude increase in the number of bound states that can be obtained with the same effort for a given system. The idea leading to the DVR was developed in 1965 [95] but has only recently been widely applied in the solution of vibration–rotation problems, mainly due to the work of Light and co-workers [77].

Using a combination of the above techniques and non-Eckart embedded Hamiltonians Tennyson and co-workers [4, 5] performed variational calculations of energy levels and transition intensities in H_3^+ (and other molecules). They have achieved excellent agreement between the observed spectra of H_3^+ and its isotopomers in the first few vibrational levels and transitions calculated using the MBB surface. This work enabled the astrophysical identification of H_3^+ [97]. They also predicted many transition frequencies and intensities and these have been useful in assigning spectra that were subsequently observed. One such calculation was of the intensity of the $(\nu_1 + \nu_2) \leftarrow \nu_2$ combination band that showed a breakdown in "normal" spectroscopic selection rules [98]. The calculated intensities were far higher than normally expected, and so an experimental search was performed and the band located [99]. This work has been reviewed by Tennyson et al. [84].

Tennyson and co-workers [100] adjusted an *ab initio* pe surface to generate an "experimental" pe surface for H_3^+ by inversion of the available spectroscopic data. They generated spectral term values from their vibration–rotation calculations, and adjusted iteratively 11 parameters from the *ab initio* surface of Lie and Frye [50] in order to bring calculated and measured term values into the best agreement. The data set used contained 243 H_3^+ energy levels containing vibrational levels up to $3\nu_2$ and rotational levels up to $J = 9$. The resulting semiempirical pe surface reproduced the observed data with a standard deviation of $0.053\ \text{cm}^{-1}$. The surface was then used in order to calculate revised estimates of the H_3^+ band origins up to $5\nu_2$.

When fundamental frequencies of the isotopomers, D_3^+, D_2H^+, and H_2D^+ were calculated from the same surface it was found that the agreement with experimental data was good for D_3^+, but poor for the mixed isotopomers, giving errors of about $0.5\ \text{cm}^{-1}$. This suggests that the errors are due to the limitations of the Born–Oppenheimer approximation, which takes no account of unequal nuclear masses. The possibility that limitations of the Born–Oppenheimer approximation would affect the accuracy of their potential was discussed by Lie and Frye [50], who estimated that the nonadiabatic correction is probably very small, but that the adiabatic correction contributes about $5\ \text{cm}^{-1}$ to the dissociation energy of H_3^+. They consider that of this about $1\ \text{cm}^{-1}$ will be due to "potential shape" effects, and note that the numbers are similar to those found in H_2 [101]. In the case of H_2^+ and its isotopomers, the limitations of the Born–Oppenheimer approximation have been very fully explored both theoretically and experimentally [10, 102] and again the magnitudes of shifts due to adiabatic corrections to the potential are consistent with the discrepancies found for H_3^+.

The semiempirical potential of Tennyson and co-workers [100] should therefore be considered an effective potential surface that is valid only for H_3^+ and different effective potential surfaces must be calculated for each isotopomer.

III. LABORATORY MEASUREMENTS OF LOW-ENERGY SPECTRA

The following sections offer an historical description of the measurements of the spectra of H_3^+ and its isotopomers in their low-energy states. Where different groups have used similar experiments, only the first have been described. The reader is referred to Table III for a complete list of published spectra.

Many compilations of observed transition frequencies of H_3^+ and its isotopomers, together with assignments and calculated line strengths (of great importance for astronomical observations) have appeared [75, 88, 89, 98, 103–110] and it was not thought necessary to give such a listing here. Unfortunately, the most recent such compilation [105] contains a substantial number of misprints [111]. A data base of up-to-date assignments, calculated energy levels, transition frequencies, and intensities is maintained by Tennyson and co-workers [111] at University College London; they will make the data available upon request.

A. The Spectroscopy of H_3^+

Oka [6] began his search for the H_3^+ spectrum in 1975, after estimating that the strongest Doppler limited absorption line of the fundamental band of H_3^+ ($\nu_2 \leftarrow 0$) would absorb several percent of an IR beam traversing 50 m through the positive column of a hydrogen glow discharge [6]. The experiment he designed relied for its success upon two recent experimental advances. The first was the use of a long positive column of glow discharge of the type first introduced to microwave spectroscopy by Woods and co-workers [112, 113]. In their experiments they successfully observed the protonated ions HCO^+ [114] and HN_2^+ [115]. Oka took this as strongly suggestive that H_3^+ would also be observable in a glow discharge. The second development that enabled the project was of a widely tunable IR laser system by Pine [116]. The IR laser system generates the difference frequency between lines from an argon-ion laser and tunable dye laser in a temperature controlled lithium niobate crystal, resulting in radiation that is tunable from 4400–2400 cm^{-1} with a power of a few microwatts. The wide frequency coverage was essential in locating the spectrum because the best theoretical predictions showed that the spectrum would extend over 500 cm^{-1} and the frequency predictions had an estimated accuracy of only 50 cm^{-1} [117].

The experiment used a 2-m length multipass discharge cell that gave a total path length through the cell of 32 m. The cell was cooled using liquid nitrogen and its ends sealed using CaF_2 Brewster angle windows. In later work on high J transitions, water cooling of the discharge was used. In order to enable lock-in amplification, the IR radiation was frequency modulated by varying the cavity length of the argon ion laser.

By 1980, 15 lines of the H_3^+ spectrum had been obtained. Figure 3.1(*a*) (see page 44) shows a stick spectrum showing the complete set of lines found, and their assignments. The assignment of the lines was achieved by Watson using a program originally written for analyzing the H_3 data of

TABLE III
Spectra of H_3^+ and Its Isotopomers Observed in the Laboratory

Year and Reference	Species	Observed	Assignment[a]	Method of Assignment
1980 Oka [17]	H_3^+	15 lines	$\nu_2(\ell = \pm 1) \leftarrow 0$	Perturbation Hamiltonian
1980 Shy et al. [18]	D_3^+	4 lines 4 lines	$\nu_2 \leftarrow 0$ unassigned	Variational calculations
1981 Shy et al. [139]	H_2D^+	7 lines 2 lines	$\nu_2 \leftarrow 0$	Assigned in 1986 by Foster et al. [148] using perturbation Hamiltonian
1981 Oka [118]	H_3^+	15 new lines	$\nu_2 \leftarrow 0$	Perturbation Hamiltonian
1982 Shy [136]	D_3^+ D_2H^+	37 new lines 31 new lines	$\nu_2 \leftarrow 0$	Assigned by Watson et al. [131, 149] using perturbation Hamiltonian and supermatrix
1982 Carrington et al. [20]	H_3^+ D_3^+ D_2H^+	First observation of IR predissociation spectra	No assignment	
1984 Amano and Watson [140]	H_2D^+	27 lines	$\nu_1 \leftarrow 0$	Perturbation Hamiltonian
1984 Watson et al. [120]	H_3^+	16 new lines	$\nu_2 \leftarrow 0$	Perturbation Hamiltonian
1984 Lubic and Amano [142]	D_2H^+	35 lines	$\nu_1 \leftarrow 0$	Perturbation Hamiltonian
1984 Bogey et al. [143] and Warner et al. [144]	H_2D^+	1 rotational line	$1_{10} - 1_{11}$	Previous experiments
1984 Carrington et al. [21]	H_3^+ H_2D^+ D_2H^+ D_3^+	27,000 lines	Predissociating states near dissociation	Classical trajectory calculations
1985 Saito et al. [146]	H_2D^+	1 rotational line	$2_{20} - 2_{21}$	Previous experiments and variational calculations
1985 Amano [141]	H_2D^+	10 new lines	$\nu_1 \leftarrow 0$	Perturbation Hamiltonian
1986 Foster et al. [148]	H_2D^+	31 lines 35 lines	$\nu_2 \leftarrow 0$ $\nu_3 \leftarrow 0$	Perturbation Hamiltonian and supermatrix calculation
1986 Jennings et al. [150]	H_2D^+	1 rotational line	$2_{20} - 2_{21}$	
1986 Foster et al. [149]	D_2H^+	54 lines 34 lines Includes data of Wing and Shy [136]	$\nu_2 \leftarrow 0$ $\nu_3 \leftarrow 0$	Perturbation Hamiltonian and supermatrix calculation
1987 Majewski et al. [123]	H_3^+	113 lines 35 additional but unassigned lines not listed	$\nu_2 \leftarrow 0$	Perturbation Hamiltonian and supermatrix calculation

TABLE III (*Continued*)

Year and Reference	Species	Observed	Assignment[a]	Method of Assignment
1987 Watson et al. [137]	D_3^+	49 lines 11 lines	$\nu_2 \leftarrow 0$ unassigned	Perturbation Hamiltonian
1989 Majewski et al. [126]	H_3^+	49 lines	$2\nu_2 \leftarrow 0$	Perturbation Hamiltonian and variational calculations
1990 Jennings et al. [150]	D_2H^+	2 rotational lines	$2_{20} \leftarrow 2_{11}$ $1_{11} \leftarrow 0_{00}$	Perturbation Hamiltonian
1990 Bawendi et al. [86]	H_3^+	14 new lines 70 lines 14 lines 21 lines 136 lines	$\nu_2 \leftarrow 0$ $2\nu_2(\ell = 2) \leftarrow \nu_2$ $2\nu_2(\ell = 0) \leftarrow \nu_2$ $\nu_1 + \nu_2 \leftarrow \nu_1$ unassigned	Variational calculations and previous experiments
1990 Xu et al. [129]	H_3^+	6 new lines 28 re-measured.	$2\nu_2(\ell = 2) \leftarrow 0$	Variational calculations and previous experiments
1990 Nakanaga et al. [125]	H_3^+	FTIR in absorption	$\nu_2 \leftarrow 0$	Previous experiments
1991 Lee et al. [130]	H_3^+	4 lines	$3\nu_2(\ell = 1) \leftarrow 0$	Variational calculations
1992 Xu et al. [99]	H_3^+	9 lines 21 lines 30 new lines 13 new lines 93 lines	$\nu_1 \leftarrow 0$ band $\nu_1 + \nu_2 \leftarrow \nu_2$ band $\nu_2 \leftarrow 0$ band $2\nu_2(\ell = 2) \leftarrow \nu_2$ unassigned	Variational calculations

[a] The assignments listed are those given in the original papers, but many have been subsequently adjusted by new fits. An up-to-date data base of assignments is kept by J. Tennyson and co-workers at University College London.

Herzberg [6]. By 1981 30 lines of the spectrum had been observed and assigned [118]. The transition frequencies used for the search for the 15 additional lines were based on predictions made from Watson's initial assignment.

The $\nu_2 \leftarrow 0$ spectrum has no obvious regularities, but Watson assigned the spectrum to the *P*, *Q*, and *R* branches of a planar symmetric rotor. The large gap in the spectrum ($\sim 100\,\text{cm}^{-1}$) at around $2600\,\text{cm}^{-1}$ indicated that the species responsible could not occupy the lowest rotational state of $J = K = 0$. This is the hallmark of a system composed of three identical fermions (see Section II.B.4), and immediately shows that H_3^+ consists of three equivalent protons, that is, its shape at equilibrium is that of an equilateral triangle. As described in Section II.B.3, the spectrum is strongly perturbed by vibration–rotation interactions and a large ℓ doubling.

In 1982, a conference abstract by Steinmetzger et al. [119] announced the observation of IR emission spectra resulting from the reaction of $H_2^+ + H_2 \rightarrow H_3^+ + H$. Possible sources of the given emission were vibrationally and rotationally excited H_3^+, neutral electronically excited H_3, and highly excited metastable H_2. These lines are *not* due to H_3^+ and are likely to be H_2 Rydberg lines [67] or H_3 Rydberg – Rydberg transitions [120].

In 1984 the $\nu_2 \leftarrow 0$ band of H_3^+ was remeasured by Watson et al. [121] who discovered 16 new lines. The IR laser–discharge technique was improved by two extensions, the use of discharge modulation (either velocity modulation [122] or amplitude modulation) and wider frequency coverage by the use of diode lasers. The assignment of the further lines was found to be problematic for high J transitions as K components were frequently closely spaced, particularly for high J values on the high-frequency side of the Q branch, where the $Q^+(J, K)$ patterns of different J values overlapped. The effective Hamiltonian used converged slowly, and extrapolations beyond $J = 5$ were thought to be uncertain. To overcome this difficulty Padé-type expressions were used in the rotational Hamiltonian and term values, and this provided a better model for all J levels. The year 1984 also saw a review by Gudeman and Saykally [122] on "velocity modulation IR spectroscopy of molecular ions" in which they mentioned the observation of 41 transitions in the $2\nu_1 - \nu_2$ band of H_3^+; these lines are *not* H_3^+ lines and are very likely to have been H_2 Rydberg lines (H_2 emission lines can be velocity modulated by momentum transfer from electrons to H_2 upon electronic excitation of H_2) [67].

In 1987 observation of the H_3^+ $\nu_2 \rightarrow 0$ spectrum in emission was achieved by use of a high pressure hollow-cathode discharge cell and a Bomem Fourier transform infrared (FTIR) spectrometer [123]. The discharge cell used had a small volume, a high current density, and could operate stably over a wide pressure range. The cell was a hollow cathode design with a 15-cm length, 5-mm diameter water-cooled copper cathode. The inner surface of the cathode was polished to enhance the collection of IR radiation for the emission measurements. The anode of the cell was a copper ring-shaped loop made of 3-mm copper tubing, and was sited at one end of the cathode. The cell was capable of dissipating up to 1 kW, and could be run with stable discharges over pressure ranges of 5–50 Torr, and currents of up to 2.5 A. A highly stable current-regulated power supply was used to power the discharge.

The Bomem spectrometer was used with a CaF_2 beam splitter and a liquid nitrogen cooled InSb detector. The radiation from the cell was collected with a CaF_2 lens, and focused into the entrance diaphragm of the spectrometer. Spectra were obtained with an apodized resolution of 0.05–0.1 cm^{-1} using data accumulation times of 1–2 h. The H_3^+ lines were distinguished from those of neutral species of means of recording the spectra at a number of different pressures. It was found that a sufficient test was to compare two spectra, one taken at 50 Torr, and one taken at

5–10 Torr, since the lines of H_3^+ are more prominent at the higher pressure.

The high temperature of the discharge used in this work enabled the measurement of lines up to $J'_{max} = 10$. Altogether 113 lines of the $\nu_2 \rightarrow 0$ band were assigned using a 29 parameter Hamiltonian that used a Padé formulation of the centrifugal distortion effects. It was found that the higher rotational levels of ν_2 were affected by Birss resonance [124] with levels of ν_1. An additional 35 lines were observed but could not be assigned unambiguously, and their transition frequencies were not reported. It was thought that the lines might belong to the hot bands $2\nu_2 - \nu_2$ and $(\nu_1 + \nu_2) - \nu_2$, but it was not certain that they were lines of H_3^+.

The FTIR spectroscopy of H_3^+ was also accomplished by Nakanaga et al. in 1990 [125]; they observed the $\nu_2 \leftarrow 0$ band in absorption. The ions were generated in a discharge of pure hydrogen in a water cooled hollow cathode cell. The discharge was sufficiently stable to allow data collection times of several hours. The cell used multipassing to achieve an optical path length of 10 m. No new transitions were reported.

The first overtone band of H_3^+, $2\nu_2 \leftarrow 0$, was analyzed almost simultaneously from measurements made in the laboratory [126] and in the atmosphere of Jupiter [97] (see below). The laboratory spectrum was recorded by Majewski and co-workers [126] in 1985 at the same time that the $\nu_2 \rightarrow 0$ emission spectrum was recorded (using FTIR measurements on a hollow cathode discharge), but at that time it could not be analyzed. The major difficulty in making assignments was the very high rotational temperature. The assignment was enabled by the observation of the $2\nu_2 - \nu_2$ band (see below) and by high quality *ab initio* predictions of the rotational levels of the $2\nu_2$ state by Miller and Tennyson [127] using the MBB [48] surface. This enabled the assignment of many of the lower J transitions of the $2\nu_2$ band. Subsequently [126], *ab initio* calculations were made for $J \leq 12$ and this enabled 49 lines to be assigned. In the fit two lines were included that had been observed in Jupiter's atmosphere. The assignments were monitored by fits to Watson's effective Hamiltonian [70]. The $2\nu_2$ state has two vibrational substates with $|\ell_2| = 0$ (A_1') and $|\ell_2| = 2$ (E'). Transitions from the $|\ell_2| = 2$ component are only allowed as a perpendicular band $E' - A_1'$ at low J, and at high J the character of the states becomes increasingly mixed. It was found that the effective Hamiltonian fits were unsatisfactory, and had little predictive power; a large number of centrifugal terms were required to reproduce the observed lines.

In 1990 Oka and co-workers [86] published their observations of three hot bands of H_3^+: 70 lines were observed in the $2\nu_2$ $(\ell = 2) \leftarrow \nu_2$ band; 14

lines were observed in the $2\nu_2$ $(\ell = 0) \leftarrow \nu_2$ band; 21 lines were observed in the $\nu_1 + \nu_2 \leftarrow \nu_1$ band and 136 lines were unassigned. The observations were made using a liquid nitrogen cooled 6 kHz ac discharge of mixtures of H_2 and He. Velocity modulation detection was used and the light source was a difference frequency laser system whose frequency coverage had been extended to 5300–1900 cm^{-1} by the use of both $LiNbO_3$ and $LiIO_3$ mixing crystals. The transition frequencies measured were in good agreement with predictions of band origins and rotational term values by Miller and Tennyson [128]. The same experiment [129] was also used for observation of 34 lines in the overtone band $2\nu_2$ $(\ell_2 = 2) \leftarrow 0$, which included 6 new lines.

The second overtone band of H_3^+ $(3\nu_2 \leftarrow 0)$ was obtained in a similar experiment [130] using a tunable diode laser operating at 1.45 μm (6897 cm^{-1}). Four transitions were measured and assigned using *ab initio* calculations on the MBB [48] pe surface. These transitions obey the selection rules $\Delta|K - \ell_2| = 3$, as do the $0 - 2\nu_2$ transitions which, combined with other measured transitions, allowed the determination of absolute rovibrational energy values for the ground state and the ν_2, $2\nu_2(\ell_2 = 2)$, $2\nu_2(\ell_2 = 0)$, and $3\nu_2(\ell_2 = 1)$ excited states.

In order to obtain information on absolute rovibrational energy values for ν_1 and $\nu_1 + \nu_2$ it was necessary to observe the $\nu_1 \leftarrow 0$ and $\nu_1 + \nu_2 \leftarrow 0$ bands that involve excitation of the totally symmetric stretch of H_3^+, and are therefore "forbidden". To this end a search was conducted by Xu et al. [99] using a difference frequency–discharge spectrometer. The $\nu_1 \leftarrow 0$ transition cannot be driven by vibrational excitation alone as it has symmetry $A_1' \leftarrow A_1'$, but intensity can be produced by the interaction between vibration and rotation. A consideration of the selection rules given above shows that for this band the selection rule will be $\Delta K = \pm 3$. The symmetry of the $\nu_1 + \nu_2 \leftarrow \nu_2$ transition is $E' \leftarrow E'$ and is therefore allowed; the intensity of the transitions is enhanced by Fermi resonance [99] (see, e.g. [63]). The selection rules are $\Delta K = \pm 1$, $\Delta \ell_2 = \mp 2$. The intensity of such transitions depends on centrifugal distortion of the ground vibrational state (which breaks the C_3 symmetry and produces a small dipole moment in the plane of the molecule) and mechanical and electrical anharmonicities in the degenerate vibrational state. The general theory of such transitions has been considered by Watson [131] and Mizushima and Venkateswarlu [132]. This theory has been used to calculate expected intensities of these bands in H_3^+ [89, 103, 104]. In general the $\nu_1 + \nu_2 \leftarrow \nu_2$ band gains intensity through Fermi resonance while the $\nu_1 \leftarrow 0$ band borrows intensity through the Birss resonance discussed above between the ν_1 and ν_2 states.

Xu et al. [99] observed 9 transitions of the $\nu_1 \leftarrow 0$ band, together with

21 transitions in the $\nu_1 + \nu_2 \leftarrow \nu_2$ band; experimental transition frequencies were in excellent agreement with those predicted by Miller et al. [98]. In the course of the work a further 30 new transitions in the $\nu_2 \leftarrow 0$ band and 13 new transitions in the $2\nu_2$ $(\ell_2 = 2) \leftarrow \nu_2$ band were observed, together with 93 unassigned transitions that were probably due to H_3^+.

B. The Spectroscopy of D_3^+

The first vibration–rotation spectrum of any molecular ion to be obtained was that of HD^+, by Wing et al. [134] in 1976. In this pioneering study, and for his work on the isotopomers of H_3^+, an ion-beam–laser spectrometer was used, which will now be described. Ions are formed in a source that is held at a potential of 1–10 kV, and extracted to earth potential, resulting in an ion beam. All ions formed in the source will be extracted provided they survive for more than a microsecond or so. The ion beam is interacted with the beam from a carbon monoxide (CO) laser, which is introduced at a near-collinear geometry. The CO laser is line tunable, over the region accessible for CO (1550–1920 cm^{-1}). Continuous frequency tuning is obtained using the Doppler effect; the speed of the ions (V) is varied by changing the beam potential through which they are accelerated, and therefore the frequency that the ions experience is changed according to the relativistic Doppler shift formula,

$$\nu_{\text{ion}} = \nu_{\text{laser}} \sqrt{\frac{1 \mp \frac{V}{c}}{1 \pm \frac{V}{c}}}$$

where the upper and lower signs denote co- and counterpropagation of the laser beam with respect to the ion beam. By using all available lines, and co- and counterpropagating ion and laser beams, a large proportion of the CO laser region can be covered for a light ion.

At resonance, when the frequency of radiation matches a transition frequency, the laser causes a population transfer between the two states involved. In general different states have different cross sections for charge exchange and this is exploited in the detection technique. After interaction with the laser beam the ion beam collides with a target gas, the parent ion beam of interest is then mass selected, and its beam current is monitored using a Faraday cup. When the ion beam velocity brings the ion into resonance with the Doppler shifted frequency of the laser light a modulation occurs in the surviving ion beam current.

Detection is enhanced by chopping the laser beam at 1 kHz and using lock-in amplification.

The use of a mass spectrometer (MS) in these studies offers certain unique advantages, which are also exploited by Carrington and co-workers [20–22] (see Section V).

1. The ratio of charge to mass of the ion of choice is known unambiguously, which greatly facilitates isotope studies.
2. Ions are formed hot in an essentially collision-free environment, enabling excited states of the ions to be probed.
3. The ion source conditions can easily be varied to obtain the most intense beam of ions; when a transition is obtained source conditions can be varied to maximise its intensity.
4. The high velocity of the ions results in "kinematic compression" [135] of the Doppler width.

From the *ab initio* prediction of the vibration–rotation Hamiltonian matrix for H_3^+ of Carney and Porter [43, 45], Wing and co-workers [18] calculated that the ν_2 vibrational band system of D_3^+ lay in the same spectral region as the emission lines of the CO laser. The D_3^+ ion was therefore an ideal candidate for detection using this experiment. The disadvantage of using a beam of D_3^+ rather than a beam of H_3^+ is that for a given rotational temperature the same number of ions are spread over many more rotational levels, thereby giving a lower population in any one level. In the experiments of Carrington and Kennedy [21] (Section V) this led to spectra of D_3^+ that were markedly weaker than those of H_3^+.

In total Wing and co-workers [18] initially observed eight transitions between vibration–rotation levels of the ground electronic state of D_3^+, having searched about 4% of their available frequency coverage. Of these eight transitions, four could be unambigously assigned to the fundamental degenerate vibrational band using the constants of Carney and Porter [19]. A subsequent adjustment to the vibrational frequencies and rotational energies by Porter led to a fit of 0.03-cm^{-1} maximum error. The remaining four transitions were attributed to transitions between high vibrational levels, but could not be assigned at the time. Shy [136] later extended these measurements to 37 more lines.

The $\nu_2 \leftarrow 0$ fundamental band of D_3^+ was also observed by Watson et al. [137] using a discharge–IR diode laser experiment. Some 60 lines were measured, and the data set was combined with 45 lines previously measured by Shy [136]; five lines were measured using both techniques. A total of 84 lines were fitted, starting from the original assignments and matrix elements of Carney and Porter, and then with an effective

Hamiltonian for the ν_2 state [70] using the Padé formulation of Polyansky [138] (the same Hamiltonian used for the analysis of the H_2D^+ and D_2H^+ described below). Some 11 lines were unassigned, and were attributed to hotter vibrational or rotational states.

C. Spectroscopy of H_2D^+ and D_2H^+

Using Wing's ion beam laser experiment Shy et al. [139] found the first IR spectrum of H_2D^+ in 1981. An assignment of the eight observed lines was not possible, but from the calculations of Carney and Porter [59] it was thought that the lines originated from high J P-branch transitions in the $\nu_3 \leftarrow 0$ and $\nu_2 \leftarrow 0$ bands.

In 1984 Amano and Watson [140] used the IR laser–discharge technique to locate 27 lines of the $\nu_1 \leftarrow 0$ band of H_2D^+. They modulated the discharge at about 17 kHz in order to modulate the concentration of the molecular ions, and detected signals using a lock-in amplifier. The search was begun using transition frequencies derived from the constants of Carney and Porter. Observed and calculated frequencies were typically in agreement within 0.01 cm^{-1}. Only $\Delta K_a = 0$ transitions were observed. Measurement of this spectrum was subsequently extended by Amano [141], who found an additional 10 lines of the band using a hollow cathode discharge and a tunable laser system. Two $\Delta K_a = \pm 2$ transitions were measured.

In an identical experiment Lubic and Amano [142] observed 35 transitions in the $\nu_1 \leftarrow 0$ band of D_2H^+. As for H_2D^+ the search was based on estimates of transition frequencies from Carney and Porters *ab initio* constants.

Given the recently detected vibrational spectrum of H_2D^+ it became obvious that it should be possible to observe pure rotational spectra of these ions and the rotational transition $1_{10} \leftarrow 1_{11}$ in H_2D^+ was detected nearly simultaneously by Bogey et al. [143] and Warner et al. [144] using near-identical experiments. Ions were generated within a liquid nitrogen cooled negative glow discharge lengthened by an axial magnetic field (300 G) of the type developed by De Lucia et al. [145]. The spectrometer consisted of a microwave source, an absorption cell, and a microwave detector. In order to avoid discharge instabilities, a large excess of Ar was used (8 mTorr), with H_2 and D_2 added in a 2:1 ratio (0.5–3 mTorr in total). The use of an axial magnetic field made the technique ion–sensitive; the behavior of a particular transition as a function of the magnetic field was used to discriminate against spurious signals from neutral species.

Both groups measured the $1_{10} \leftarrow 1_{11}$ transition frequency at 372, 421.3 MHz (12.42 cm^{-1}), lying between the frequency predicted by

Amano from his analysis of the ν_1 band of H_2D^+ (372,383 MHz) and that of Porter, which was calculated *ab initio* (372,681 MHz).

In 1985 a second microwave transition of H_2D^+ was discovered by Hirota and co-workers [146], who determined the transition frequency of $2_{20} \leftarrow 2_{21}$ to be 155,987.185 MHz. The experiment used microwave absorption of a liquid nitrogen cooled discharge, which was of a different design to that described above. The ions were generated in a glass discharge cell of 10 cm diameter and 1.1 m length, whose ends were vacuum sealed using Teflon Lenses, which also enabled collimation and focusing of the microwave beam. The discharge was maintained in a hollow cathode made of stainless steel, of 60 cm length, which could be cooled to liquid nitrogen temperature. The signal was observed using a discharge current of 200 mA, using partial pressures of 26 and 34 mTorr for H_2 and D_2, and an integration time of 320 s. Application of a magnetic field of 50 G decreased the line intensity by a factor of 5, in agreement with other observations of Hirota that such a field decreases the ion production efficiency of a hollow cathode. Attempts to improve the intensity of the signal by applying a field to the negative glow region outside the cathode, after De Lucia et al. [145], were unsuccessful as it caused the discharge to become unstable.

In 1986, Foster et al., observed and analyzed the $\nu_2 \leftarrow 0$ and $\nu_3 \leftarrow 0$ bands of H_2D^+ [148] and D_2H^+ [149]. Twelve lines were measured but not assigned: these are are listed in Table II of [110]. The spectra were measured using discharge–IR laser techniques. Some 66 lines of the H_2D^+ spectrum were observed. These lines were fitted using an effective Hamiltonian employing a Padé representation of a conventional *A*-reduced centrifugal Hamiltonian for each vibrational level, together with Coriolis and higher rotational interactions between ν_2 and ν_3. In addition a "supermatrix" model was employed in which the matrix of the untransformed Hamiltonian was set up in a large vibration–rotation basis and diagonalized directly. In this model most of the matrix elements that were off-diagonal in vibration were constrained to values derived from Carney's *ab initio* calculations, while the ν_1 parameters were fitted to lines observed by Amano (described above). As a result of these analyses it was also possible to assign seven of the nine lines reported by Shy et al. [139]. For D_2H^+ 72 new lines were observed, together with a remeasurement of 5 that had previously been measured by Shy [136]. The analysis was made in a similar manner to that for the H_2D^+ data. The analysis made it possible to assign 11 other lines observed by Shy [136], giving a total of 88 assignments. The data were subsequently reanalyzed by Polyansky and McKellar [110] using a simultaneous least-squares analysis with the ν_1 band data and rotational transitions. Two of the rotational transitions included were the $2_{20} \leftarrow 2_{11}$ and $1_{11} \leftarrow 0_{00}$ transitions measured by Evenson and co-workers [150] at 1,370,051.6 MHz and 1,476,605.5 MHz, but to my knowledge never published separately. They were able to achieve a standard deviation for the ν_2/ν_3 data of $0.0046\ \mathrm{cm}^{-1}$, using fewer parameters. It was found necessary to reassign five transitions.

D. Discussion

Since the first measurements of the H_3^+ spectrum by Oka in 1980, H_3^+ spectroscopy has entered a renaissance. The interplay between experiment and theory has been a characteristic of all aspects of the spectroscopy. Higher and higher overtone and combination bands of H_3^+ and its isotopes have been measured, often as a result of predictions of intensity and frequency made from first principle calculations using highly accurate pe and dipole surfaces [151]; this will surely continue. The experimental spectroscopy of the "low"-energy regime of H_3^+ is rapidly approaching the barrier to linearity, which is believed to drive the chaotic behavior of the molecule observed near dissociation by Carrington and co-workers [20–22]. Theoretical calculations of states close to the barrier show that it is no longer easy to assign particular energy levels to overtones or combinations of the normal modes, which can be calculated near the equilibrium configuration [152].

The theoretical calculation of the energy levels of states that are close to or above the barrier to linearity requires the use of vibration–rotation Hamiltonians framed in a coordinate system not dependent on a Taylor series expansion about the normal-mode coordinates, that is, a departure from the Eckart separation of the motion of the center of mass of the molecule and its internal nuclear motions. The Watson Hamiltonian is no longer valid in these regions because the Jacobian of the Eckart transformation is zero when the molecule accesses linear geometries. For this reason the Watson Hamiltonian has gradually fallen out of favor for variational calculations of energy levels of H_3^+.

Sutcliffe and Tennyson (reviewed in [85]) formulated general methods of generating vibration–rotation Hamiltonians corresponding to any particular choice of internal coordinates, but such methods are difficult to apply. Nevertheless, using one such Hamiltonian Tennyson and co-workers [94] calculated energy levels of H_3^+ up to dissociation at zero angular momentum, finding 848 bound states (including the forbidden A_1 states).

IV. ASTROPHYSICAL MEASUREMENTS OF H_3^+ SPECTRA

The H_3^+ ion and its isotopomers play a crucial role in our understanding of the chemistry of the ISM, and the detection of H_3^+ within the ISM has therefore been desirable since H_3^+ chemistry was first proposed by Herbst and Klemperer [153] and Watson [154]. The whole field of molecular astrophysics has been reviewed by Herzberg [155]. Searches [156, 157] for H_3^+ in the ISM have succeeded only in establishing upper limits for the column density, for which theoretical oscillator strengths are required [158]. Due to the astrophysical importance of H_3^+ there have been many calculations and compilations of theoretical line positions and intensities (see Section III for references), as well as partition functions and equilibrium constants [159, 160] relevant to the reactions of formation and the deuterium fractionation discussed below.

There are now tentative observations of H_2D^+ in the ISM [161] and of H_3^+ in the dark cloud L1457 [158]. The H_3^+ ion has been observed in other astrophysical environments as well. The first such observation was of H_3^+ in the Jovian aurora [97]. The H_3^+ ion has subsequently been observed in other planetary auroras; the mechanisms leading to such intense emission from H_3^+ are outside the scope of this chapter. The second astrophysical identification of H_3^+ (plausible, but unproven) is of spectral lines recorded in the expanding envelope of Supernova 1987a [162].

A. Astrophysical Importance of H_3^+

Radioastronomers observed many polyatomic molecules in the interstellar medium in dense molecular clouds [163]. In 1973 it was recognized that ion–molecule reactions were an important mechanism by which many such molecules could be formed [153, 154]. Ion–neutral reactions occur much more rapidly than neutral–neutral reactions in the ISM (typically by 3–5 orders of magnitude) for two principal reasons: first, an ion can polarize a neutral species causing a charge/induced-dipole attraction between them, and second, because reactions between ions and neutral species often do not have an activation energy; very little energy is available within the ISM due to its low temperatures. The H_3^+ ion was recognized as one of the most important molecular ions in the chemistry of the ISM because of its ability to protonate neutrals by the reaction $H_3^+ + X \rightarrow HX^+ + H_2$. For example, the ions HCO^+ and HN_2^+, which are observed by radioastronomers are believed to be produced by this reaction from neutral CO and N_2.

A further important phenomenon in the ISM involving H_3^+ is that of deuterium fractionation; it is well known that in exchange reactions different isotopes may participate at significantly different rates. The ratio of H/D was determined in the hot-big-bang at about 10^5, but two exothermic isotopic substitution reactions lead to the concentration of D in H_2D^+ [158]:

$$H_3^+ + HD \rightarrow H_2D^+ + H_2 + 232\,K$$

$$H_3^+ + D \rightarrow H_2D^+ + H + 509\,K$$

The H_2D^+ ion can introduce either H^+ or D^+ into the ion–molecule reaction sequence. The first reaction is exothermic for two reasons: the zero-point energy of H_2D^+ lies below that of H_3^+ by more than that of HD below H_2 and the ground state of H_3^+ is $J = 1$ (see above). As H_2D^+ does not consist of three identical particles, $J = 0$ is allowed, so formation of H_2D^+ from H_3^+ and HD can release the rotational energy of H_3^+. Cold interstellar clouds may well have temperatures of about 10 K, in which case the ratio of H_2D^+ to H_3^+ will become close to 1:10. In the case of cold dark dust clouds ($T \approx 5 - 10$ K) the apparent ratio of DCO^+/$HCO^+ > 1$ [164].

B. The H_3^+ Ion in the ISM

The H_3^+ ion does not have an electrical dipole allowed pure rotation spectrum, and therefore attempts at astrophysical detection have concentrated on observing vibrational (IR) transitions in H_3^+ [165]. It should also be possible to locate the spectrum of H_3^+ by the indirect mechanism of observing the visible emission spectrum of H_3 caused by electron, H_3^+ recombination. A tentative identification of the H_3 emission spectrum was made in 1984 [166] but was later withdrawn [167]. Dalgarno et al. [168] pointed out that H_2D^+ had a considerable dipole moment, and hence some rotational (microwave) transitions that would be accessible to astronomers. The ion–molecule reaction mentioned above allows for a strong deuteration effect only when kinetic temperatures in a molecular cloud are below about 30 K, therefore in searching for H_2D^+ cool dense clouds are favored. The laboratory observations [143, 144] of the $1_{10} - 1_{11}$ rotational transition in H_2D^+ located the frequency very accurately, and stimulated astrophysical searches. For the astrophysical detection to be possible, unless there is a suitable light source behind the cloud, there must be a sufficient temperature for the 1_{10} energy level to be populated if the $1_{10} - 1_{11}$ transition is to be seen in emission.

The deuterium fractionation reaction is known to lead to a strong enhancement of H_2D^+, and therefore also of DCO^+. Measurements of DCO^+ have been made [169, 170] and provide information on the likely degree of deuteration of H_3^+ as a function of position within molecular clouds, allowing the prediction of regions of the clouds that would be likely candidates for astronomical searches for H_2D^+.

1. *Observation of H_2D^+?*

As a result of the above considerations, several searches were made for the spectra of H_2D^+ in the ISM. Early searches were performed before the laboratory measurements of the transition frequency, and were unsuccessful [171, 172]. After the laboratory frequency of the $1_{11} \rightarrow 1_{10}$ transition became available, Phillips et al. [173] mounted a search for the transition in the molecular cloud core regions of NGC 2264 and TMC-1, which they selected on the basis of the DCO^+ observations of Wootten et al. [169] and Guelin et al. [170]. A line was detected at the expected wavelength in emission in NGC 2264, but not in TMC 1. The velocity shift and width of the line measured were consistent with those for DCO^+, and the strength was consistent with a partial thermalization of H_2D^+ between its ortho and para states. It is not possible to exclude the observation as being due to a coincidence with a transition frequency of some other molecule, but no other is known at the same frequency. Searches for the $1_{11} - 1_{10}$ transition of H_2D^+ were subsequently made by Pagani et al. [174] and van Dishoeck et al. [175]. The objects searched for the transition included NGC 2264, but no confirmation of the initial detection could be made, despite an increase in sensitivity of up to two orders of magnitude [175] over the first observation.

In 1993 Boreiko and Betz [161] reported the observation of a transition at 1370 GHz (the frequency of the $1_{10} - 0_{00}$ transition in H_2D^+) in the IRc2 region of Orion (M42). The transition was observed in absorption on two occasions separated by almost 3 months. Calculations using chemical models showed that if the line were a transition of H_2D^+ that the spectrum of H_3^+ should be directly observable in absorption towards IRc2. Such an observation would be very exciting, and probably the best way of confirming the observed transition as being due to H_2D^+.

2. *Observation of H_3^+?*

After the observation of the H_3^+ emission spectrum in Jupiter (see below), it was suggested [5] that it might be possible to detect the IR spectrum of H_3^+ in emission in regions of the ISM that have been heated to temperatures similar to that of Jupiter's aurora. Such regions might be created by shock waves passing through the ISM, for example, in regions where young stars are forming, or in regions affected by supernova explosions. Tennyson and co-workers [158] conducted two such searches, concentrating on the emission lines in the Q and R branches of H_3^+ in the K and L windows. Objects searched included the proto-planetary nebula CRL 618, parts of the supernova remnant IC 443, and the dark cloud L1457, which is thought to contain a hard X-ray source [176].

The first search (February 1992) shows a tantalizing observation of an emission feature (at a 2σ level) in the dark cloud L1457. The feature could be attributed to a slightly red shifted H_3^+ $\nu_2 \rightarrow 0$ $R(3)$ transition at a wavelength of 3.534 μm. Poor weather prevented the confirmation of this observation.

C. Measurements of H_3^+ in Planetary Atmospheres

The first confirmed observation of H_3^+ in a nonterrestrial environment was that of Drossart et al. [97] with the serendipitous discovery of the H_3^+ spectrum emitted from Jupiter's aurora.

1. *The Spectrum of the Jovian Aurora*

The H_3^+ ion was detected *in situ* in the magnetosphere of Jupiter in 1980 [177] by Voyager 2. In 1989, a high-resolution spectrum at a wavelength of 2 μm (5000 cm^{-1}, within the "K-window") was taken of the southern auroral region of Jupiter, recorded at the Canada–France–Hawaii telescope. The aim was to examine the polar regions for evidence of an aurora, by monitoring quadrupole transitions of H_2 [5] using an FT spectrometer. The quadrupole transitions were detected, but many other lines of comparable intensity were found [178]. It was proposed that these lines were due to H_3^+, and Watson recognized the spectrum as due to the first overtone transition of H_3^+ $[2\nu_2(\ell_2 = 2) \leftarrow 0]$. This band had recently been observed in the laboratory [86]. *Ab initio* calculations by Tennyson and co-workers [97] were used to assign the high J transitions. Calculations and spectra were matched with 23 transitions fitted to an excitation temperature of 1099($\pm$100) K, together with a column density of ions in the $2\nu_2(\ell_2 = 0)$ state of $1.39(\pm 0.13) \times 10^9$ cm^{-2} and a ratio of ortho/para H_3^+ of 0.58 ($\pm$0.03). The determination of the ratio between ortho and para H_3^+ is of interest because detailed symmetry considerations show [179] that para H_2 can only form para H_3^+, but ortho H_2 forms $\frac{2}{3}$ ortho H_3^+ and $\frac{1}{3}$ para H_3^+. The equilibrium ortho/para ratio of H_2 is a sensitive function of temperature [180], para H_2 dominating at low temperatures. The temperature of a molecular cloud should therefore be reflected in the ortho-para ratio of H_3^+, if no equilibration of H_3^+ takes place after its formation.

The observation of the $2\nu_2 \rightarrow 0$ overtone transition of H_3^+ led to a natural expectation that the fundamental band $\nu_2 \rightarrow 0$ should be observable in the same environment, and a search for this transition was initiated by Oka and Geballe [181] using the United Kingdom infrared telescope (UKIRT) with a grating spectrometer (CGS2). The band was observed on September, 14–19, 1989. Ten emission lines were seen of the

Q branch of the fundamental in the 4-μm (2500 cm^{-1}) region. Oka and Geballe [18] also attempted to observe the previously detected 2μm band on September 9–10, but were unsuccessful. This is clear evidence of a strong temporal variation of the Jovian plasma. The fundamental band was subsequently observed in its entirety in the Jovian aurora using an FTIR spectrometer [182] against a virtually zero background; this beautiful spectrum is shown in Fig. 3.1(*b*).

The 4-μm band of H_3^+ was observed at much higher intensity than the 2-μm band, which suggested that the emission is due to H_3^+ in thermal equilibrium at temperatures of 1100 [97] to 670 K [181], excited into high vibrational states by collision with H_2 and He. This was also suggested by subsequent observation of both the 2- and 4-μm emissions [183]. However, more recent measurements show that H_3^+ temperatures and densities are subject to considerable temporal and spatial variations [5], which are not suggestive of equilibrium.

The emission spectrum of H_3^+ in the Jovian auroras has subsequently been employed as a sensitive probe of the auroras themselves. This spectrum has been used to image the auroras [184] and to study their morphology by using IR cameras [184, 185]. The images show that the intensity of the auroral emission can vary on a time scale of 1 h, and have a different spatial distribution to the auroral activity inferred from observations in the ultraviolet (UV) and longer wavelength IR (which may therefore be controlled by different processes).

2. *The Spectrum of the Uranian Aurora*

The H_3^+ fundamental spectrum was located in Uranus on April 1, 1992 [5] by Trafton et al., [186], who used the UKIRT telescope with CGS4 spectrometer. They found 11 emission features, which were assigned to lines from the Q branch of the $\nu_2 \rightarrow 0$ band. An analysis indicated a rotational temperature of 740 ± 25 K, a disk-averaged H_3^+ column abundance of 6.5×10^{10} molecules cm^{-2} ($\pm 10\%$), and an ortho/para ratio of 0.51 ± 0.03. As a result of the temperature measurements, it was found that H_3^+ was a major cooling agent in the atmosphere and estimates of the amount of energy deposition onto Uranus had to be revised upwards by a factor of 6 to allow for the energy being reradiated.

3. *The Spectrum of the Saturnian Aurora*

Geballe et al. [187] detected the H_3^+ $\nu_2 \rightarrow 0$ fundamental band in emission in Saturn's polar regions. The emission is weaker that that of Jupiter's polar regions by two orders of magnitude. The observations were made on UKIRT on July 18–19, 1992 using a grating spectrometer (CGS4). Three vibration–rotation transitions were observed and assigned. Com-

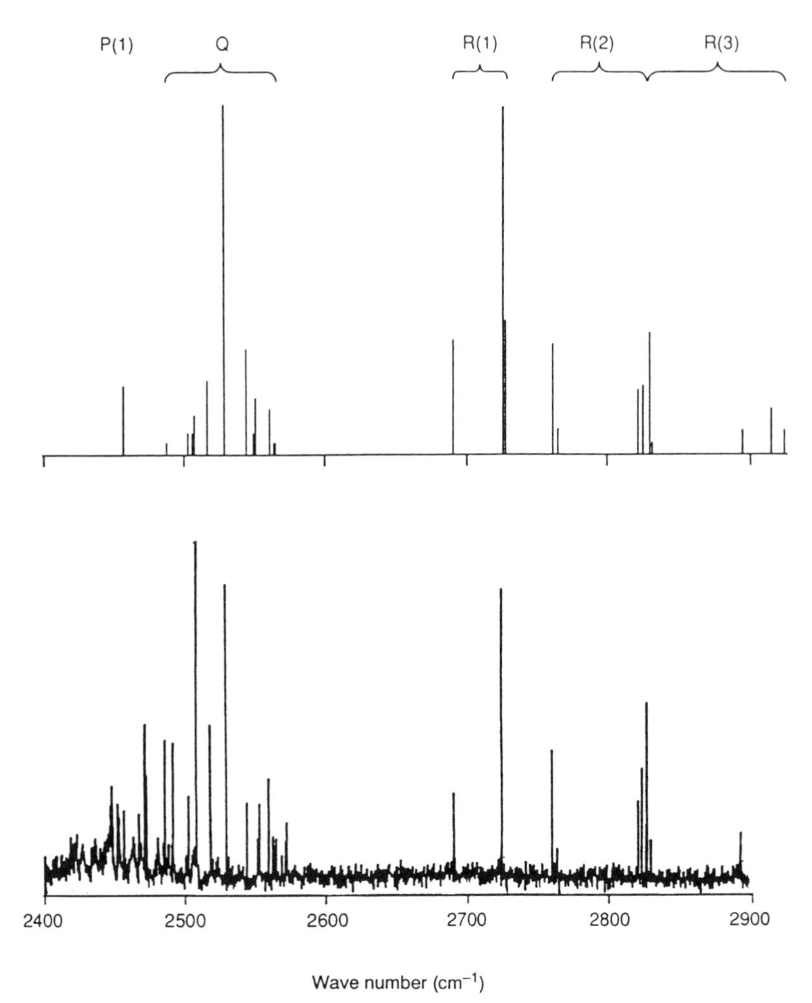

Figure 3.1. (*a*) Shows a stick spectrum showing the complete set of lines found by Oka [17] in 1980, together with the assignments of Watson. It took $4\frac{1}{2}$ years to locate the spectrum, and 54 days of scanning to locate all the lines shown [67, 118]. (*b*) Shows the beautiful FTIR spectrum of the same band recorded from emission of Jupiter's atmosphere [182], the spectrum was recorded using a total integration time of 72 min! (Reprinted with permission from the Astrophysical Journal, 1990.) The difference in line intensities between the two spectra is due to the differing temperatures of the Jovian aurora and a discharge tube.

parison with the data from Uranus indicated that the emission from Saturn is a few times weaker, and this is also consistent with the finding of Trafton et al. [186] that the H_3^+ line intensity from Uranus is a few percent of that from Jupiter.

D. Measurements of H_3^+ in Supernova 1987a?

The supernova explosion detected on February 23, 1987 (designated SN 1987a) in the Large Magellenic Cloud was intensely studied. One aspect of the study was the recording of IR spectra in the L-window region. It was found that the spectra between 2.95–4.15 μm were dominated by three hydrogen recombination lines [188]. However, after day 110 other strong previously unidentified features became apparent at 3.41 and 3.53 μm: these reached a peak of intensity at day 192. These are the wavelengths at which H_3^+ was detected in the Jovian aurora [181, 182].

A $\nu_2 \rightarrow 0$ H_3^+ spectrum was computed using observed transition wavelengths and calculated transition probabilities, with an ortho/para ratio of 1, and fit to the observed data from day 192 between 3.53 and 3.6 μm [162] (3390–2410 cm^{-1}). In the fit the mass of emitting H_3^+ (M_e), the velocity width at one-half maximum (V_w), and the temperature (T), were adjusted. The fit gave the values $M_e = 5.1 \times 10^{-8}$ solar masses, $T = 2050$ K, and $V_w = 3200$ km s^{-1}.

The high velocity of the expanding shell causes considerable Doppler broadening of the observed spectra, and rotational fine structure is therefore lost. However, the agreement between the calculated and observed spectrum is impressive. As a check of the fit the H_3^+ spectrum was calculated throughout the full region of the L window, using the fitted values. It was found that at wavelengths longer than 3.318 μm the observed spectrum was consistent with the fit, but that at shorter wavelengths two predicted features (at 3.2 and 3.02 μm) were absent. However, these wavelengths correspond to regions of water absorption in the atmosphere, which may account for the discrepancy.

The possibility that other emitters might also contribute to the spectra was allowed for by reducing the temperature of a second fit to 1000 K. In this case the calculated discrepancy between the computed and observed spectra around 3.02 μm was removed, and the intensity of the predicted feature at 3.2 μm was greatly reduced.

Chemical modeling of the processes expected in the supernova yielded abundances of H_3^+ at day 200, which were in reasonable agreement with those deduced from the spectrum at day 192. The observation of H_3^+ placed an upper limit on the temperature of the envelope of 3000 K (above which molecules are rapidly destroyed), and showed that mixing of the two expanding shells (the outer of light elements, the inner of heavier elements) was negligible; the presence of heavy elements would have prevented the formation of H_3^+ by destroying H_2 in endothermic reactions.

The identification of H_3^+ in supernova 1987a is not conclusive but is strongly supported by circumstantial evidence: No known atomic spectrum can explain the observed features, plausible chemical models can account for the fitted mass, and the identification of the CO overtone bands [189] in the spectrum shows that molecules are present. Oka [8] points out that the rotational temperature deduced from the fit (1000–2050 K) badly mismatches the high kinetic temperature expected from the expansion velocity of 3200 km s^{-1}. He suggests that it may be possible to reconcile the results if IR and far-IR emissions through "forbidden" rotational transitions are fast enough to lower the rotational temperature significantly. Further observations of H_3^+ are no longer possible in the remnants of Supernova 1987a and confirmation must await further observations in future supernovas.

E. Discussion

The observation of H_3^+ in extraterrestrial environments is becoming commonplace! Calculations indicate that current instruments should be on the edge of the required sensitivity to detect H_3^+ in the ISM, as do the tantalizing observations of Tennyson et al. [158] and Boreiko and Betz [161]. Where it has been observed in the atmospheres of other planets H_3^+ is becoming an invaluable source of information of the conditions in their ionosphere, able to probe both temporal and spatial variations in energy output (and hence input), which also place important constraints on models of the atmospheres. The use of H_3^+ as an astronomical probe has recently been discussed by Oka and Jagod [71].

V. THE H_3^+ PREDISSOCIATION SPECTRUM

In 1982 Carrington et al. [20] discovered a remarkable IR spectrum of H_3^+ involving energy levels within 1100 cm^{-1} of the first dissociation limit. The spectrum was incredibly dense and displayed no regular features. At the time of its discovery there had been no theoretical prediction that such a spectrum might exist and there was no theory available that could account for its properties. The spectrum was examined in detail by Carrington and Kennedy [21] and their observations were published in 1984. The study covered the region from 874 to 1094 cm^{-1} using a continuous wave (CW) carbon dioxide (CO_2) laser. Approximately 27,000 absorption lines were recorded; each line was detected by measuring an increase in the number of fragment protons produced on resonance. Subsequent theoretical investigations by several different groups, described in Section V.C, provided some understanding of the molecular physics involved, and led to further predictions about the spectrum. Some of these predictions were tested by Carrington et al. [22] in 1993 and the results of these tests, together with other new findings, are described in Section V.D.

Carrington and Kennedy [21] established a number of important features of the H_3^+ predissociation spectrum, which acceptable theoretical models must accommodate. A full appreciation of some of these features requires an understanding of the technical aspects of the experiment that is now described.

A. Experimental Details

The experiment of Carrington and co-workers [21] is similar to that of Wing and co-workers [134] (described in Section III.B), in that it is an ion beam–laser experiment, and enjoys the same advantages over "traditional" spectroscopic arrangements. The ion beam machine is a modified tandem MS and its main features are illustrated in Fig. 5.1. The experiment is designed to obtain IR spectra of molecular ions in levels close to their dissociation limits. The specificity of the experiment for levels close to dissociation arises because transitions between levels are detected by monitoring fragmentation of the parent ion beam, thus any process that enables a change in the number of fragments to be observed when a transition is in resonance, will enable detection of the transition for example,

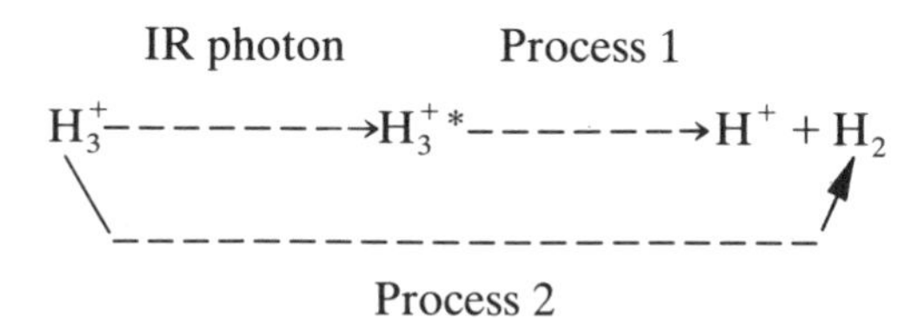

Provided the rates for process 1 and 2 are different, a transition due to the absorption of an IR photon may be observed as a change in the number of fragment ions produced. Spectroscopic transitions are driven by IR radiation from a line tunable CW carbon dioxide laser operated with $^{12}CO_2$ or $^{13}CO_2$, which gives coverage from 874 to 1094 cm^{-1}.

The use of fragmentation as a detection mechanism gives a specificity for the uppermost levels supported by a potential surface. Single photon transitions will only be observed when occurring within one IR photon of the lowest dissociation limit. This specificity allows unique experimental information to be determined; most spectroscopic experiments probe

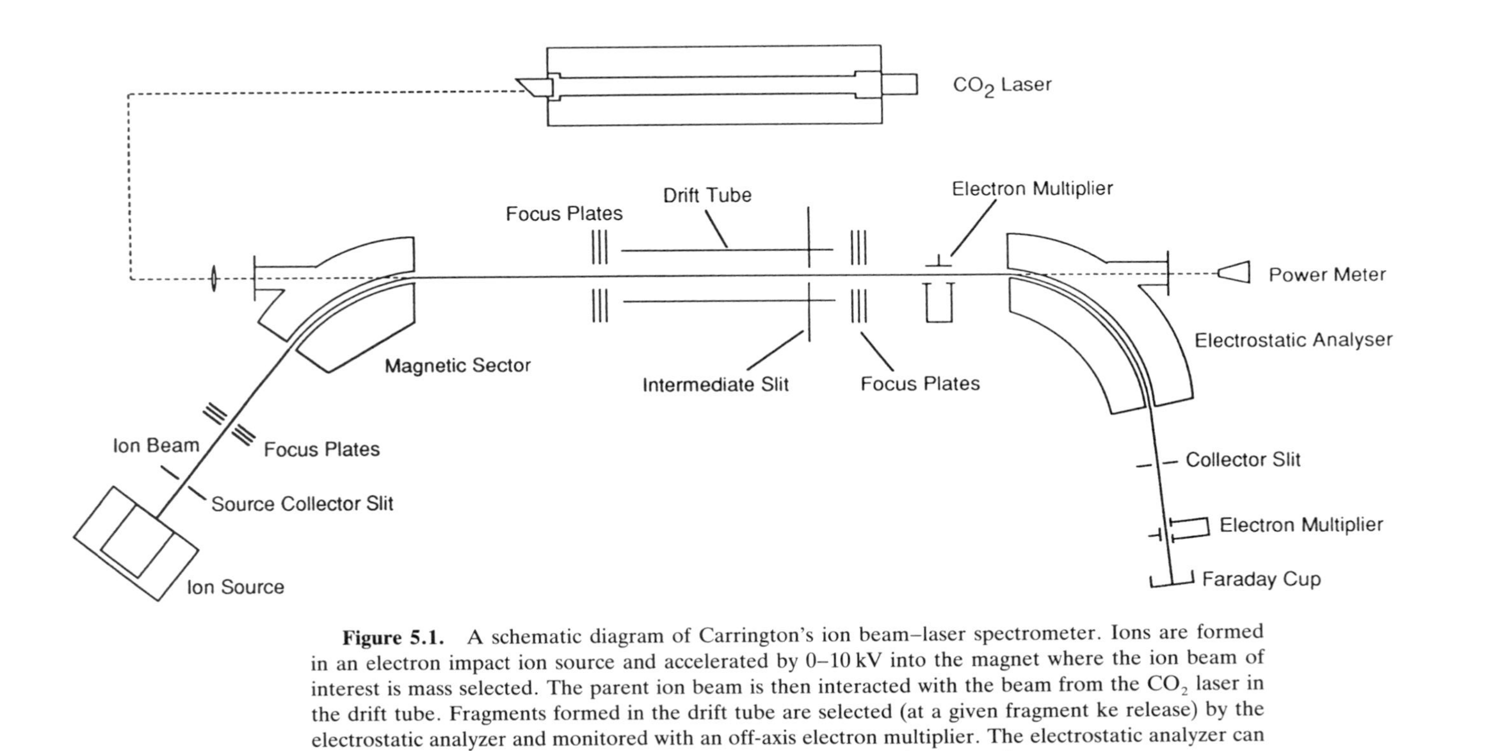

Figure 5.1. A schematic diagram of Carrington's ion beam–laser spectrometer. Ions are formed in an electron impact ion source and accelerated by 0–10 kV into the magnet where the ion beam of interest is mass selected. The parent ion beam is then interacted with the beam from the CO_2 laser in the drift tube. Fragments formed in the drift tube are selected (at a given fragment ke release) by the electrostatic analyzer and monitored with an off-axis electron multiplier. The electrostatic analyzer can also be used to record ke spectra of resonant or nonresonant fragmentation.

energy levels close to the bottom of the potential energy well but this experiment necessarily probes those levels within 1094 cm^{-1} of the top of the potential well. That these states should be populated may seem surprising, but the process of ionization of a neutral species and formation of an ion by chemical reactions in the ion source both lead to highly vibrationally and/or rotationally excited ions.

Carrington et al. [190] used many types of fragmentation process to enable this spectroscopy with different molecules, that is, tunneling through a rotational barrier, subsequent excitation to a repulsive electronic state [191], and electric field dissociation [192]. The fragmentation is monitored against background fragmentation that results from processes such as collision induced dissociation with residual gas in the analyzer. The transfer in population at resonance results in an increase or decrease in fragmentation depending on the initial populations of the states involved. The interaction of the ion and laser beams occurs in a 40-cm length metal tube, which is called the drift tube. The use of a bias voltage ensures that only fragments created in the drift tube are monitored.

Fragment ions formed in the drift tube by a predissociation process possess a centre-of-mass ke, whose magnitude depends on the energy of the predissociating level with respect to the energy of the dissociation channel. The fragment ions are detected using an electrostatic analyzer (ESA). The ESA is a ke analyzer and so the ke of the fragments themselves can be measured. While monitoring fragments the ESA can be used in two different ways.

1. As a window that transmits only fragment ions whose ke lies within a predetermined range; this is achieved by setting the ratio of the ESA voltage to the ion source voltage.
2. The ESA voltage may be scanned and the fragment ion intensity recorded; this procedure yields a fragment ion scattering pattern, recorded in the laboratory coordinate system. A simple transformation yields the center-of-mass ke release.

Detection of resonant transitions is enhanced by the use of lock-in amplification. One of two methods is used: (1) amplitude modulation of the laser beam that yields normal absorption line shapes, or (2) frequency modulation of the laser beam via the Doppler effect, which yields first derivative line shapes. For frequency modulation an ac potential of up to ± 10 V, with frequencies up to 10 kHz is applied to the drift tube. The resulting velocity modulation of the parent ion beam corresponds to frequency modulation of the laser frequency because of the Doppler effect. An effective frequency modulation of some 10–20 MHz is obtained. It should be noted that the velocity modulation technique is far more sensitive than amplitude modulation of the laser beam because it discriminates against background photodissociation, whereas amplitude modulation does not. However, the amplitude of the frequency modulation should ideally be optimized for each transition (to the natural line width of the transition) and the use of one frequency modulation amplitude throughout the whole spectrum results in some lines being under- or overmodulated, with consequent distortion of line shapes and intensities.

B. The Predissociation Spectrum

The H_3^+ ion was found to have an extremely strong and dense IR spectrum throughout the region that could be scanned (872–1094 cm^{-1}). The spectra were detected by means of monitoring H^+ fragments; no photofragment H_2^+ ions could be detected. The observation was unexpected, and at the time there was no theoretical explanation for it. The experiment has an intrinsically high resolution and the complete recording of the spectrum took over 2500 h; over 26,500 lines of signals/noise (S/N) greater than 2:1 were recorded with a laser power of 7 ± 3 W, with an output time constant of 1 s, and a drift tube modulation amplitude of 7 V. As will be seen to be significant, the ESA was adjusted to collect H^+ fragments with zero ke release. However, all ion-beam slits were open to their maximum aperture giving a ke window of approximately $\pm 250\ cm^{-1}$ about 0. A typical spectrum is shown in Fig. 5.2(*a*).

Carrington and co-workers [20–22] chose to define line intensities in terms of S/N ratios. The noise is the statistical noise of the background proton fragment peak; if the ke window is set at 0 cm^{-1}, spectra are recorded against the maximum background noise, but with a ke window centered at 3000 cm^{-1} the recording is on the wings of the background fragment peak, where the noise is much lower. It would be difficult to define genuine line intensities of the experiment but it is believed that the S/N ratios provide realistic measures of line intensities; at a ke window centered at 3000 cm^{-1}, for example, the background noise is smaller but the signal protons are correspondingly reduced because the collimating slit before the ESA selects only a small fraction of the fragments, which are scattered over the full 4π solid angle.

The nature of the spectrum was at first unknown, and could have arisen from four possible mechanisms that led to fragmentation.

1. Enhanced collision induced dissociation of the upper state on resonance (as in Wing's experiment). This mechanism was eliminated when the addition of hydrogen gas to the analyzer region *reduced* signal intensity, rather than enhancing it.

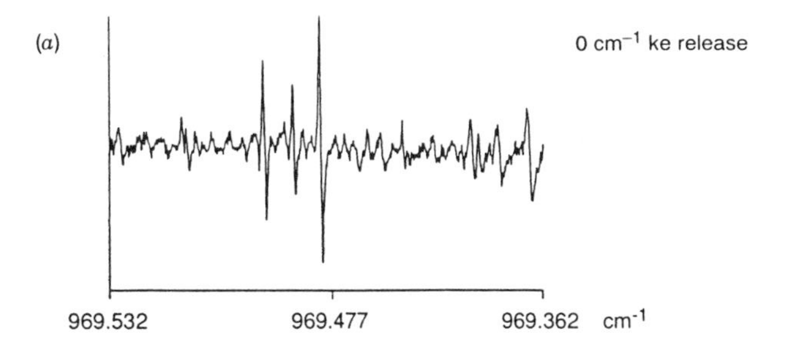

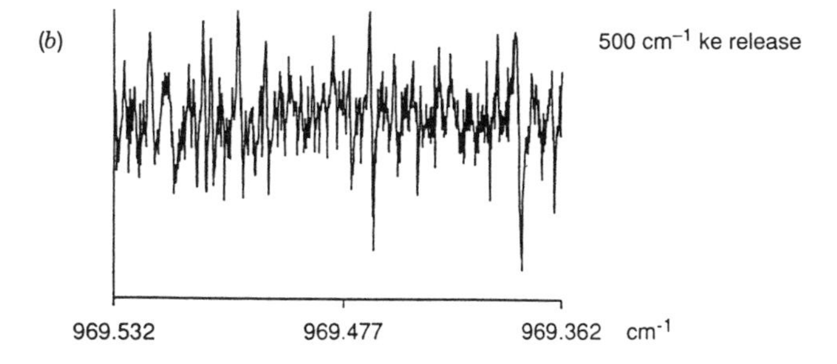

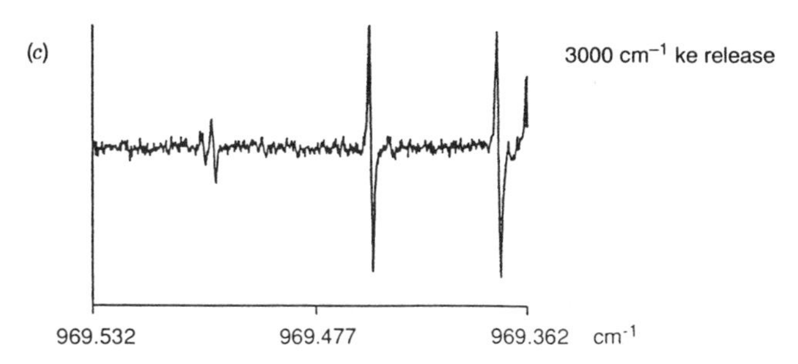

Figure 5.2. Typical predissociation spectra of H_3^+ recorded using frequency modulation (hence first-derivative line shapes): (*a*) at $0\,\mathrm{cm}^{-1}$ ke release many strong lines are seen (there are at least 37 shown), (*b*) at $500\,\mathrm{cm}^{-1}$ ke release many weak lines are found, and (*c*) at $3000\,\mathrm{cm}^{-1}$ ke release a few intense lines are found against a low noise background.

2. Direct photodissociation of the upper state on resonance. This mechanism was eliminated when it was shown that addition of off-resonance laser power did not increase the intensity of the spectrum.
3. Multiphoton excitation processes from bound levels. Detailed measurements of many line intensities at differing laser powers

showed a linear power dependence (until the onset of saturation). Strong lines saturate at powers greater than 2 W, and the spectrum is considerably less dense at low powers. Multiphoton processes do not contribute to the spectrum.

4. Some form of predissociation process.

These experiments indicated that predissociation was the mechanism by which the excited states decomposed, and further strong evidence of this was the fact that line widths vary greatly throughout the spectrum. When the spectrum was recorded using amplitude modulation it was found that the line widths varied from the Doppler limited value of about 10 MHz to several hundred megahertz. The variation in line width is a characteristic of a predissociation spectrum involving energy levels of varying lifetimes, reflecting the natural line width ($\Delta E \cdot \Delta T \geq h/2\pi$). The mechanism for production of the spectrum is therefore:

$$H_3^+ \xrightarrow{h\nu} H_3^{+*} \xrightarrow{\text{Predissociation}} H^+ + H_2(v, J)$$

and it should be noted that a very large family of dissociation limits exists corresponding to production of the H_2 fragment molecules in different vibration–rotation (v, J) levels (see Fig. 2.1). The experiment could not distinguish through which channel the dissociation occurred.

1. *The Experimental Lifetime Constraints*

One of the major constraints that all theories of the observed predissociation spectrum of H_3^+ must take into account is the observational constraints on the lifetimes of the initial and final states observed. The lifetime of an initial level must be sufficiently long to retain a significant population of a state that has been populated in the ion source and accelerated through the instrument to the drift tube region; this implies that the initial states should have lifetimes of order 10^{-6} s. A similar constraint exists on the final state lifetimes. For predissociation from a final state to be observed, it must fragment within the time of passage through the drift tube. An H_3^+ ion accelerated by 5 kV travels at a velocity of 5.68×10^5 m s^{-1}; it therefore passes through the 40-cm drift tube in 7×10^{-7} s, and must fragment with a characteristic lifetime smaller than this. Conversely, the maximum width of a line that is likely to be detected in scans when using a drift tube modulation of 10 V amplitude (which discriminates against broad lines) is approximately 0.005 cm^{-1}, giving a lower limit on the final state lifetime of 7×10^{-9} s.

2. *Kinetic Energy of Fragmentation*

The H^+ ke release for any given resonance may be determined but to do so for all observed resonances would require a prohibitive time (~1 h per resonance). However, sections of the spectrum were examined in this manner and it was found that in many resonances the fragment H^+ ke release is between 2000 and 3000 cm^{-1}. As the input energy of the photons is at most 1094 cm^{-1}, and as multiphoton processes are not involved this proves that many of the transitions must involve an initial level that lies at least 2000 cm^{-1} above the appropriate dissociation limit.

3. *The Pseudo-Low-Resolution Spectrum*

Although no pattern (e.g., combination differences) could be readily discerned in the H_3^+ predissociation spectrum, it was felt that a low-resolution recording would be informative. Unfortunately, as the experiment was intrinsically high resolution, such a spectrum could not be directly obtained but a pseudo-low-resolution spectrum was constructed using a computer convolution of the high resolution data assuming a 4-cm^{-1} wide Gaussian line shape for each measured line. For a convolution including only the 1934 strongest lines in the spectrum it was apparent that there was structure, and the result is shown in Fig. 5.3, which shows four "clumps" of high intensity at 876, 928, 978, and 1034 cm^{-1}, with an uncertainty of about 1 cm^{-1} in the peak positions.

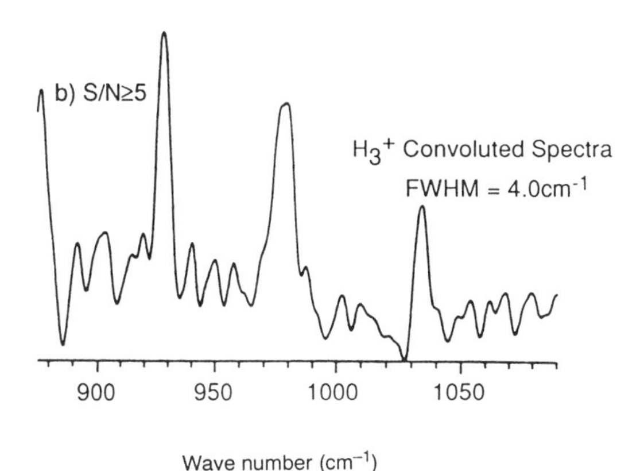

Figure 5.3. The pseudo-low-resolution convolution spectrum of H_3^+ obtained when only the 1934 most intense lines are included. Four "clumps" of high intensity were found centered at 876, 928, 978, and 1034 cm^{-1}, with an uncertainty of about 1 cm^{-1} in the peak positions.

4. *Isotope Studies*

One advantage of obtaining spectra in an MS is the ease with which isotope studies can be performed. Infrared predissociation spectra of D_3^+, D_2H^+, and H_2D^+ were also investigated with the ESA set to transmit fragments at zero ke release. In each case it was found that photofragment diatomic ions were not detectable, but spectra could be obtained by monitoring atomic ions. The D_3^+ spectrum was found to be weaker and more dense than that of H_3^+, but was not studied in detail. For the heteronuclear ions H_2D^+ and D_2H^+ the possibility existed of detecting ions by either monitoring H^+ or D^+ fragments. For H_2D^+ the MS has insufficient resolution to separate out D_2^+ ions, and consequently at low ke release the main contributor to D^+ fragments was direct excitation from high-lying vibration–rotation levels of D_2^+ ($1s\sigma_g$) to the repulsive electronic state ($2p\sigma_u$) [10]. For D_2H^+, no such complication arose, and spectra were recorded at zero ke release monitoring H^+ and D^+ in turn. The spectra had different transition wavenumbers and the spectrum recorded using D^+ fragments was of lower intensity than that recorded using H^+ fragments. As with H_3^+, the mixed-isotope species and D_3^+ were found to have some resonance lines with fragment kinetic energies greater than 2000 cm^{-1}.

5. *Summary of the Predissociation Spectrum Features*

The main features of the H_3^+ IR predissociation spectrum established by Carrington and Kennedy [21] were as follows.

1. At zero ke release approximately 27,000 lines with S/N ratios of 2:1 or greater were observed over the range 874–1094 cm^{-1}, using an average CW laser power of 7 W.
2. Line widths vary from about 10 MHz to several hundred megahertz. Lines of greater width were undermodulated, and therefore not detected.
3. In addition to protons produced by the predissociative IR transitions, background protons also arose from spontaneous predissociation, collision-induced dissociation, and a weak continuous photodissociation. These background protons determined the noise level against which the IR resonances were detected.
4. It was not possible to determine any pattern in the 27,000 resolved lines but convolution of the strongest 1934 lines revealed the presence of a coarse-grained structure consisting of four clumps centered at 876, 928, 978, and 1034 cm^{-1}.
5. When the ESA was set to transmit protons of large center-of-mass

ke release (3000 cm^{-1} or greater), a much smaller number of lines were observed. The observation of these lines proved the existence of long-lived levels lying at least 2000–3000 cm^{-1} above the lowest dissociation limit.

6. Experiments designed to test for multiphoton transitions gave negative results; the spectrum arose from single-photon excitations.
7. Predissociation spectra of the isotopic species D_2H^+ and H_2D^+ were also observed at zero ke release. Spectra recorded by monitoring H^+ fragments were different and more intense than those observed when monitoring D^+ fragments.
8. Because of the instrumental geometry lifetime constraints were imposed on the spectrum. All predissociating states had a lifetime between 10^{-7} and 10^{-9} s, and all initial states had lifetimes greater than 1 μs.

C. Theoretical Understanding of the H_3^+ Predissociation Spectrum

The predissociation spectrum of H_3^+ provided a challenge to theoreticians that was vigorously met. Prior to its discovery no such spectrum had been anticipated. The existence of states above the dissociation limit with lifetimes of microseconds was unaccounted for, as was the complexity of the spectrum and the tantalizingly simple structure in the low-resolution spectrum and the channel dependence of the mixed-isotope spectra. The following section describes some of the many calculations that were performed in order to account for the observations.

1. *Rigid-Rotor Model*

One intriguing aspect of the H_3^+ predissociation spectrum is the coarse-grained spectrum, which shows four distinct clumps. Carrington and Kennedy [21] noted that the frequencies of these four clumps are in close coincidence with the $N = 5 \leftarrow 3$ rotational transitions of H_2 in the $v = 0$, 1, 2, and 3 states. A model of H_3^+ in the high-energy regime was therefore postulated of a hydrogen molecule interacting with a proton. This rigid-rotor approximation is similar in nature to the description commonly used for van der Waals' molecules. Its consequences were investigated in a series of papers by Child and co-workers [193–195] and by Drolshagen et al. [196].

Child and co-workers [193–195] established the properties of the rigid-rotor model. Using the SDL [47] *ab initio* pe surface for H_3^+ it was found that final states with lifetimes in the experimental window could be accounted for at total angular momenta of $J = 30$–45. Approximately 600 such states were found within the required energy range and the long lifetimes of the states were accounted for by trapping behind a series of centrifugal barriers. These lifetimes were calculated using an internal scattering model. The density of states may seem too small to account for the observed 27,000 lines, but it should be noted that in the high-power limit, and in the absence of any good quantum numbers, only $(27{,}000)^{1/2} = 165$ states would be needed in each of the initial and final energy ranges. The existence of three good quantum numbers puts the required number of states to about 600!

In the rigid-rotor model, transitions within the experimental wavenumber region were found to be due to transitions between states of different angular momentum of the H^+ orbiting H_2. The model did not predict any spectral intensity above 900 cm^{-1} unless the spectrum was assumed to be fully saturated. Nothing in the model suggested a possible origin of the coarse-grained spectrum.

Lifetimes of states within the rigid-rotor model were investigated further by Drolshagen et al. [196] who examined the H_2/H^+ collision model without rotational excitation ($J = 0$). They found that interchannel vibrational couplings were insufficient to account for the lifetimes of the states observed in the spectrum.

2. *Nature of States Involved in the Spectrum*

Following similar ideas used by Chesnavich [197] and Rynefors and Nordholm [198], Pollak derived a classical Hamiltonian, which showed that the total angular momentum (J) of a triatomic molecule could lead to angular momentum barriers behind which classical infinitely long-lived states could be trapped above the dissociation limit [199]. In quantum theory the lifetimes of such states are finite and governed by the rate of tunneling through the rotational barrier. Chambers and Child [200] subsequently showed that this Hamiltonian is formally identical with that used in Child's earlier work on the quantal rigid-rotor model. To account for the observed predissociation spectrum the long-lived states must fulfill three criteria:

1. That states above the dissociation limit could decay by quantal tunneling through the rotational barriers within the lifetime window of the experiment.
2. That 880–1100 cm^{-1} below these long-lived states there should be states sufficiently long lived to have reached the drift tube after formation in the ion source.
3. That a sufficient density of states should exist to account for the observed number of lines.

The nature of the states trapped behind such barriers was found to be chaotic [201], or more properly ergodic (that all of the accessible volume of phase space was visited if a trajectory was integrated for infinite time). This was found to be a general feature of simple molecules, dependent only on a sufficiently deep pe well. This finding has profound implications for molecular spectroscopy; almost any strongly bound simple molecule will have chaotic spectra in the high-energy regions close to or above dissociation.

Pollak's [199] investigation of the effect of angular momentum on the H_3^+ Hamiltonian showed that there was a maximal J value beyond which the potential will not support any additional bound states and following his suggestion Miller and Tennyson [202] therefore performed variational calculations on four pe surfaces to find the highest J values that could support a bound state. The highest bound J values were found to be $J = 46$ for the SDL, MarB [203], and MBB [48] surfaces, while the DIM [204] surface gave a maximum $J = 50$. The number of bound states was found to depend strongly on the value of the dissociation energy, but was relatively insensitive to details of the surface topology.

Gomez-Llorente and Pollak [205] used a semiclassical prescription for estimating the resonance lifetimes, and found that with this prescription the observed range of lifetimes could easily be accounted for. In ergodic systems time averages can replace averages over initial conditions (i.e., different trajectories) because one trajectory visits all of accessible phase space. Time averages over one trajectory were therefore computed in order to calculate tunneling probabilities at a given energy and angular momentum. The lifetime τ of a particular state was calculated by assigning a particular probability P_i of tunneling through the potential barrier to each collision with the barrier. The total tunneling lifetime was therefore calculated using

$$\tau = \lim_{T \to \infty} \left(\frac{T}{\sum_i P_i} \right)$$

where T is the time for which the trajectory was integrated. Individual tunneling probabilities (P_i) were calculated using a 1-D sudden approximation; all coordinates except the tunneling coordinate were frozen, and the tunneling probability along that coordinate calculated using semiclassical formulas (10% accuracy was claimed).

Another explanation for long-lived states in H_3^+ above dissociation has been put forward by Quack [66]. He points out that vibrational levels with A_2^- symmetry have nuclear spin $I = \frac{3}{2}$, and that if such levels were to decompose to yield a H_2 fragment with angular momentum of zero, that is,

$$H_3^+ \rightarrow H_2(v, j=0) + H^+$$

that a nuclear spin flip and a symmetry change would be required. These processes should require long lifetimes, of the order of microseconds, even in the absence of a centrifugal barrier. However, it is hard to imagine that a ke release of 4000 cm^{-1} could result from a state of zero angular momentum, and this mechanism is unlikely to account for the observed isotope spectra. The direct calculation of lifetimes arising in this manner is said to be possible if accurate wave functions are known, but to my knowledge no such calculations have yet been performed. This analysis shows that even for large centrifugal barriers, lifetimes will be greatly influenced by symmetry considerations.

3. *Correspondence between Kinetic Energy Release and Angular Momentum*

In the calculations of lifetimes it was noted that the probability of a proton tunneling through a barrier scaled exponentially with the energy displacement from the top of the barrier [106] (as expected for heavy particles). It was found that the experimental lifetime constraints on the levels showed that only energy levels close to the top of the centrifugal barrier could contribute to the observed predissociation spectrum at any given angular momentum. The centrifugal barriers are far out in the exit channel where the system may be considered as a proton–diatom (to a good approximation). The tunneling lifetime is therefore mainly determined by the energy available for the separation of the proton and the diatom: For the tunneling probability to be large, the fragment must have a maximum momentum in the direction of dissociation, and for a given centrifugal barrier this dependence is exponential in energy.

The result of these calculations is that fragmenting H_3^+ ions that contribute to the observed spectrum will have most of the available energy in ke of the fragments rather than internal energy of the diatomic fragment. As a result, the total ke of the fragments will scale with the total angular momentum of the fragmenting complex.

4. *Formation of Highly Energetic States*

Although it was calculated that long-lived states of H_3^+ could exist trapped behind centrifugal barriers, it was not known if such states could be populated by the usual reaction of formation: $H_2^+ + H_2 \rightarrow H_3^+ + H$. This problem was addressed by Eaker and co-workers [207–210] who investigated whether the assumed mechanism of formation of H_3^+ in the ion source, by the reaction

$$H_2 + H_2^+ \ (v = 0\text{–}18) \rightarrow H_3^+ + H$$

could account for the formation of long-lived highly energetic states of H_3^+ (all vibrational levels of H_2^+ are populated when it is formed by electron impact ionization due to favorable Franck–Condon factors with the ground vibrational state of H_2 [10]). They performed classical trajectory calculations on the H_3^+ surface, which was parameterized using a diatomic in molecules function. It was found that very highly excited states of H_3^+ could be formed from reaction with H_2^+ in the higher vibrational levels (10–17) when up to 95% of the vibrational energy became internal energy of the H_3^+.

That most of the available energy in vibrationally excited H_3^+ becomes internal energy of H_3^+ may be understood when it is realized that in the high vibrational states H_2^+ spends most of its time close to the outer turning point where it looks very like a hydrogen atom and a proton. The $H_2^+ + H$ reaction dynamics at high v are therefore very similar to a proton combining with H_2. In that case, all of the available energy becomes internal energy if a long-lived complex is formed.

All trajectories of H_3^+ formed by highly excited $H_2^+ + H_2$ were found to be chaotic. The H_3^+ motions were followed for 0.36 ps, which corresponded to about 15 periods of the lowest frequency motion. On examination of those molecules that survived for longer than this it was discovered that many levels did not undergo dissociation within 30 ps, these were described as quasistable. It was not feasible to simulate H_3^+ motions on a time scale directly relevant to the experiments, but it was found that quasistable complexes were not weakly coupled $H_2 \cdots H^+$ orbiting complexes. The dynamics were of chaotic vibrational motion characterized by strong collisions that led to rapid exchange between loosely and strongly bound atoms in H_3^+. Despite this many quasistable molecules were found above any rigorous centrifugal barrier, and this suggested that partial decoupling of vibrational and orbital motions inhibits the vibrational redistribution that could lead to dissociation. It was noted that none of the H_3^+ complexes that they studied would have the hydrogen free-rotor spectrum suggested by Carrington and Kennedy [21]. From these calculations there is no doubt that the reaction of $H_2^+ + H_2$ can form H_3^+ in the states relevant to the H_3^+ predissociation spectrum.

5. *Isotope Effects*

One of the intriguing aspects of Carrington and Kennedy's data was the observation that spectra obtained from H_2D^+ and D_2H^+ were different if different fragmentation channels were used for the observation, for example, the spectrum of D_2H^+ observed was different according to whether D^+ or H^+ fragmentation was used as the probe. Pollak, Schlier and their co-workers [211, 212] suggested a mechanism for this effect that was subsequently confirmed by Chambers and Child [200] using a slightly different method. The difference in the spectra is due to the nature of the angular momentum barriers for the two dissociation channels. The D_2H^+ ion will be specifically discussed, but similar arguments apply to H_2D^+, and the conclusions are listed for both.

In the dissociation of D_2H^+, the fragmentation route can take either of two channels, with the dissociating states being either $D_2 \cdots H^+$ (1) or $HD \cdots D^+$ (2). The total angular momentum barriers can then be approximated by the orbital angular momentum alone, with the appropriate classical pe expression $V(R) + L^2/2\mu R^2$, where μ is now the reduced mass appropriate to (1), $m(D_2)m(H^+)/m(D_2H^+)$, or (2), $m(HD)m(D^+)/m(D_2H^+)$. As the barrier in the deuteron channel is always lower than in the proton channel, classically one expects to observe spectra only by monitoring the H^+ channel. However, quantum mechanics demands that two further effects be considered: the difference in zero-point energies between D_2 and HD, and the difference in ionization potential between H and D. Allowing for these differences shifts the barriers to the positions shown in Fig. 5.4. It is seen that there will be a predominant dissociation to H^+ at low angular momentum (low ke release), a brief region in which both channels are open, and then a predominant dissociation to D^+ at high angular momentum (high ke release). The specific conclusions for D_2H^+ and H_2D^+ are as follows:

For H_2D^+, protons should have ke less than 1190 cm^{-1} and deuterons should have ke greater than 540 cm^{-1}

For D_2H^+, protons should have ke less than 680 cm^{-1} and deuterons should have ke greater than 286 cm^{-1}

This accounted for the lower intensity of lines in the D_2H^+ spectrum found at zero ke release when monitoring D^+ rather than H^+.

6. *The Origin of the Clumps*

Many spectra that originate in classically chaotic regions of energy are known, for example, the stimulated emission pumping (SEP) spectrum of ethyne [213, 214], the quadratic Zeeman effect spectra of the hydrogen atom in a strong magnetic field [215–217] and the SEP spectrum of Na_3 [218]. For all these systems, clumps of high intensity in low-resolution spectra or characteristic frequencies found from limited FTs of the spectra

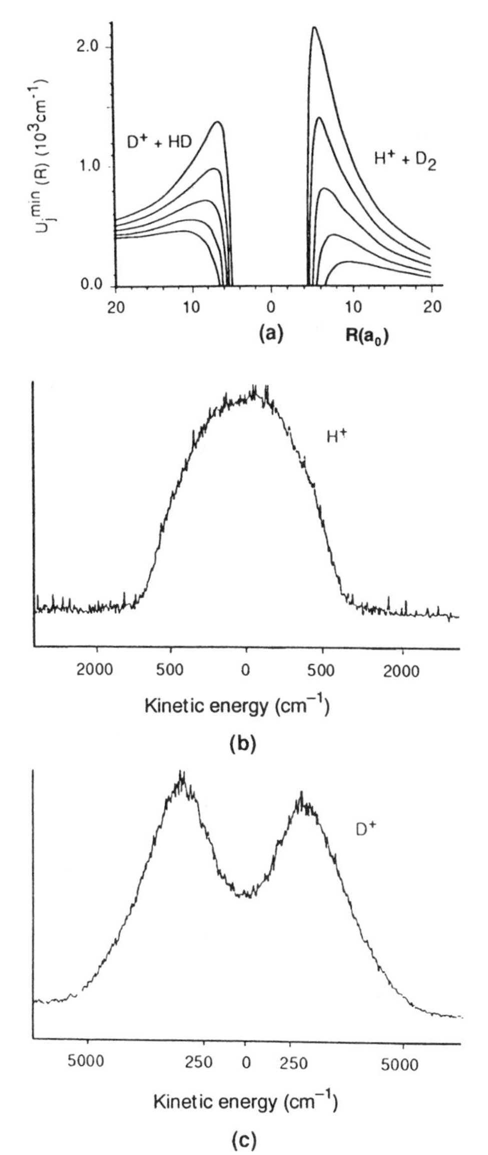

Figure 5.4. Channel specific predissociation in D_2H^+ showing the correspondence between observed ke release of both fragments and the calculated ke release and angular momentum barriers: (*a*) Angular momentum barriers, for (left) $HD + D^+$ and (right) $D_2 + H^+$, at total angular momenta of $J = 20$, 25, 30, 35, and 40. (*b*) The observed off-resonance H^+ fragment ke distribution. (*c*) The corresponding off-resonance D^+ fragment ke distribution.

can be accounted for by wave functions related to remnants of stable or unstable periodic orbits [219, 220]. This phenomenon was first investigated by Heller [221] using the model system of the stadium billiard that is well known to given classical chaos. It was found that although trajectories were chaotic, there were isolated trajectories that were stable for long periods of time. The full quantum solutions for the same system showed that the wave function had large amplitude along these trajectories, and thus a remnant of order survived into the classical chaos, and this remained true in the quantum solution. Heller referred to this as "scarring" of the wave function. The connections between periodic orbits and quantum mechanical eigenfunctions and spectra in the chaotic region has been reviewed by Founargiotakis et al. [222] and Zakrzewski [223].

As Pollak [224] has shown the normal modes of periodic orbits (which can be determined from a linear stability analysis) play a similar role in the interpretation of high-energy spectra to the expansion about the normal modes of a molecule at low energies. The normal mode analysis is based on an expansion about the minimum of the pe surface. At increased energies, the motions of the molecule move far away from this minimum but at high energies the classical dynamics still has invariant objects, these are the periodic orbits of the system.

Berblinger et al. [225] and Gomez-Llorente and Pollak [226] investigated whether remnants of a periodic orbit could be responsible for the clumping in the predissociation spectrum of H_3^+. One periodic orbit was found that survived in the regions of energy and angular momentum associated with the H_3^+ predissociation spectrum; a large amplitude bending motion of quasilinear H_3^+, with a frequency of 972 cm^{-1}. This was nicknamed a horseshoe orbit due to the appearance of its trajectory when plotted in configuration space (Fig. 5.5), and may be visualized as a proton oscillating through and perpendicular to the bond of vibrating H_2. The orbit was found to be stable over a wide range of angular momentum and energy above the dissociation limit.

Heller used the term *scar* only in relation to unstable periodic orbits, and the horseshoe orbit is stable, and therefore strictly speaking cannot cause a scar. Despite the fact that the horseshoe orbit occupied an insufficient volume of phase space to account for any quantum states in the usual EBK [227] semi classical quantization scheme the small regular region surrounding it suggested that it would still produce a substantial quantum localization.

The horseshoe orbit provides a possible (and I believe correct) explanation for the origin of the clumped spectrum of H_3^+. The normal modes of the horseshoe orbit correspond to 972 cm^{-1} for a bending motion and 643 cm^{-1} for the antisymmetric stretch [205]. The rotational constant of the orbit (calculated on the DIM surface) was found to be 30 cm^{-1} for angular momentum directed along the C_{2v} symmetry axis. The clumping in the predissociation spectrum was assigned as the antisymmetric stretch of the horseshoe state (centered at 643 cm^{-1}) with a rotational constant (B) of 30 cm^{-1}. The specific transitions in the clumped

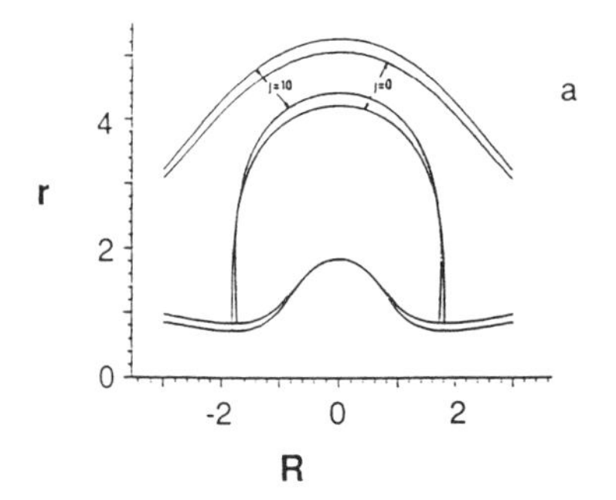

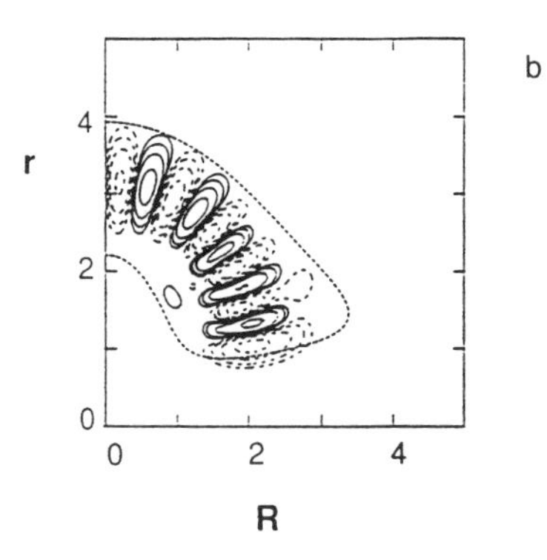

Figure 5.5. "Horseshoe" periodic orbits; the motion is along the center line of (*a*) and corresponds to a wide amplitude bending motion of quasilinear H_3^+. The parameters r, R are the Jacobi coordinates of Fig. 2.2 (in bohr) with θ fixed at 90°. (*a*) Shows two classical horseshoes calculated above the dissociation limit by Gomez-Llorente and Pollak [206] with equal action, and $J = 0$ and 10, together with the classical turning points at energies 0 and 2982 cm^{-1} [shown above and below the horseshoe, and corresponding to the dotted line in (*b*)]. (*b*) Shows a quantum horseshoe (state 101) calculated below dissociation at $J = 0$ by Tennyson and Henderson [93]; the dotted outer contour gives the classical turning point.

spectrum were assigned as $R(3)$, $R(4)$, $R(5)$, and R(6). The angular momentum of the levels involved in the strong transitions that result in the low-resolution clumping must therefore be small; consequently the ke of the fragments used to measure these lines must also be small if the model is correct, and each clump should be due to one J value.

With hindsight it is not surprising that the rigid-rotor model proposed by Carrington and Kennedy [21] should have failed to account for the clumping in the low-resolution spectrum. The horseshoe orbit constantly visits a region of phase space completely unexplored in the model, that is, the collinear configuration.

Taylor and co-workers [228–230] used classically guided quantum calculations on the DIM surface to investigate further the horseshoe orbits. The quantum effects of the periodic orbits were studied by laying down a Gaussian basis set along the periodic orbit, and diagonalizing the Hamiltonian. It was found that quantum mechanical resonances were formed about the periodic orbit, despite the failure of semiclassical quantization, and it was suggested that the stable periodic orbit identifies a relatively decoupled region of phase space. This approach to the interpretation of chaotic molecular spectra has been reviewed by Taylor and Zakrzewski [231]. Further calculations in a Feschbach formalism identified the resonance as an intense state (P), which interacts with the chaotic continuum background (Q) leading to a Lorentzian peak in the coarse grained spectrum. It was found that states localized around the rotating horseshoe and that the semiclassical quantized energy and the stabilized eigenvalues of the quantum resonance were in good agreement. The widths of the states were found to be 10–20 cm^{-1}, so that states should be resolved in the low-resolution spectrum as observed. Calculations using the MBB [48] surface gave a rotational spacing (2B) of 50 cm^{-1}, in good agreement with the average experimental separation between the clumped peaks of 52.6 cm^{-1}.

A reexamination of the Carrington and and Kennedy [21] data showed that the clumping in the spectrum can be accounted for by just the 100 most intense lines, and the horseshoe orbits investigated by Taylor et al. [228–230] show a sufficient density of states to account for 100 transitions between the horseshoe states. These intense lines are superimposed upon the chaotic background, and the intensity comes when the phase-space trajectory at a particular energy and angular momentum lies close to a periodic orbit of the system. The chaotic background borrows amplitude from the remnants of periodic motion.

In a combined study, Tennyson, Brass and Pollak [232, 233] investigated full quantum solutions of H_3^+ to see if these calculations could also locate the "horseshoe" states. In agreement with the expectations of the classical and classically guided quantum studies, the horseshoe orbits were found to persist strongly at high energies. It is clear from this study that although the volume of phase space occupied by the horseshoe orbits is too small to give rise to EBK quantization [227] there *is* nevertheless a localization of quantum states around the horseshoe orbit as was expected. A further "inverted hyperspherical" periodic orbit was found that persisted to the region of interest, but which was much less stable than the horseshoe orbit. A study of the full configuration space again located the horseshoe orbit. The periodic orbits were investigated at increasingly high energies, and were found to persist, but with decreased amplitude. This is contrary to the findings of other studies on model potentials [222, 221, 234], where it has been found that the intensity of the wave function becomes localized more strongly about the periodic orbit as the energy of the system increases. The inverted hyperspherical modes were found to be largely unstable in classical calculations of the full configuration space but persisted in the quantum calculations. It is seen that a full quantum solution is preferable where possible as it automatically gives all possible information on the systems behavior.

One consequence of the upper state "lifetime window" is that the lower state in any transition is necessarily within a closely specified energy below the top of the centrifugal barrier. The explanation for the low-resolution spectrum in terms of clumping of energy levels around the horseshoe periodic orbits requires that the experiment is observing states that have sequential J values (i.e., $J'' = 3$–6). A criticism of this model is that if one J'' value results in a state that undergoes a transition resulting in an upper state within the "lifetime window" of the experiment, then the next J'' values are unlikely to do so because the vibrational separation is calculated at approximately 643 cm^{-1}. The theory above resolves this difficulty by coupling the horseshoe orbits to a large background of states in a Feschbach resonance picture and this idea is strongly supported by recent quantum calculations by Le Sueur et al. [235]. They found that the "horseshoes . . . leak intensity into the nearby bath states" and "the coupling to the bath also causes the horseshoe scar to be spread over several states. Naming a particular state as THE horseshoe state is therefore largely a subjective matter."

7. *Summary of the Theoretical Predictions*

To my best knowledge the current consensus is that:

1. The H_3^+ predissociation spectrum is in an energy region with classically chaotic dynamics.
2. High angular momentum creates quasibound states in H_3^+ lying far above the lowest dissociation limit, and this is expected to be a general feature of molecules with a deep pe well.
3. Long-lived states of H_3^+ also occur with low angular momentum, and it is these states that are believed to give rise to the strongest lines in the predissociation spectrum causing the clumping in the spectra by leaking intensity to nearby states.
4. Total angular momentum differentiates between predissociation products for mixed isotopes of H_3^+. This is expected to be a general phenomenon.
5. The clumps in the pseudo-low-resolution spectrum are due to a localization of intensity in the wave function attributable to the horseshoe periodic orbit, which consists of a large amplitude bending motion of quasilinear H_3^+.
6. The H_3^+ predissociation spectrum obtained is limited by the lifetime window inherent in the experiments. There is probably a very large region of frequencies and lifetimes over which such a spectrum might be observable by other methods.

Three major theoretical predictions arose from the work of Pollak, Schlier and their co-workers [2, 199, 201, 204–206, 211, 212, 224–226] (described above) that could be investigated experimentally:

1. For the coarse-grained spectrum of H_3^+, the lines that contribute maximum intensity to the pseudo-low-resolution convolution should have nearly zero ke release.
2. Any molecular ion with a sufficiently deep pe well should support many bound levels in the continuum, and hence display spectra similar to that of H_3^+. This has not yet been tested.
3. For D_2H^+ and H_2D^+, spectra taken monitoring D^+ fragments should be much more intense at high ke releases than those of H^+ fragments, which should disappear at a fairly well-defined threshold.

The theories developed to account for the predissociation spectrum of H_3^+ met the challenge provided by experiment and returned it by providing predictions that could be tested experimentally. The following section describes the experiments of Carrington et al. [22], which were conducted between 1984 and 1992 specifically to test these predictions.

D. Further Measurements of the Predissociation Spectrum

One of the most important aspects of the H_3^+ predissociation spectrum to emerge from the theoretical studies was the relationship between frequencies, line intensities, and proton ke release. It was therefore important to reexamine in more detail the spectrum in the region of one of the peaks previously observed in the low-resolution convolution; an important conclusion of the theory was that the convoluted peaks should arise from the most intense lines with small ke releases.

The experimental verification of this prediction was problematic. It might at first be thought that it would be sufficient to obtain ke spectra of the most intense lines recorded by Carrington and Kennedy [21], but this would be tautological; the original data was obtained with the ke release centered at $0\,\text{cm}^{-1}$, and therefore the most intense lines known were necessarily those with low ke release. In order to make a meaningful measurement, Carrington et al. [22] rerecorded the predissociation spectrum over the range $964.00–991.60\,\text{cm}^{-1}$ (centered on the 978-cm^{-1} clump), using the ESA as a ke window, set to transmit protons whose center-of-mass ke lay within selected ranges.

A typical example of the spectra obtained is shown in Fig. 5.2. With the ke window set at $0\,\text{cm}^{-1}$ we observed lines with a range of intensities and widths; approximately 17 lines/cm^{-1} have S/N ratios of 10 or greater, and there is no identifiable pattern. With a window centered at

500 cm^{-1} a very large number of weak lines of varying intensities and widths is observed (approximately 60 lines/cm^{-1}), these might be mistaken for noise but are highly reproducible. Much more interesting spectra are found when we use a window of 3000 cm^{-1};these consist of a low-intensity background and a number of strong lines well separated from each other. The strong lines show no obviously regular pattern, except that they appear to occur either as single lines, or in small clusters of up to five lines. The single lines are the strongest observed. The simplicity and sparseness of the high ke spectrum suggests that at high ke release (high total angular momentum) the spectrum is once again regular. This is consistent with the explanation of the low J (low ke) H_3^+ spectrum arising in a region of classical chaos. The chaos at low J is driven by the anharmonicity in the potential of the saddle point at the barrier to linearity; when the saddle point is energetically accessible, chaotic dynamics are found. The anharmonicity due to dissociation is far weaker than that of the saddle point, but is important at high J. Nevertheless, as the total angular momentum increases, the depth of the potential well is decreased and the linear geometry becomes increasingly accessible to the point where bound and quasibound states are no longer chaotic [225, 111]. The predissociation spectrum of H_3^+ at high ke may therefore ultimately be assignable in the traditional spectroscopic sense, with each line corresponding to two sets of nearly good, and good quantum numbers. However, for each ke window it would require several years to record the H_3^+ spectrum over the available wavenumber range!

1. Correspondence of High-Intensity Lines and Low Kinetic Energy Release

Because of the importance of the low ke release spectrum, recordings of the region centered at 978 cm^{-1} were made at a number of ke offsets. The data for the 408 strongest lines as a function of ke release is shown in Fig. 5.6(*a*); 371 lines with S/N ratios of 10:1 or greater were measured. These 371 lines were reexamined using ke windows of 0–25, 25–50, 50–100, and greater than 100 cm^{-1}, and the results are shown in Fig. 5.6(*b*). This diagram very clearly shows the concentration of intense lines with very small energy releases.

2. Convolutions of the New High-Resolution Data

A number of convolutions of the new data were made in the manner of Carrington and Kennedy [21]. For the zero ke centered window it was found that a convolution of the new data using a full width at half-maximum of 2 cm^{-1} and ke $\leq$50 cm^{-1} showed the original 978-cm^{-1} clump is split into an apparent 1:3:3:1 quartet. The original and the new

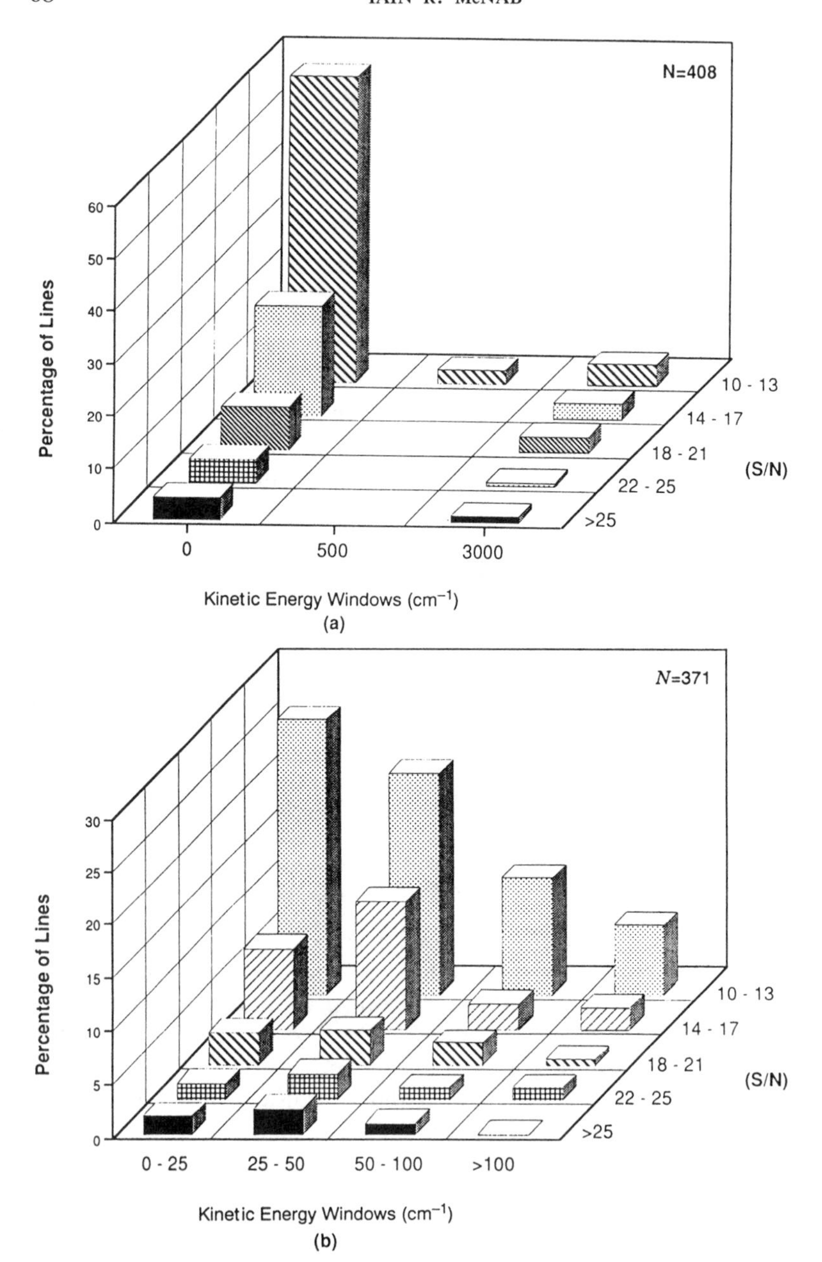

Figure 5.6. Showing the correspondence between lines observed in the region 964.0–991.6 cm^{-1} (centered on the 978-cm^{-1} clump) and low ke release. (*a*) shows data for the 408 strongest lines at ke windows centered at 0, 500, and 3000 cm^{-1}. (*b*) Shows the distribution of the 371 strongest lines for small ke releases.

data sets were consistent with each other, the main difference being that the lower laser power used in the later work (2 W, compared with 7 W previously) led to fewer strong lines in the center region around 980 cm^{-1}. Further analysis of the original data of Carrington and Kennedy [21] for the three other clumps, now convoluted with $\Gamma = 2$ cm^{-1} and including all lines with $S/N \geq 2$, revealed a poorly resolved doublet structure in each case.

In summary, the prediction of Pollak that the overall 978-cm^{-1} clump arose from lines with very small ke releases is confirmed by the experiments; the additional structure observed in the convolutions is presumably significant but as yet no explanation for it has been provided.

3. *Isotope Effects*

As mentioned above, isotope studies are particularly easy in an MS, and experimentally verifiable theoretical conclusions had been reached (summarized in Section V.C.5) that linked the intensity of H^+ and D^+ fragmentation of the mixed-isotopomers H_2D^+ and D_2H^+ to the ke release at which spectra were measured. Carrington et al. [22] examined these conclusions.

Two observations were made upon the mixed-isotopic species, fragmentation was examined both on- and off-resonance.

a. Resonant Transitions in H_2D^+ and D_2H^+. The observations on-resonance were of the transitions and their associated fragment ke releases. Spectra arising from the detection of H^+ and/or D^+ were recorded using ke windows of 0, 750, 1200, 2000, and 3000 cm^{-1} for H_2D^+, and 0, 300, 750, 1500, and 3000 cm^{-1} for D_2H^+. Both ions exhibited IR predissociation spectra throughout the CO_2 laser region that were considerably more dense, and weaker, than the H_3^+ spectrum.

It is important to understand that the limitations on ke resolution mean a strong line in, for example, the 300-cm^{-1} window will contribute much reduced intensity in the adjacent ke windows. The effective resolution increases rapidly as the ke release decreases, so that a strong line with close to zero energy release will not contribute to intensity in the next window. The limitations arise because the relatively slow predissociation process leads to fragment ions being scattered over the full 4π solid angle.

It was found that the measured intensities of H^+ and D^+ fragments from the predissociation spectra of H_2D^+ and D_2H^+ agreed very well with the theoretical predictions, which were

1. Lines that can be observed only through H^+ fragments should have ke releases less than or equal to 540 cm^{-1}.

2. Lines that can be observed through the observation of either H^+ or D^+ fragments should have ke releases between 540 and 1190 cm^{-1}.
3. Lines that can be observed only through D^+ fragments should exhibit ke releases greater than 1190 cm^{-1}.

For H_2D^+ 186 lines were measured via H^+ fragments and all had ke releases less than or equal to 1190 cm^{-1}, 201 lines were measured via D^+ fragments, and 179 of these had ke releases greater than or equal to 540 cm^{-1}.

Similar results were found for D_2H^+, that is, 197 lines were measured via D^+ fragments and 190 lines were measured via H^+ fragments. It was found that D^+ spectra showed ke releases that were maximized in the range 300–750 cm^{-1} but were still abundant in the 1500- and 3000-cm^{-1} windows. Spectra measured with H^+ fragments were concentrated in the 0- and 300-cm^{-1} windows, and none were observed in the 3000-cm^{-1} window.

b. Studies of the Nonresonant Background Dissociation. All of the isotopic species studied showed atomic or diatomic fragmentation in the absence of laser irradiation and, for the atomic fragments, these were believed to arise mainly from spontaneous predissociation. The fragments were readily observed at the lowest pressures that could be achieved in the drift tube region, and all of the peaks showed structure when recorded by scanning the ESA voltage. Introduction of off-resonance laser power also resulted in an increase in atomic fragment ion intensity, but with loss of structure.

A further confirmation of the explanation of the channel dependence of the mixed-isotope spectra was provided by examining the ke distribution of fragments formed from the dissociation observed in the absence of laser irradiation. The H^+ fragment peak from D_2H^+ [Fig. 5.4(*b*)] shows a ke distribution that is maximized at zero-energy release, with a half-width of 647 cm^{-1}. Conversely, the D^+ fragment peak [Fig. 5.4(*c*)] shows two maxima, which indicates that the preferred ke release is approximately 700 cm^{-1}; at this ke value, the H^+ fragment peak intensity is close to zero. The results for H_2D^+ are similar; the H^+ peak showed evidence of maxima at $\pm$140 cm^{-1}, while the D^+ peak exhibited a very clear splitting of approximately 1100 cm^{-1}.

These results are consistent with the observations of the resonance lines. The triatomic ions are formed in metastable states with a large range of lifetimes, and therefore the observation of spontaneous predissociation was not surprising, not was the fact that the observed ke values of fragments due to measurements both on-resonance and without laser irradiation were consistent with one another.

The diatomic fragment ion peaks were much less intense, and less interesting; no off-resonance or on-resonance photofragmentation was observed. The lowest $H_2^+ + H$ dissociation limit lies almost 2 eV above the $H^+ + H_2(v = 0, N = 0)$ limit. If H_3^+ states lying above the $H_2^+ + H$ limit are populated in the ion source, they are either short lived or predissociate preferentially through the lower energy channel.

4. *Lifetimes of the Predissociating States*

a. Direct Measurement of Initial State Lifetimes. It will be recalled that lifetime limits exist on the initial and final states, due to instrumental geometry. To confirm these estimates of lifetimes, which are so important in theoretical considerations of the spectra, Carrington et al. [22] measured the lifetimes of a few states directly. They assumed that one or both of the states involved in a given transition was metastable, in which case a change in the time interval between formation (in the ion source) and observation (in the drift tube) could lead to a consequent change in the observed line intensity. Many regions of the H_3^+ spectrum can be observed at different accelerating potentials provided different laser lines (or orientation) are used. At high beam potentials the ion beam is accelerated to a greater velocity, and consequently takes less time to reach the drift tube than at a low accelerating potential. Most lines in the spectrum were unaffected by the small changes in flight time that could be realized. After a small proportion of the spectrum was scanned, eight examples were found where at low voltages (longer flight times) the initial state had decayed significantly by the time the ions reached the drift tube, which yielded a weaker observed line intensity.

By assuming first-order kinetics to determine the lifetime of the initial level involved in the transition, that is, by assuming that the observed intensity of the transition was proportional to the population in the initial level when the molecule reached the drift tube, and from the measurements of intensity of high- and low-source potentials (velocities), initial state lifetimes were calculated. All the lifetimes measured in this manner were similar, the range extending only between 1.1 and 2.3 μs These measurements reinforce the validity of the lifetime constraints on the observed spectrum.

b. Lifetimes of Upper States. As the spectrum arises through predissociation, lifetimes of the upper states involved in any transition can be determined directly from the observed line widths. The region 1025–1045 cm^{-1} (centered on the 1034-cm^{-1} clump) was rerecorded, using a zero ke window only, at low laser powers (2 W—to avoid saturation and line broadening). The spectra were obtained using laser chopping for signal modulation, rather than drift tube voltage modulation, and consequently the line shapes, widths, and intensities were not subject to distortions produced by modulation. Because absorption rather than first-derivative line shapes were obtained, peak areas were also tabulated.

The data obtained shows one interesting feature. A three-dimensional (3-D) plot exhibiting the line width variations as a function of frequency shows that the upper predissociating states have lifetimes τ in the range $5 \times 10^{-10} < \tau < 2 \times 10^{-8}$ s, as expected, but the maximum intensity region in the clump arises from lines with the smallest widths. The histogram is shown in Fig. 5.7. The outer frequency regions (1025–1030 cm^{-1} and 1040–1045 cm^{-1}) have a much higher proportion of broader lines than the more intense center region. This shows that the states contributing to the center of the clump have longer lived upper states than those contributing to the edges.

This observation is consistent with the horseshoe model, and the calculations of Le Sueur et al. [235]. One expects that there will be a correlation between the degree of coupling between the horseshoe state and the nonresonant background and the lifetime of the state; this

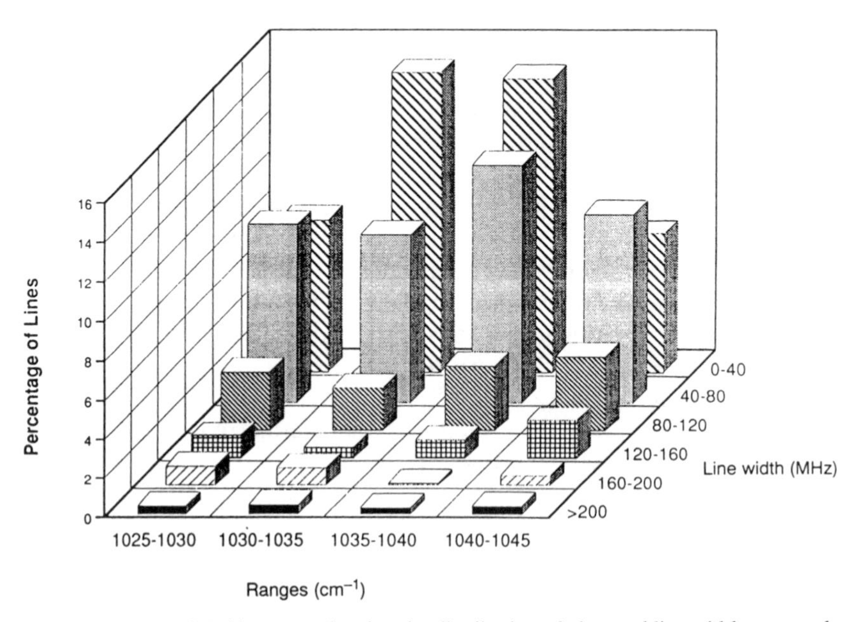

Figure 5.7. A 3-D histogram showing the distribution of observed line widths across the high wavenumber clump. It can be seen that there are more lines with narrow widths in the center of the clump, although a roughly uniform density of lines across the range is found.

coupling will decrease with increasing energy difference between the horseshoe and the background.

5. *Multichannel Dissociation*

As the H_3^+ spectrum contains so many lines, it would not be surprising to discover examples of predissociation occurring through two or more channels at similar rates (i.e., to two or more dissociation limits $H^+ + H_2(v, N)$ corresponding to differing internal excitation of the H_2 fragment). A search for such lines was made and several were observed.

Because of the high density of lines in the H_3^+ spectrum, it is difficult to be certain that the multiple scattering patterns, indicative of a multiple dissociation channel, do not arise accidentally from two different transitions with closely similar frequencies. In order to test this the ke spectra were rerecorded at a range of different source pressures, which might have changed the relative populations of two different predissociating levels, and over a range of slightly different ion beam potentials, which would discriminate between two different transitions with slightly different frequencies. Such tests were consistent with genuine multichannel dissociations.

The difference in ke release between two channels must correspond to a particular vibration–rotation level separation in H_2. If these could be measured with sufficient accuracy it would be possible to locate the position of the predissociation level with respect to two different dissociation limits $H_2(v, N) + H^+$. Unfortunately the ke resolution of Carrington's experiment falls short of these requirements.

6. *Summary of the Additional Features*

The reexamination of the H_3^+ IR predissociation spectrum by Carrington et al. [22] established the following additional features:

1. As predicted by theory, there is a correlation between the most intense lines in the H_3^+ spectrum and low ke release (low J according to the theory). This correlation is strong evidence for the explanation of the clumping in the spectrum being due to transitions between horseshoe orbits.
2. As predicted by theory, there is a correlation between the propensity of the mixed isotopomers H_2D^+ and D_2H^+ to dissociate to H^+ and D^+ fragments and the ke of the fragments. This finding demonstrates a channel specific dissociation controlled by total angular momentum.
3. Convolutions of the high-resolution data show a 1:3:3:1 quartet

within the 1033-cm^{-1} clump. No theoretical explanation has been offered to account for this observation.

4. An apparently nonchaotic H_3^+ spectrum exists when recorded from fragments at high ke release (high angular momentum). This is consistent with theoretical expectations.
5. Several examples of multichannel dissociation were found.
6. In the high wavenumber clump (1034 cm^{-1}) it was found that at the center of the clump upper states had long predissociation lifetimes, while at the edges shorter lifetimes were observed. This finding is consistent with theoretical expectations.

E. Discussion

The measurements of the predissociation spectrum reviewed in Section V.D substantially confirmed the conclusions of the theories developed since the first observations. It is believed that the nature of the predissociation spectrum (recorded at low ke release) is chaotic, and arises due to excitation to long-lived levels trapped behind total angular momentum barriers.

One theoretical conclusion that has been proven unambiguously is that the channel specific predissociation of the mixed-isotopomers H_2D^+ and D_2H^+ is strongly controlled by the ke of the fragmentation. This finding proves the correlation between fragment ke release and total angular momentum that was first deduced theoretically.

The low-resolution clumping of the data is believed to be due to remnants of horseshoe periodic orbits localizing probability density at certain energies. Carrington's further experiments lend support to this hypothesis, in particular the low ke, which is associated with the intense lines in the spectrum bears out the predictions of Pollak. Quantum calculations have the advantage of automatically locating all states, and we look to future quantum calculations to prove or disprove the assignment of the spectrum to the horseshoe orbits.

The more general conclusions raised by the H_3^+ predissociation spectrum are

1. One should expect that tightly bound molecules will display chaotic spectra in high-energy regions.
2. Channel-specific dissociation processes can be controlled by total angular momentum.

These general conclusions have arisen from detailed classical and quantum calculations in a region of energy that is classically chaotic. In addition to the confirmation of theoretical predictions, further experimental data were obtained that raise more questions and possibilities.

There has been much discussion in the literature about the nature of chaotic molecular spectra, and the feasibility of performing traditional spectroscopic analysis of such spectra. The consensus up to a few years ago was that the most that could be extracted from such spectra was statistical information such as spectral rigidity and level spacing statistics [236].

Considerable effort has been devoted to obtaining such data on real molecules. Random matrix fluctuations were reported for NO_2 in 1988 [237], and subsequently in the stimulated emission pumping spectrum of ethyne [238]. It is almost certain that the statistical analysis of the energy level spacing for the observed H_3^+ predissociation spectrum would follow the same pattern, but unfortunately it is not yet possible to test this assertion; the level spacing analysis relies upon the fact that the analysis should be carried out only for levels that have the same good quantum numbers in common. In the case of the H_3^+ predissociation spectrum it is not yet possible to assign these quantum numbers.

The predissociation spectrum was first "assigned" using classical trajectory calculations, and the coarse grained peaks found by Carrington and Kennedy [21] were shown to correspond to a "normal-mode" assignment based around the horseshoe periodic orbit, rather than around the equilibrium geometry (as is the case of the low-energy spectra). This type of analysis is expected to be of general utility in understanding the gross features of coarse grained molecular spectra obtained in regions of high energy. The analysis is similar to the perturbation Hamiltonian approach used at low energies [224], in that it has a highly physical interpretation.

It has been argued that low-resolution spectra of chaotic regions (clumped data) contain all the important information about the dynamics of the molecule, and that the chaotic background on which this is superimposed is of purely statistical interest. In a 1988 review of the quantum chaos problem, Elyutin stated [239] "If the positions of levels vary in a complex way upon the parameters of the Hamiltonian, there is no physical interest in a detailed description of a system of levels." This is contradicted by the work of Standard et al. [240] who argue that there is no reason in principle why the traditional approach to spectroscopic analysis of labeling transitions and refining a Hamiltonian (which may include the pe surface) cannot be extended to chaotic spectra. They investigated such a fitting process for a model system starting in a region where the spectra were not chaotic, and gradually including more of the complex data set in an iterative manner—a "bootstrap" approach. In order to achieve the fitting it is only necessary that energy levels be labeled, not that they be labeled with good quantum numbers! They concluded that chaotic spectra had a very high information content, and allowed parameters in a Hamiltonian to be determined very accurately.

Variational calculations offer the best hope of achieving at least a partial assignment of the H_3^+ predissociation spectrum, involving the most intense lines, which are widely separated. Henderson and Tennyson [241] have now calculated all the $J = 0$ bound states of H_3^+ supported by both the MBB, SDL, and DIM potentials, finding that between 800 and 900 states are supported. The complete quantum calculation of the predissociation spectrum of H_3^+ begins to appear possible. However, before such a calculation can be made there are still fundamental problems that need to be addressed. Although the MBB dipole and energy surfaces have been used to calculate extremely accurate vibration–rotation energies and intensities for low-lying levels, the surfaces are not well behaved at dissociation, the region of interest for the predissociation spectrum. The publication of a surface that is accurate to dissociation is eagerly awaited. A further problem is the calculation of the population of individual levels created in the ion source, although it may be sufficient to consider only transition intensities, and assume a roughly equal population of states close to and above dissociation. The vibration–rotation problem has essentially been solved for bound states but it remains to extend the theory to calculate energies and widths of quasibound states. With such an extension, and with pe and dipole surfaces that are accurate at dissociation, there is a real possibility that the most intense lines in the low ke H_3^+ predissociation spectrum could be assigned, together with the sparse spectrum that was observed at high ke release. However, the nature of the spectroscopic assignment for the most intense lines in the low ke predissociation spectrum may not satisfy traditional spectroscopists. Such an assignment would consist of a 2:1 correspondence of calculated energy levels and spectral lines. The energy levels will have only three good quantum numbers, parity, total angular momentum, and permutation symmetry. It is possible that further nearly good quantum numbers might be used to label transitions: the degree of excitation of "normal modes" of periodic orbits (such as the horseshoe orbit), but such assignments may not exist for all states involved in the spectrum.

VI. EXCITED ELECTRONIC STATES OF H_3^+

Experimental evidence of excited electronic states has been provided by only two groups. Peart and Dolder [242] measured excitation of H_3^+ in electron impact causing dissociation to protons, and found excitation to two excited states with energies 15 and 19.2 eV above the ground state.

The states were assigned with the use of Kawaoka and Borkman's 1971 calculations [53] to be the $^3E'$ and $^1E'$ equilateral triangle geometry repulsive excited states. Bae and Cosby [243] observed bound-free photodissociation in H_3^+ at photon energies between 2.4 and 4.5 eV to form H_2^+, that is,

$$H_3^+ + h\nu \rightarrow H_2^+ + H$$

they interpreted this observation, using the pe curves of Talbi and Saxon [54], as a single-photon absorption from highly excited levels of the H_3^+ ground state into the continuum of the first excited singlet electronic state, for example, from $1^1A' \rightarrow 2^1A'$ (C_s symmetry). No spectroscopy of excited electronic states has yet been accomplished.

VII. CONCLUSION

This chapter considered both the low-energy H_3^+ spectra obtained in the laboratory and in extraterrestrial environments. It also considered the predissociation spectrum, together with the different theoretical frameworks in which the spectra have been analyzed.

The assignment of the H_3^+ spectra at low energies is a textbook example of the utility of fitting observed line positions to effective Hamiltonians. It is also an illustration of the problems associated with this approach; the molecule is so anharmonic that even close to equilibrium a large number of constants are required to fit observed transitions. The resulting effective Hamiltonians have little predictive power in regions of higher energy. Conversely, first-principle variational calculations of line positions and intensities using suitably embedded Hamiltonians have proved of great utility in assigning spectra in all energy regions. The extension of perturbation Hamiltonians into the regions of energy probed by variational calculations is of interest [244] because the magnitude of constants in perturbation Hamiltonians gives physical insight into the relative importance of the processes that give rise to the observed structure of energy levels.

The low-energy spectra of H_3^+ provide a beautiful demonstration of how accurate *ab initio* calculations have become for a molecule when accurate pe and dipole surfaces are available. The low-energy spectra are well understood, and the agreement between theoretical calculations of line positions and intensities is outstanding. Since the MBB pe and dipole surfaces became available, calculated spectra have enabled experimentalists to locate transitions with a minimal search problem, leading to ever higher energy states being probed. This is an on-going process: so far levels of H_3^+ have been probed up to $3\nu_2$, but accurate predictions now exist for levels up to $6\nu_2$ and laboratory observations will surely follow.

Since the first observation of the fundamental band of H_3^+ by Oka in 1980 [17] there have been many searches for the H_3^+ spectrum in extraterrestrial environments. Such searches have at last born fruit, and emission spectra of H_3^+ have been definitely located in three planetary ionospheres. Plausible detections exist of emission from heated regions of a dark molecular cloud and the expanding envelope of Supernova 1987a, while H_2D^+ may have been detected in the IRc2 region of Orion.

Where it has been observed, the emission from H_3^+ is providing a valuable tool for monitoring atmospheric and other processes, and has already led to revisions of models of planetary conditions. Hopefully, the spectrum of H_3^+ will be located in the ISM in the next few years, and such observations will become commonplace and serve as a useful tool for astronomical observations.

We have seen that both the low-energy and predissociation spectra of H_3^+ are assignable in principle in terms of good and nearly good quantum numbers through the use of variational calculations of vibration–rotation energies and transition intensities. The basis for the nearly good quantum numbers is different in the two energy regions. In the low-energy region, the nearly good quantum numbers are normal modes of the equilibrium geometry, while in the high-energy region they are normal modes of particular periodic orbits (in the intermediate region, local mode behavior is found [245]). As more molecular spectra are obtained at intermediate and high energies, one expects that spectroscopists will become ever more aware of the limitations of using rigid rotor and anharmonic oscillator models to describe molecular spectra. As a community, spectroscopists are moving out of the bottom of the molecular potential energy well and being forced to modify and extend traditional techniques of analysis and interpretation. The reduction of spectroscopic data to a set of parameters in an effective Hamiltonian based on near-equilibrium geometries may no longer by meaningful or worthwhile for spectra obtained in regions of high energy—if the number of constants required to describe a set of transitions is of the same order of magnitude as the original data set, then it may be preferable to communicate data sets directly.

The predissociation spectrum of H_3^+ is understood theoretically in terms of classical mechanical models based on the remnants of periodic motion causing regions of high intensity in the chaotic background, and the experimental evidence supports this. Quantum calculations by Tennyson and co-workers [235] show that the "horseshoe" states persist up to dissociation and these calculations provide a unifying framework in which both the low-energy spectra and the predissociation spectra may be

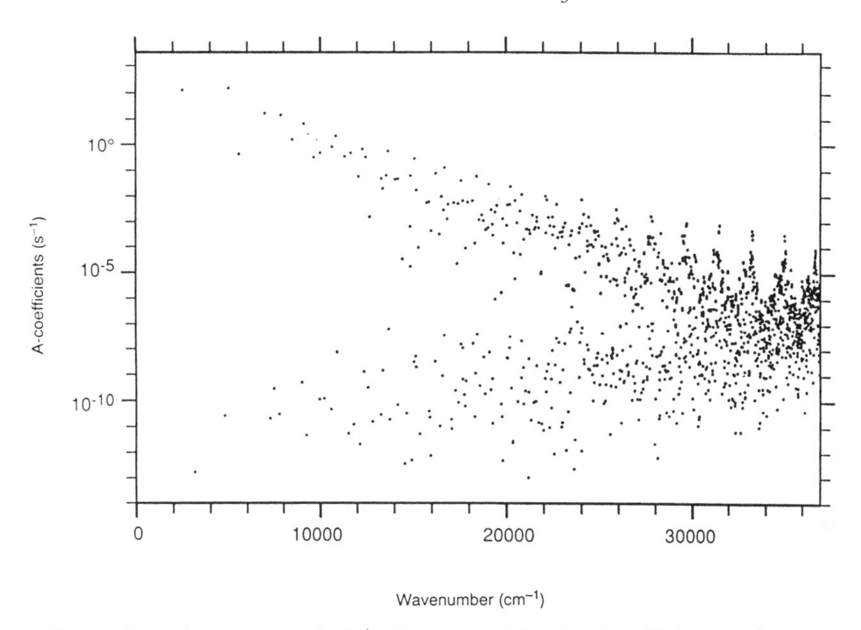

Figure 6.1. Gateway states in H_3^+, showing the Einstein A-coefficients to the ground state calculated by Le Sueur et al. [235]. It is seen that intense transitions are expected even up to 2000 cm^{-1} below dissociation, close to the energy range of the observed predissociation spectrum.

understood. Experiment has yet to probe H_3^+ in the energy region between the two, although the barrier to linearity is rapidly being approached by observation. Experiments that probe the intermediate energy regions would enable a pe surface for H_3^+ to be completely determined from experiment.

In recent work, Tennyson and co-workers [235] examined the transition intensities between the ground state of H_3^+ and highly excited states, which extended to dissociation. The found extreme variations in intensity, with the most intense transitions (which could provide a "*gateway*" to upper levels) corresponding to excitation to the "horseshoe" states. The Einstein A-coefficients for these transitions are shown in Fig. 6.1. The calculated transition intensities are such that these "gateway" states should be observable and may provide an experimental connection between the low- and high-energy regimes of the H_3^+ spectrum. Experiments to observe these states are currently under consideration [111].

ACKNOWLEDGMENTS

I am indebted to Professor A. Carrington, F.R.S., for his inspiration and guidance and to Dr. Y.D. West; together we worked upon the H_3^+

predissociation spectrum for many years. I have benefitted greatly from conversations and correspondence with Professor M.S. Child, F.R.S., Professor A.S. Dickinson, Professor P. Jensen, Professor E. Pollak, Professor H.S. Taylor, Dr. J. Tennyson, Professor T. Oka and Professor J.K.G. Watson. I thank the Research Committee of the University of Newcastle upon Tyne and the Science and Engineering Research Council for their support of my experimental work.

REFERENCES

1. R. N. Porter, *Ber. Bunsenges. Phys. Chem.*, **86**, 407 (1982).
2. E. Pollak and C. Schlier, *Acc. Chem. Res.*, **22**, 223 (1989).
3. H. S. Taylor, *Acc. Chem. Res.*, **22**, 263 (1989).
4. J. Tennyson, S. Miller, and B. T. Sutcliffe, *J. Chem. Soc. Faraday Trans. 2*, **84**, 1295 (1988).
5. S. Miller and J. Tennyson, *Chem. Soc. Rev.*, **21**, 281 (1992).
6. T. Oka, in *Molecular Ions: Spectroscopy, Structure and Chemistry*, T.A. Miller, and V.E. Bondybey, Eds., North-Holland, Oxford, 1983, p. 73.
7. A. Carrington and I. R. McNab, *Acc. Chem. Res.*, **22**, 218 (1989).
8. T. Oka, *Rev. Mod. Phys.*, **64**, 1141 (1992).
9. A. Carrington, *J. Chem. Soc. Faraday Trans. 2*, **82**, 1089 (1986).
10. A. Carrington, I. R. McNab, and C. A. Montgomerie, *J. Phys. B.*, **22**, 3551 (1989), and references cited therein.
11. J. J. Thompson, *Philos. Mag.*, **24**, 209 (1912).
12. R. E. Christoffersen, *J. Chem. Phys.*, **41**, 960 (1964).
13. M. J. Gaillard, D. S. Gemmell, G. Goldring, I. Levine, W. J. Pietsch, J. C. Poizat, A. J. Ratkowsi, J. Remillieux, Z. Vager, and B. J. Zabransky, *Phys. Rev. A*, **17**, 1797 (1978).
14. G. Herzberg, *Trans. R. Soc. Can.*, **5**, 3 (1967).
15. G. Herzberg, *Trans. R. Soc. Can.*, **20**, 151 (1982), and references cited therein.
16. G. Herzberg, *Ann. Rev. Phys. Chem*, **38** 27 (1987).
17. T. Oka, *Phys. Rev. Lett.* **45**, 531 (1980).
18. J.-T. Shy, J. W. Farley, W. E. Lamb, and W. H. Wing, *Phys. Rev. Lett.*, **45**, 535 (1980).
19. G. D. Carney and R. N. Porter, *Phys. Rev. Lett.*, **45**, 537 (1980).
20. A. Carrington, J. Buttenshaw, and R. A. Kennedy, *Mol. Phys.*, **45** 753 (1982).
21. A. Carrington, and R. A. Kennedy, *J. Chem. Phys.*, **81**, 91 (1984).
22. A. Carrington, I. R. McNab, and Y. D. West, *J. Chem. Phys.*, **98**, 1073 (1993).
23. M. Born and K. Huang, *Dynamical Theory of Crystal Lattices*, Oxford, 1954, Appendix VIII.
24. J. M. Combes and R. Seiler, in *Quantum Dynamics of Molecules*, R. G. Woolley, Ed., Plenum, New York, 1980, p. 435.
25. B. T. Sutcliffe, *J. Chem. Soc. Faraday Trans.*, **89**, 2321 (1993).

26. Quoted in: G. Handler and J. R. Arnold, *J. Chem. Phys.*, **27**, 144 (1957).
27. J. Hirschfelder, H. Eyring, and N. Rosen, *J. Chem. Phys.*, **4**, 121 (1936).
28. J. Hirschfelder, H. Eyring, and N. Rosen, *J. Chem. Phys.*, **4**, 130 (1936).
29. J. Hirschfelder, H. Diamond, and H. Eyring, *J. Chem. Phys.*, **5**, 695 (1937).
30. D. Stevenson and J. Hirschfelder, *J. Chem. Phys.*, **5**, 933 (1937).
31. J. O. Hirschfelder, *J. Chem. Phys.*, **6**, 795 (1938).
32. J. O. Hirschfelder and C. N. Weygandt, *J. Chem. Phys.*, **6**, 806 (1938).
32a. R. A. Kennedy, Ph.D. Thesis, University of Southampton, Southampton, U.K., 1984.
33. C. A. Coulson, *Proc. Cambridge Philos. Soc.*, **31**, 244 (1935).
34. See, for example, P. J. Knowles, in *Supercomputational Science*, R. G. Evans and S. Wilson, Eds., Plenum, New York, 1990, p. 211, and references cited therein.
35. J. N. Murrell, S. Carter, S. C. Farantos, P. Huxley, and A. J. C. Varandas, *Molecular Potential Energy Functions*, Wiley, London, 1984.
36. J. B. Anderson, *J. Chem. Phys.*, **96**, 3702 (1991).
37. R. K. Preston and J. C. Tully, *J. Chem. Phys.*, **54**, 4297 (1971).
38. F. O. Ellison, *J. Am. Chem. Soc.*, **85**, 3540 (1963).
39. W. Ketterle, H.-P. Messmer, and H. Walther, *Europhys. Lett.*, **8**, 333 (1989).
40. C. W. Bauschlicher, Jr., S. V. O'Neill, R. K. Preston, H. F. Schaefer III, and C. F. Bender, *J. Chem. Phys.*, **59**, 1286 (1973).
41. F. O. Ellison, N. T. Huff, and J. C. Patel, *J. Am. Chem. Soc.*, **85**, 3544 (1963).
42. See, for example, G. C. Schatz, *Rev. Mod. Phys.*, **61**, 669 (1989).
43. G. D. Carney and R. N. Porter, *J. Chem. Phys.*, **60**, 4251 (1974).
44. G. Simons, R. G. Parr, and J. M. Finlan, *J. Chem. Phys.*, **59**, 3229 (1973).
45. G. D. Carney and R. N. Porter, *J. Chem. Phys.*, **65**, 3547 (1976).
46. C. F. Giese and W. R. Gentry, *Phys. Rev. A*, **10**, 2156 (1974).
47. R. Schinke, M. Dupuis, and W. A. Lester, Jr., *J. Chem. Phys.*, **72**, 3909 (1980).
48. W. Meyer, P. Botschwina, and P. G. Burton, *J. Chem. Phys.*, **84**, 891 (1986).
49. S. Miller and J. Tennyson, *J. Mol. Spectrosc.*, **126**, 183 (1987).
50. G. C. Lie and D. Frye, *J. Chem. Phys.*, **96**, 6784 (1992).
51. D. Frye, A. Preiskorn, G. C. Lie, and E. Clementi, *J. Chem. Phys.*, **92**, 4948 (1990).
52. M. J. Bramley, J. R. Henderson, J. Tennyson, and B. T. Sutcliffe, *J. Chem. Phys.*, **98**, 10104 (1993).
53. K. Kawaoka and R. F. Borkman, *J. Chem. Phys.*, **54**, 4234 (1971).
54. D. Talbi and R. P. Saxon, *J. Chem. Phys.*, **89**, 2235 (1988).
55. L. J. Schaad and W. V. Hicks, *J. Chem. Phys.*, **61**, 1934 (1974).
56. R. Ahlrichs, C. Votava, and C. Zirz, *J. Chem. Phys.*, **66**, 2771 (1977).
57. P. E. S. Wormer and F. de Groot, *J. Chem. Phys.*, **90**, 2344 (1989).
58. G. D. Carney, *Mol. Phys.*, **39**, 923 (1980).
59. G. D. Carney and R. N. Porter, *Chem. Phys. Lett.*, **50**, 327 (1977).
60. G. D. Carney, *Chem. Phys. Lett.*, **78**, 200 (1981).
61. G. D. Carney, *Can. J. Phys.*, **62**, 1871 (1984).
62. R. S. Mulliken, *J. Chem. Phys.*, **23**, 1997 (1955).

63. G. Herzberg, *Molecular Spectra and Molecular Structure, II. Infrared and Raman Spectra of Polyatomic Molecules*, Krieger, Florida, 1991.
64. E. Teller, in *Hand. Jahrb. Chem. Phys. 9II*, 43 (1934).
65. See, for example, G. Herzberg, *The Spectra and Structure of Simple Free Radicals*, Dover, New York, 1988, p. 144.
66. M Quack, *Phil. Trans. R. Soc. London A*, **332**, 203 (1990).
67. T. Oka, private communication.
68. R. S. Mulliken, *Phys. Rev.*, **43**, 279 (1933).
69. See, for example, P. R. Bunker *Molecular Symmetry and Spectroscopy*, Academic, London, 1979.
70. J. K. G. Watson, *J. Mol. Spectrosc.*, **103**, 350 (1984).
71. T. Oka and M.-F. Jagod, *J. Chem. Soc. Faraday. Trans.*, **89**, 2147 (1993).
72. See, for example, H. Kroto, *Molecular Rotation Spectra*, Wiley, London, 1975, p. 36.
73. See, for example, H. H. Nielsen, *Rev. Mod. Phys.*, **23**, 91 (1951).
74. J. K. G. Watson, *Mol. Phys.*, **15**, 479 (1968).
75. S. Miller and J. Tennyson, *J. Mol. Spectrosc.*, **136**, 223 (1989).
76. J. Tennyson, *J. Chem. Soc. Faraday Trans.*, **88**, 3271 (1992).
77. Z. Bačić and J. C. Light, *Ann. Rev. Phys. Chem.*, **40**, 469 (1989).
78. C. Eckart, *Phys. Rev.*, **47**, 552 (1935).
79. B. T. Sutcliffe, in *Quantum Dynamics of Molecules*, R. G. Woolley, Ed., Plenum, London, 1980, p. 1.
80. J. K. G. Watson, *Mol. Phys.*, **19**, 465 (1970).
81. E. B. Wilson and J. B. Howard, *J. Chem. Phys.*, **4**, 260 (1936).
82. B. T. Darling and D. M. Dennison, *Phys. Rev.*, **57**, 128 (1940).
83. R. J. Whitehead and N. C. Handy, *J. Mol. Spectrosc.*, **55**, 356 (1975).
84. J. Tennyson, S. Miller, J. R. Henderson, and B. T. Sutcliffe, *Phil. Trans. R. Soc. Lond A*, **332**, 329 (1990).
85. B. T. Sutcliffe, in *Methods of Computational Chemistry*, Vol. 4, Chapter 2, S. Wilson, Ed., Plenum, New York, 1992.
86. M. G. Bawendi, B. D. Rehfuss, and T. Oka, *J. Chem. Phys.*, **93**, 6200 (1990).
87. V. Špirko, P. Jensen, P. R. Bunker, and A. Cejchan, *J. Mol. Spectrosc.*, **112**, 183 (1985).
88. P. Jensen, V. Špirko, and P. R. Bunker, *J. Mol. Spectrosc.*, **115**, 269 (1986).
89. P. Jensen and V. Špirko, *J. Mol. Spectrosc.*, **118**, 208 (1986).
90. J. Tennyson and B. T. Sutcliffe, *Mol. Phys.*, **51**, 887 (1984).
91. J. Tennyson and B. T. Sutcliffe, *Mol. Phys.*, **58**, 1067 (1986).
92. See, for example, R. N. Zare, *Angular Momentum*, Wiley-Interscience, New York, 1988.
93. J. Tennyson and J. R. Henderson, *J. Chem. Phys.*, **91**, 3815 (1989).
94. J. R. Henderson, J. Tennyson, and B. T. Sutcliffe, *J. Chem. Phys.*, **98**, 7191 (1993).
95. D. O. Harris, G. O. Engerholm, and W. Gwinn, *J. Chem. Phys.* **43**, 1515 (1965).
96. Deleted in Proof.
97. P. Drossart, J.-P. Maillard, J. Caldwell, S. J. Kim, J. K. G. Watson, W. A. Majewski,

J. Tennyson, S. Miller, S. K. Atreya, J. T. Clarke, J. H. Waite Jr., and R. Wagener, *Nature (London)*, **340**, 539 (1989).
98. S. Miller, J. Tennyson, and B. T. Sutcliffe, *J. Mol. Spectrosc.*, **141**, 104 (1990).
99. L.-W. Xu, M. Rösslein, C. M. Gabrys, and T. Oka, *J. Mol. Spectrosc.*, **153**, 726 (1992).
100. B. M. Dinelli, S. Miller, and J. Tennyson, *J. Mol. Spectrosc.*, **163**, 71 (1994).
101. W. Kołos, K. Szalewicz, and H. J. Monkhorst, *J. Chem. Phys.*, **84**, 3278 (1990).
102. A. Carrington and R. A. Kennedy, *Gas Phase Ion Chemistry*, Ed., M. T. Bowers, Academic, London, 1984, p. 393.
103. F.-S. Pan and T. Oka, *Astrophys. J.*, **305**, 518 (1986).
104. S. Miller and J. Tennyson, *Astrophys. J.*, **335**, 486 (1988).
105. L. Kao, T. Oka, S. Miller, and J. Tennyson, *Astrophys. J. Suppl. Ser.*, **77**, 317 (1991).
106. S. Miller, J. Tennyson, and B. T. Sutcliffe, *Mol. Phys.*, **66**, 429 (1989).
107. J. Tennyson and B. T. Sutcliffe, *Mol. Phys.*, **54**, 141 (1985).
108. P. Jensen, *J. Chem. Soc., Faraday Trans. 2*, **84**, 1315 (1988).
109. I. N. Kozin, O. L. Polyansky, and N. F. Zobov, *J. Mol. Spectrosc.*, **128**, 126 (1988).
110. O. L. Polyansky and A. R. W. McKellar, *J. Chem. Phys.*, **92**, 4039 (1990).
111. J. Tennyson, private communication.
112. R. C. Woods, *Rev. Sci. Instrum.*, **44**, 282 (1973).
113. T. A. Dixon and R. C. Woods, *Phys. Rev. Lett.*, **34**, 61 (1975).
114. R. C. Woods, T. A. Dixon, R. J. Saykally, and P. G. Szanto, *Phys. Rev. Lett.*, **35**, 1269 (1975).
115. R. J. Saykally, T. A. Dixon, T. G. Anderson, P. G. Szanto, and R. C. Woods, *Astrophys. J. Lett.*, **205**, 101 (1976).
116. A. S. Pine, *J. Opt. Soc. Am.*, **64**, 1683 (1974); **66**, 97 (1976).
117. R. N. Porter, private communication to T. Oka referenced in [17].
118. T. Oka, in *Laser Spectroscopy V*, A. R. W. McKellar, T. Oka, and B. P. Stoicheff Eds., Springer-Verlag, New York, 1981, p. 320.
119. U. Steinmetzger, A. Redpath, and A. Ding, *Ber. Bunsenges. Phys. Chem.*, **86**, 468 (1982).
120. J. K. G. Watson, private communication.
121. J. K. G. Watson, S. C. Foster, A. R. W. McKellar, P. Bernath, T. Amano, F. S. Pan, M. W. Crofton, R. S. Altman, and T. Oka, *Can. J. Phys.* **62**, 1875 (1984).
122. C. S. Gudeman and R. J. Saykally, *Ann. Rev. Phys. Chem.*, **35**, 387 (1984).
123. W. A. Majewski, M. D. Marshall, A. R. W. McKellar, J. W. C. Johns, and J. K. G. Watson, *J. Mol. Spectrosc.*, **122**, 341 (1987).
124. This name was used by J. K. G. Watson in his analysis of the ν_2 band [123] after F. W. Birss who considered the general case of perturbations of the form q^2J^2: F. W. Birss, *Mol. Phys.*, **31**, 491 (1976).
125. T. Nakanaga, F. Ito, K. Sugawara, H. Takeo, and C. Matsumura, *Chem. Phys. Lett.*, **169**, 269 (1990).
126. W. A. Majewski, P. A. Feldman, J. K. G. Watson, S. Miller, and J. Tennyson, *Astrophys. J.*, **347**, L51 (1989).
127. S. Miller and J. Tennyson, *J. Mol. Spectrosc.*, **136**, 223 (1989).

128. S. Miller and J. Tennyson, *J. Mol. Spectrosc.*, **128**, 530 (1988).
129. L.-W. Xu, C. Gabrys, and T. Oka, *J. Chem. Phys.*, **93**, 6210 (1990).
130. S. S. Lee, B. F. Ventrudo, D. T. Cassidy, T. Oka, S. Miller, and J. Tennyson, *J. Mol. Spectrosc.*, **145**, 222 (1991).
131. J. K. G. Watson, *J. Mol. Spectrosc.*, **40**, 536 (1971).
132. M. Mizushima and P. Venkateswarlu, *J. Chem. Phys.*, **21**, 705 (1953).
133. M. R. Aliev and V. M. Mikhailov, *Acta Phys. Hung.*, **55**, 293 (1984).
134. W. H. Wing, G. A. Ruff, W. E. Lamb, and J. E. Spezeski, *Phys. Rev. Lett.*, **36**, 1488 (1976).
135. S. L. Kaufman, *Opt. Commun.*, **17**, 309 (1976).
136. J. T. Shy, Ph.D. Thesis, University of Arizona, Tuscon, Arizona 1982; No. 8305995. University Microfilms International, Ann Arbor, Michigan, USA.
137. J. K. G. Watson, S. C. Foster, and A. R. W. Mckellar, *Can. J. Phys.*, **65**, 38 (1987).
138. O. L. Polyansky, *J. Mol. Spectrosc.*, **112**, 79 (1985).
139. J.-T. Shy, J. W. Farley, and W. H. Wing, *Phys. Rev. A*, **24**, 1146 (1981).
140. T. Amano and J. K. G. Watson, *J. Chem. Phys.*, **81**, 2869 (1984).
141. T. Amano, *J. Opt. Soc. Am. B*, **2**, 790 (1985).
142. K. G. Lubic and T. Amano, *Can. J. Phys.*, **62**, 1886 (1984).
143. M. Bogey, C. Demuynck, M. Denis, J. L. Destombes, and B. Lemoine, *Astron. Astrophys.*, **137**, L15 (1984).
144. H. E. Warner, W. T. Conner, R. H. Petrmichl, and R. C. Woods, *J. Chem. Phys.*, **81**, 2514 (1984).
145. F. C. De Lucia, E. Herbst, G. M. Plummer and G. A. Blake, *J. Chem. Phys.*, **78**, 2312 (1983).
146. S. Saito, K. Kawaguchi, and E. Hirota, *J. Chem. Phys.*, **82**, 45 (1985).
147. Deleted in Proof.
148. S. C. Foster, A. R. W. McKellar, I. R. Peterkin, J. K. G. Watson, F. S. Pan, M. W. Crofton, R. S. Altman, and T. Oka, *J. Chem. Phys.*, **84**, 91 (1986).
149. S. C. Foster, A. R. W. McKellar, and J. K. G. Watson, *J. Chem. Phys.*, **85**, 664 (1986).
150. D. A. Jennings, C. Demuynck, M. Banek, and K. M. Evenson, private communication to Polyansky and McKellar, referenced in [110].
151. S. Miller and J. Tennyson, *J. Mol. Spectrosc.*, **128**, 530 (1988).
152. B. M. Dinelli, S. Miller, and J. Tennyson, *J. Mol. Spectrosc.*, **153**, 718 (1992).
153. E. Herbst and W. Klemperer, *Astrophys. J.*, **185**, 505 (1973).
154. W. D. Watson, *Astrophys. J.*, **183**, L17 (1973).
155. G. Herzberg, *Spectrochim. Acta Part* A, **89**, 63 (1990).
156. T. R. Geballe and T. Oka, *Astrophys. J.*, **342**, 855 (1989).
157. J. H. Black, E. F. van Dishoeck, S. P. Willner, and R. C. Woods, *Astrophys. J.*, **358**, 459 (1990).
158. J. Tennyson, S. Miller, and H. Schild, *J. Chem. Soc. Faraday Trans.*, **89** 2155 (1993).
159. K. S. Sidhu, S. Miller and J. Tennyson, *Astron. Astrophys.*, **255**, 453 (1992).
160. S. Chandra, V. P. Gaur, and M. C. Pande, *J. Quant. Spectrosc. Radiat. Transfer.*, **45**, 57 (1991).

161. R. T. Boreiko and A. L. Betz, *Astrophys. J.*, **405**, L39 (1993).
162. S. Miller, J. Tennyson, S. Lepp, and A. Dalgarno, *Nature* (*London*), **355**, 420 (1992).
163. For a recent review, see J. Lequeux and E. Roueff, *Phys. Rep.*, **200**, 241 (1991).
164. B. E. Turner and B. Zuckerman, *Astrophys. J.*, **225**, L79 (1978).
165. T. Oka, *Phil. Trans. R. Soc. Lond A*, **303**, 543 (1981).
166. C. J. Pritchet and C. J. Grillmair, *Publ. Astron. Soc. Pac.*, **96**, 349 (1984).
167. G. Gussie and C. Pritchet, *J. R. Astron. Soc. Can.*, **82**, 69 (1988).
168. A. Dalgarno, M. Oppenheimer, and R. S. Berry, *Astrophys. J.*, **183**, L21 (1973).
169. A. Wootten, R. B. Loren, and R. L. Snell, *Astrophys. J.*, **225**, 160 (1982).
170. M. Guelin, W. D. Langer, and R. W. Wilson, *Astron. Astrophys.*, **107**, 107 (1982).
171. P. Angerhofer, E. Churchwell, and R. N. Porter, *Astrophys. J. Lett.*, **19**, 137 (1978).
172. T. G. Phillips, P. M. Solomon, P. J. Huggins, and G. Blair, unpublished results, reported in [173].
173. T. G. Phillips, G. A. Blake, J. Keene, R. C. Woods, and E. Churchwell, *Astrophys. J.*, **294**, L45 (1985).
174. L. Pagani, P. G. Wannier, M. A. Frerking, T. B. H. Kuiper, S. Gulkis, P. Zimmermann, P. J. Encrenaz, J. B. Whiteoak, J. L. Destombes, and H. M. Pickett, *Astron. Astrophys.*, **258**, 472 (1992).
175. E. F. van Dishoeck, T. G. Phillips, J. Keene, and G. A. Blake, *Astron. Astrophys.*, **261**, L13 (1992).
176. J. P. Halpern and J. Patterson, *Astrophys. J.*, **312**, L31 (1987).
177. D. C. Hamilton, G. Gloeckler, S. M. Krimigis, C. O. Bostrom, T. P. Armstrong, W. I. Axford, C. Y. Fan, L. J. Lanzerotti, and D. M. Hunten, *Geophys. Res. Lett.*, **7**, 813 (1980).
178. L. Trafton, D. F. Lester, and K. L. Thompson, *Astophys. J.*, **343**, L73 (1989).
179. M. Quack, *Mol. Phys.*, **34**, 477 (1977).
180. See, for example, G. Herzberg, *Molecular Spectra and Molecular Structure, Vol I—Spectra of Diatomic Molecules*, Krieger, Malabar, 1989, pp. 139–140.
181. T. Oka and T. R. Geballe, *Astrophys. J.*, **351**, L53 (1990).
182. J.-P. Maillard, P. Drossart, J. K. G. Watson, S. J. Kim, and J. Caldwell, *Astrophys. J.*, **363**, L37 (1990).
183. S. Miller, R. D. Joseph, and J. Tennyson, *Astrophys. J.*, **360**, L55 (1990).
184. R. Baron, R.D. Joseph, T. Owen, J. Tennyson, S. Miller, and G. E. Ballester, *Nature* (*London*), **353**, 539 (1991).
185. S. J. Kim, P. Drossart, J. Caldwell, J.-P. Maillard, T. Herbst, and M. Shure, *Nature* (*London*), **353**, 536 (1991).
186. L. M. Trafton, T. R. Geballe, S. Miller, J. Tennyson, and G. E. Ballester, *Astrophys. J.*, **405**, 761 (1993).
187. T. R. Geballe, M.-F. Jagod, and T. Oka, *Astrophys. J.*, **408**, L109 (1993).
188. W. P. S. Meikle, D. A. Allen, J. Spyromilio, and G.-F. Varani, *Mon. Not. R. Astron. Soc.*, **238**, 193 (1989).
189. J. Spyromilio, W. P. S. Meickle, R. C. M. Learner, and D. A. Allen, *Nature* (*London*), **344**, 327 (1988).
190. A. Carrington, I. R. McNab, and C. A. Montgomerie, *Chem. Phys. Lett.*, **149**, 326 (1988).

191. A. Carrington, I.R. McNab, and C. A. Montgomerie, *J. Chem. Phys.*, **87**, 3246 (1987).
192. A. Carrington, I. R. McNab, and C. A. Montgomerie, *Phys. Rev. Lett.*, **61**, 1573 (1988).
193. M. S. Child, *J. Chem. Soc., Faraday Trans. 2*, **82**, 1143 (1986).
194. M. S. Child, *J. Phys. Chem.*, **90**, 3595 (1986).
195. R. Pfeiffer and M. S. Child, *Mol. Phys.*, **60**, 1367 (1987).
196. G. Drolshagen, F. A. Gianturco, and J. P. Toennies, *Isr. J. Chem.*, **29**, 417 (1989).
197. W. J. Chesnavich, *J. Chem. Phys.*, **77**, 2988 (1982).
198. K. Rynefors and S. Nordholm, *Chem. Phys.*, **95**, 345 (1985).
199. E. Pollak, *J. Chem. Phys.*, **86**, 1645 (1987).
200. A. V. Chambers and M. S. Child, *Mol. Phys.*, **65**, 1337 (1988).
201. J. M. Gomez-Llorente and E. Pollak, *Chem. Phys. Lett.*, **138**, 125 (1987).
202. S. Miller and J. Tennyson, *Chem. Phys. Lett.*, **145**, 117 (1988).
203. B. Martire and P. G. Burton, *Chem. Phys. Lett.*, **121**, 479 (1985).
204. Ch. Schlier and U. Vix, *Chem. Phys.*, **95**, 401 (1985).
205. J. M. Gomez-Llorente and E. Pollak, *Chem. Phys.*, **120**, 37 (1988).
206. J. M. Gomez-Llorente and E. Pollak, *J. Chem. Phys.*, **90**, 5406 (1989).
207. C. W. Eaker and G. C. Schatz, *J. Phys. Chem.*, **89**, 2612 (1985).
208. C. W. Eaker and G. C. Schatz, *Chem. Phys. Lett.*, **127**, 343 (1986).
209. G. C. Schatz, J. K. Badenhoop, and C. W. Eaker, *Int. J. Quantum Chem.*, **XXXI**, 57 (1987).
210. J. K. Badenhoop, G. C. Schatz, and C. W. Eaker, *J. Chem. Phys.* **87**, 5317 (1987).
211. M. Berblinger, J. M. Gomez-Llorente, E. Pollak, and Ch. Schlier, *Chem. Phys. Lett.*, **146**, 353 (1988).
212. M. Berblinger, Ch. Schlier, and E. Pollak, *J. Phys. Chem.*, **93**, 2319 (1989).
213. J. P. Pique, Y. Chen, R. W. Field, and J. L. Kinsey, *Phys. Rev. Lett.*, **58**, 475 (1987).
214. Y. Chen, S. Halle, D. M. Jonas, J. L. Kinsey, and R. W. Field, *J. Opt. Soc. Am. B*, **7**, 1805 (1990).
215. A. Holle, G. Wiebush, J. Main, B. Hager, H. Rottke, and K. W. Welge, *Phys. Rev. Lett.*, **56**, 2594 (1986).
216. J. Main, G. Wiebush, A. Holle, and K. H. Welge, *Phys. Rev. Lett.*, **57**, 2789 (1986).
217. J. Main, A. Holle, G. Wiebush, and K. H. Welge, *Z. Phys. D*, **6**, 295 (1987).
218. M. Broyer, G. Delacretaz, G. Q. Ni, R. L. Whetten, J. P. Wolf, and L. Woste, *J. Chem. Phys.*, **90**, 4260 (1989).
219. D. Wintgen and A. Hönig, *Phys. Rev. Lett.*, **63**, 1467 (1989).
220. J. M. Gomez-Llorente and H. S. Taylor, *J. Chem. Phys.*, **91**, 953 (1989).
221. E. J. Heller, *Phys. Rev. Lett.*, **53**, 1515 (1984).
222. M. Founargiotakis, S. C. Farantos, G. Contopoulos, and C. Polymilis, *J. Chem. Phys.*, **91**, 1389 (1989).
223. J. Zakrzewski, *Acta Phys. Pol.*, **A77**, 745 (1990).
224. E. Pollak, *Philos. Trans.* R. Soc. *London A*, **332**, 343 (1990).
225. M. Berblinger, E. Pollak, and Ch. Schlier, *J. Chem. Phys.*, **88**, 5643 (1988).
226. J. M. Gomez-Llorente and E. Pollak, *J. Chem. Phys.*, **88**, 1195 (1988).

227. See, for example, M. S. Child, *Semiclassical Mechanics with Molecular Applications*, Chapter 7. Oxford University Press, Oxford, 1991.
228. H. S. Taylor, J. Zakrzewski, and S. Sinai, *Chem. Phys. Lett.*, **145**, 555 (1988).
229. J. M. Gomez-Llorente, J. Zakrzewski, H. S. Taylor, and K. C. Kulander, *J. Chem. Phys.*, **89**, 5959 (1988).
230. J. M. Gomez-Llorente, J. Zakrzewski, H. S. Taylor, and K. C. Kulander, *J. Chem. Phys.*, **90**, 1505 (1989).
231. H. S. Taylor and J. Zakrzewski, *Phys. Rev. A*, **38**, 3732 (1988).
232. J. Tennyson, O. Brass, and E. Pollak, *J. Chem. Phys.*, **92**, 3005 (1990).
233. O. Brass, J. Tennyson, and E. Pollak, *J. Chem. Phys.*, **92**, 3377 (1990).
234. P. O'Connor, J. Gehlen, and E. J Heller, *Phys. Rev. Lett.*, **58**, 1296 (1987).
235. C. R. Le Sueur, J. R. Henderson, and J. Tennyson, *Chem. Phys. Lett.*, **206**, 429 (1993).
236. For excellent reviews see Th. Zimmerman, L. S. Cederbaum, H.-D. Meyer, and H. Koppel, *J. Phys. Chem.*, **91**, 4446 (1987); M. Carmeli, *Statistical Theory and Random Matrices*, Marcel Dekker, New York, 1983.
237. Th. Zimmermann, H. Köppel, L. S. Cederbaum, G. Persch, and W. Demtröder, *Phys. Rev. Lett.*, **61**, 3 (1988).
238. R. L. Sundberg, E. Abrahamson, J. L. Kinsey, and R. W. Field, *J. Chem. Phys.*, **83**, 466 (1985).
239. P. V. Elyutin, *Sov. Phys. Usp.*, **31**, 597 (1988).
240. J. M. Standard, E. D. Lynch, and M. E. Kellman, *J. Chem. Phys.*, **93**, 159 (1990).
241. J. R. Henderson and J. Tennyson, *Chem. Phys. Lett.*, **173**, 133 (1990).
242. B. Peart and K. T. Dolder, *J. Phys. B*, **7**, 1567 (1974).
243. Y. K. Bae and P. C. Cosby, *Phys. Rev. A*, **41**, 1741 (1990).
244. O. L. Polyansky, S. Miller, and J. Tennyson, *J. Mol. Spectrosc.*, **157**, 237 (1993).
245. D. A. Sadovskii, N. G. Fulton, J. R. Henderson, J. Tennyson, and B. I. Zhilinskii, *J. Chem. Phys.*, **99**, 906 (1993).

SUPERCOOLED LIQUIDS

UDAYAN MOHANTY

Eugene F. Merkert Chemistry Center, Department of Chemistry, Boston College, Chestnut Hill, MA 02167, USA

CONTENTS

Advances in Chemical Physics, Volume LXXXIX, Edited by I. Prigogine and Stuart A. Rice.
ISBN 0-471-05157-8

I. ENTROPIC THEORIES

A. General Considerations[1]

When simple, molecular or polymeric liquids are cooled under appropriate conditions of pressure and temperature, thermodynamics requires that such a liquid should be crystalline below the freezing point, while on the other hand, one finds that they can be supercooled [1–5]. A supercooled liquid is dynamically metastable with respect to both the crystalline and the glassy states. However, its structure can be determined, and quantities such as diffusion coefficient, viscosity, thermal conductivity, structure factor as well as specific heat and entropy can be measured [1–5].

A vast amount of experimental data indicates that supercooled states are characterized by universal features [1–5]. The viscosity of glass-forming liquids as a function of temperature shows deviations from an Arrhenius behavior and can be described by the empirical Vogel–Tammann–Fulcher (VTF) or William–Landel–Ferry (WLF) laws [2]. The frequency dependence of shear viscosity and dielectric relaxation are reasonably well accounted for by the Barlow–Erginsav–Lamb (BEL) or the Cole–Cole empirical relations [1b]. There are discontinuities in the specific heat, the thermal expansion coefficient, and the isothermal compressibility at the glass transition T_g, which is defined as the temperature at which the shear viscosity has a value of 10^{13} p [1a]. These discontinuities are rather similar to a second-order transition [1a]. In fact, Doolittle [3] observed that the Ehrenfest relation for a second-order phase transition is satisfied for a number of the glasses studied [1a]. Hysteresis effects in various physical properties as a supercooled liquid is

[1] This introduction is based on a revised version of unpublished notes written in collaboration with S. Yip and in part with T. Keyes.

cooled and then heated through the glass transition region clearly reveal the influence and the importance of nonequilibrium factors [1, 4].

Kauzmann [5] observed that if one extrapolates the equilibrium data from above T_g to lower temperatures, thermodynamic "catastrophies" would occur [5–9]. For example, one obtains negative values for configurational entropy [5–9]. Based on a mean field but statistical mechanical theory of amorphous polymer molecules, Gibbs and DiMarzio (GD) resolved the Kauzmann paradox [6]. They predicted a second-order transition at a temperature T_2 below T_g where the configuration entropy vanishes. Although the results are controversial, an explicit solution to the Flory–Huggins combinatorial problem of packing of semiflexible and flexible chains as well as rods on lattice, has opened the possibility of generalizing the GD theory [7]. Arguments based on an inherent structure formulation suggest that an ideal glass transition would not occur in substances which have low molecular weights [8]. More recently, Montero et al. [9] showed that for polymeric liquids, vibrations lower the GD temperature to some finite temperature. Their arguments, however, are not based on the mean-field Flory–Huggins approximation [9].

A theoretical challenge of potential significance is a qualitative as well as quantitative description of how relaxation time is connected with the topology of the potential energy hypersurface in configuration space. An important step was taken in this direction several years ago by Goldstein [10b] and more recently by Stillinger and Weber (SW) [10a]. Stillinger and Weber [10a] investigated the many-body dynamics of a liquid by separating packing configurations from anharmonic vibrations about these molecular packings. The packing configurations correspond to the local minima in the potential energy hypersurface [10]. The free-energy barrier that needs to be surmounted for relaxation to occur is found to increase logarithmically with scale length [10a]. At low temperatures, the dynamics consist of oscillations within a local potential minimum [10–12]. The oscillatory motions are interrupted by jumps over saddle point to another nearby local minimum [10–12]. The time scale for vibrational motions is much larger than the time to jump over a barrier region [11, 12]. Based on this picture Zwanzig [11] and Mohanty [12] proposed an analysis of the Stokes–Einstein relation. These ideas have lead to an explicit relation between the size of a cooperatively rearranging region and the configuration entropy of the melt [13, 14]. The temperature dependence of relaxation behavior in supercooled liquids is attributed by Adam–Gibbs to variations in the size of a cooperative rearranging region [14]. A theory for diffusion in supercooled liquids has also been developed based on a synthesis of normal mode analysis of liquid state dynamics with hopping rate among the various local minima [15]. The unstable modes in the configuration averaged density of vibrational states play an important role in the diffusive properties of the system [15].

Cooperative dynamics in supercooled liquids have been investigated by a variety of models [16, 17]. The Potts glass has a Kauzmann paradox [18]. A phase transition in these models leads to a resolution of the paradox [18]. The free energy barriers between different structures of glass have been analyzed by using density functional theory in classical fluids [19]. The connection of spin glass theories to laboratory glass, as well as the relevance of fixed points in describing metastable liquid states, are all topics in need of further study [20].

A provoking experiment by Walker et al. [21] indicates the validity of the Stokes–Einstein law in highly viscous 2-ethylexylbenzoate. Whether the Stokes–Einstein and Debye laws hold for other supercooled liquids is an issue that has not yet been unambiguously resolved [22, 23]. A.C. temperature oscillation measurements have yielded considerable insights on enthalpy relaxation time τ_h isothermally [24–26]. The temperature dependence of τ_h for glycerol, as well as its spectral width, is in agreement with mechanical and dielectric relaxation studies [24–25]. The viscosity of *o*-terphenyl, as well as other "fragile" liquids, returns to an Arrhenius-like behavior for times larger than 10^{-6} s [13, 25]. In contrast, enthalpy relaxation is non-Arrhenius to time scale of the order of several hundred seconds [13, 25]. Consequently, the viscous modes can decouple from the slower configurational modes [13, 25]. This leads to the proposal that it is these slow modes that exhibit non-Arrhenius temperature dependence [13, 25].

The molecular level understanding of supercooled liquids has been a major impetus for theoretical studies of the dynamics of dense fluids, quasielastic neutron and light scattering experiments, and simulations of fluids in metastable states.

The discovery that a self-consistent application of the mode-coupling approximation [27, 28] can lead to a freezing transition at which the inverse shear viscosity vanishes, has raised the possibility of whether one can achieve a dynamical model of the liquid-to-glass transition [29]. The original formulation which considered only coupling to density fluctuations referred to as the primitive model, has been generalized to include coupling to current fluctuations, using two very different formalisms [30, 31]. Both formalisms show a cutoff of the ergodic-to-nonergodic transition [30, 31]. Aside from the question of describing the highly viscous state, there are nontrivial theoretical predictions [29] of nonlinear relaxation effects and scaling relations that can be tested against simulation and laboratory experiments.

Molecular dynamics studies have shown that a change in thermal expansivity or compressibility of a fluid occurs upon cooling [32] or compressing [33] into the metastable region of temperature or density. Because of the microscopic time scales in simulations, the relevance of this behavior to glass transition is not clear. A concern here is whether structural arrest has set in when one extrapolates to times scale of laboratory measurements. In simulations one finds that the self-diffusion coefficient show a cross-over behavior in the density range where the compressibility change occurs; its numerical values, however, are too large for the systems to be regarded as being kinetically arrested.

The cross-over behavior is quite general; it is seen in the temperature variations of the shear viscosity data for a number of "fragile" glass-forming liquids [34, 25]. If a characteristic temperature, denoted as T_x [25], is assigned to the cross-over region, then one finds that it lies above the glass transition temperature. Thus there is experimental evidence for regarding transition-like behavior in thermodynamic and transport properties that is distinct from structural arrest [25].

Molecular dynamics simulations are well suited for studying space–time correlation functions, and hence can provide tests of mode-coupling predictions [33, 35]. For fluids interacting through the Lennard–Jones potential, numerical comparisons of the temporal decay of the density correlation function [36] indicate that the primitive model predicts too strong a freezing effect. On the other hand, recent results based on the extended model of fluctuating hydrodynamics are found to be in better agreement with simulations [37]. This result does not necessarily confirm the validity of the mechanism for cutting off the "ideal glass transition" predicted by the primitive model. In any case, one does not expect to see a sharp transition since present mode-coupling approximations do not satisfactorily account for activated state dynamics [25]. At sufficiently low temperatures particle hopping motions must dominate [25]. How to incorporate this type of dynamical process without being empirical is a major theoretical challenge [25].

Neutron [38] and light scattering [39] measurements have also provided valuable insights on the temporal decay of density fluctuations [40]. A nonexponential behavior that is well known in relaxation studies [41] has been observed, and the results interpreted in terms of mode-coupling model descriptions of α and β relaxations [29]. The experiments appear to confirm that mode-coupling theory is capable of treating the relaxation kinetics in supercooled liquids rather well; however, its usefulness as a dynamical model of the glass transition is still an open question.

It is generally believed, though not universally, that β relaxations are a result of localized molecular motions that occur both above and below T_g

[42–44]. The β relaxations are an inherent property of the equilibrium liquid above T_g [41, 44]. However, these molecular motions involve a wide distribution of relaxation times [41, 44]. The concept of heterogeneity has been evoked in a number of theories to explain the nature of α relaxation [45–52]. Johari and Goldstein [42–44] were the first to recognize that such heterogeneity is also responsible for β relaxation [50a]. This idea was implemented by Cavaille et al. [51] with use of the Palmer et al. [52] hierarchical constraint model. Recently, more complicated relaxation processes have been observed in poly(propenylene glycol) by Bergman et al. [40], for example. How such complex relaxation behavior can be described by existing techniques is also an open question. The phenomenological models for α/β relaxations by Kivelson and co-workers [50c] may be of considerable value in analyzing such relaxation kinetics.

A new phenomena called retrograde vitrification has been discovered by Condo et al. [53]. These authors, based on a lattice theory and the GD configuration entropy idea, predicted that a liquid glass transition would occur with increasing temperature [53]. Chamberlin [54] showed that the relaxation behavior of glass-forming liquids could be described by assuming a Gaussian distribution of independently rearranging regions or clusters. The non-Arrhenius temperature dependence of relaxation rate is attributed to variations of the cluster size with temperature [13, 14, 54]. Kob and Schilling [55] explicitly studied the dynamics of a chain of particles in a one-dimensional potential that has a large number of metastable states. The surprising result is that the system is ergodic at all temperatures as depicted by various correlation functions [55]. Various models of dynamical motions past entropic barriers have been investigated in detail by Zwanzig [56, 57]. Mohanty et al. [58] showed how dynamical motions past entropic barriers in configuration space lead to nonequilibrium generalization of the Adam–Gibbs (AG) theory.

The thermal conductivity of glassy materials above 1 K has a characteristic plateau, while the specific heat C_p rises faster than T^3 [1, 59, 60]. As the temperature is raised further the thermal conductivity increases. In contrast, both C_p/T^3 and the vibrational density of states $g(\omega)/\omega^2$ develop a peak, where ω is the frequency [1, 59, 60]. This is the so-called boson peak, which can be measured by Raman and neutron scattering techniques [61]. The vibrational modes in glasses as well as the tunneling and the relaxation modes coexist with sound waves [59]. The various anomalies in undercooled liquids at low temperatures have been accounted for by a soft-potential model developed by Karpov et al. [62] and by Buchenau [59].

A host of experiments by Frauenfelder and co-workers [63, 64] and theoretical analysis by Wolynes and co-workers [65, 66] established that proteins, glasses, and spin glass share many features in common [67, 68]. Conformation motions show stretched exponential relaxation while the temperature dependence of the rate coefficient obeys either the Ferry or the VTF expressions [2, 66]. These dynamical properties have been discussed in terms of the topology of the potential energy landscape [10, 65–68].

Calorimetric experiments provided valuable insights into the thermodynamics of globular proteins [69, 70]. These experiments not only indicate the two-state nature of denaturation process but also reveal a number of universal features. For example, Privalov [69] found that entropy and enthalpy changes for a number of globular proteins extrapolate to almost the same value at or near 110 °C [71]. The heat capacity difference between the native and the unfolded states is almost independent of temperature up to 80 °C [69, 71]. The free energy of denaturation, $\Delta G(T)$, for chymotrypsinogen and ribonuclease has a parabolic temperature dependence, that is, $\Delta G(T) = \mathrm{A} + \mathrm{B}T + \mathrm{C}T^2$, where A, B, and C are constants [72–74]. This expression for $\Delta G(T)$ suggests that proteins can also unfold at low temperatures [72–74]. The phenomena is called cold denaturation and has been experimentally observed [72, 73]. Although analysis of protein stability have been carried out by a number of authors [71, 74–78], further theoretical and experimental understanding of cold denaturation is desperately needed. It may be fruitful to exploit a similarity between the statistical mechanics of cold denaturation and the AG theory of cooperative relaxation in glass-forming liquids [79].

The emphasis of this article is to survey particular new developments in equilibrium and dynamical properties of supercooled liquids. Certain topics like mode-coupling theories, β relaxation, low-temperature anomalies in glasses, and protein folding will not be discussed. What is the status of configurational entropic theories? Is there an underlying phase transition below T_g as some models predict? What is the entropic basis of the VTF relation? What are the signatures of glass transition? Is there a characteristic length scale that diverges? Are there nanoscale inhomogeneities that are responsible for nonlinear relaxation? How does one exploit the topology of the potential energy surface to analyze equilibrium and dynamical properties of glass-forming liquids? How does one construct simple models for viscosity and diffusion in undercooled liquids? Are Stokes–Einstein–Debye relations valid in supercooled liquids? These are some of the questions that will be addressed.

B. Configurational Entropy Model

The temperature dependence of relaxation time in the AG model is described in terms of the size of a cooperatively rearranging region [14].

A cooperative rearranging region is defined to be a subsystem that can rearrange into configurations independent of its environment [14]. The average transition probability that allows cooperative rearrangements, $W(z^*, T)$, is expressed in terms of a critical number[2] of monomer segments z^* [14]

$$W(z^*, T) = A \exp(-z^* \Delta\mu / k_B T) \tag{1.1}$$

where $\Delta\mu$ is the molar enthalpy, k_B is Boltzmann's constant, T is the absolute temperature, and the variable A is weakly temperature dependent compared with the exponential factor. The derivation of Eq. (1.1) is based on two main assumptions [12–14]. First, the rearranging region interacts weakly with its environment [12–14]. Second, the ensemble is close to an equilibrium distribution [12–14].

The decisive step taken by AG is to relate the critical size z^* to the molar configurational entropy $S_c(T)$ of the undercooled melt [13, 14]

$$z^*(T) = s^* N_A / S_c(T) \tag{1.2}$$

$$S_c(T) = \Delta C_p(T_g) \ln(T/T_2) \tag{1.3}$$

where N_A is Avogadro's number, s^* is the critical configuration entropy, and $\Delta C_p(T_g)$ is the specific heat difference between the equilibrium liquid and the glassy state at T_g. Observe that the temperature T_2 is obtained from the equilibrium properties of the amorphous phase [13, 14]. On substituting Eqs. (1.2) and (1.3) in Eq. (1.1), we obtain the transition probability in terms of the configurational entropy of the undercooled melt [14]

$$W(z^*, T) = A \exp\{-C/TS_c(T)\} \tag{1.4}$$

where $C = s_c^* \Delta\mu / k_B T$. An important characteristic of $W(T)$ is that it has an essential singularity as $T \to T_2$ [12, 13].

Based on Eq. (1.4), AG theory predicted a correlation between T_2 and T_g [12–14]. The ratio T_2/T_g was found to be in agreement with calorimetric and viscometric data [80–83]. The significance of these results is that it implies not only an underlying thermodynamic transition at T_2, but that the thermodynamic properties of an undercooled melt govern the temperature dependence of relaxation time within 100 °C of the glass transition [12–14].

[2] It is the smallest size that allows cooperative rearrangements [14]. The average transition probability refers to the fact that the transition probability has been averaged over all possible values of z from a lower limit z^* to an upper limit of infinity [12–14].

C. Generalization of Adam–Gibbs Model

The enormous success of the AG model is partly due to its simplicity and the idea of relating the critical size z^* to the configuration entropy of the melt.[3] This state of affairs has been questioned by Mohanty [83, 84] who recently showed that many of the predictions in the AG model are obtained without evoking relations (1.2) and (1.3). We now discuss Mohanty's formulation in detail.

1. *Topological Features*

It is useful to write the critical size as $z^*(T) = \alpha_1(T)/f(T)$, where $f(T)$ is an arbitrary function of temperature.[4] The parameter α_1 is assumed to be weakly temperature dependent in comparison to $f(T)$. The basic hypothesis by Mohanty [83, 84] is that there exists a temperature T_2 at which the critical size $z^*(T_2)$ diverges. Hence, $f(T_2)$ vanishes.

The average transition probability $W(T)$ is inversely proportional to the relaxation time $\tau(T)$. A logarithmic shift factor $\log[\tau(T)/\tau(T_s)]$ is introduced

$$-\log a_T = \log[\tau(T)/\tau(T_s)] = (\Delta\mu/2.303k_B)[z^*(T_s)/T_s - z^*(T)/T] \tag{1.5}$$

which can be rewritten as [83, 84]

$$-\log a_T = a_1(T_s)(T - T_s)/(a_2(T) + (T - T_s)) \tag{1.6}$$

where [84]

$$a_1(T_s) = \Delta\mu\ \alpha_1/2.303k_B T_s f(T_s) \tag{1.7}$$

$$a_2(T) = T_s f(T_s)/[f(T_s) + (T/(T - T_s))(f(T) - f(T_s))] \tag{1.8}$$

Equation (1.6) is a relation resembling the WLF equation. However, the parameters in Eq. (1.6) are not universal [83, 84, 14]. The relevant

[3] See Eqs. (1.2) and (1.3).

[4] For justification of this form see the renormalization group arguments in Section I.4.1. By construction, the function $f(T)$ is dimensionless. Consequently, $\alpha_1(T)$ is also dimensionless. By assumption the critical size diverges at T_2. This temperature, however, may not necessarily be the same as either the AG temperature T_2 or the VTF temperature T_0; see Section II. For simplicity of notation we have avoided introducing as yet another temperature.

question, therefore, is under what conditions does Eq. (1.6) yield the "universal" WLF parameters $c_1 = 8.86$ and $c_2 = 101.6\,°\mathrm{C}$ [14, 83, 84] ?[5]

2. *WLF and VTF Parameters*

To answer the question posed at the end of Section I.C.1, we define a dimensionless function $F(T)$ via $f(T) = \alpha_0(T)F(T)$. In terms of $F(T)$ and $\alpha(T) = \alpha_1/\alpha_0(T)$, we rewrite $a_1(T_s)$ and $a_2(T)$ as [84]

$$a_1(T_s) = \Delta\mu\, \alpha(T_s)/2303k_B T_s F(T_s) \tag{1.9}$$

$$a_2(T) = T_s F(T_s)/\{(1 + F(T_s)) + [\alpha_0(T)F(T) - \alpha_0(T_s)F(T_s) - ((T - T_s)/T)\alpha_0(T_s)]T/((T - T_s)\alpha_0(T_s))\} \tag{1.10}$$

On approximating $a_2(T)$ by $a_2(T_s)$, one obtains from Eq. (1.10) a relation due to Mohanty [83]

$$\alpha_0(T)F(T) = \alpha_0(T_s)F(T_s) - ((T - T_s)/T)\alpha_0(T_s) \tag{1.11}$$

Let us now choose T_g as the reference temperature. Since the critical size $z^*(T_2)$ diverges, relation (1.11) leads to an explicit expression for the ratio T_g/T_2 [83, 84]

$$T_g/T_2 = 1 + c_2'/(T_g - c_2') = 1 + F(T_g) \tag{1.12}$$

$$T_g - T_2 = c_2' \tag{1.13}$$

The WLF parameters c_2' and c_1' correspond to the choice $T_s = T_g$.[6] The significance of Eqs. (1.12) and (1.13) is that it shows how the kinetic glass transition temperature is correlated with the thermodynamic temperature T_2 [83, 84]. Note that in deriving Eqs. (1.12) and (1.13) we have not made use of configurational entropy or free-volume concepts.

We summarize some of the results obtained by Mohanty [83, 84]. First, the magnitude of a_1 and a_2 are governed by the term $T_s F(T_s)$. Second, validity of WLF parameters depends on whether or not $\Delta\mu\, \alpha(T_s)/2.303k_B$ is constant from one substance to the next [14, 85]. Third, the VTF empirical temperature T_0 is identical to T_2 if the reference temperature is taken to be T_g [83, 84]. Fourth, the VTF parameter B

[5] The WLF equation is of the form $-\log a_T = c_1(T - T_s)/(c_2 + (T - T_s))$. The values $c_1 = 8.86$ and $c_2 = 101.6\,°\mathrm{C}$ are an accurate representation of the experimental data if T_s is appropriately chosen [1, 2, 14]. One finds that T_s is approximately 50 °C below T_g [1, 2, 14].

[6] These WLF parameters are not universal [14, 83, 84].

equals $[\Delta\mu\ \alpha/k_B T_g]T_2$ [84]. Fifth, the ratio T_g/T_2 is remarkably constant for widely different materials [83, 84]. The mean values for T_g/T_2 are $1.28 \pm 5.78\%$ [83, 84]. Finally, as $T \to T_s$, the shift factor becomes Arrhenius [83, 84].

3. *Nonlinearity of Relaxation*

There is a host of experiments that unambiguously indicate the nonlinear nature of relaxation in the glassy state [4]. By nonlinear one means that relaxation time depends on structure [4, 84]. The mean relaxation time for shear compliance and volume relaxations during annealing of glasses obeys the empirical form suggested by Narayanaswamy [4]

$$\tau = \tau_0 \exp[x\ \Delta h^*/RT + (1-x)\ \Delta h^*/RT_f] \tag{1.14}$$

where Δh^*, τ_0, and x are constants, and R is the gas constant. The value of x lies between 0 and 1. It reflects the nonlinearity of the relaxation process [4]. The parameter Δh^* governs the rate at which the fictive temperature T_f changes with rate of cooling [4]. T_f is the temperature at which the structure would be at equilibrium [4]. Hence, for a nonequilibrium glass T_f may be less or greater than T. For an equilibrium liquid, T_f and the actual temperature T are identical [4, 84].[7]

The phenomenological approach by Narayanaswamy [4] has several shortcomings. First, the relation is empirical. Second, for an equilibrium liquid Eq. (1.14) predicts an Arrhenius behavior that is in conflict with WLF and VTF relations [4, 84]. Third, it does not explain the observed inverse correlation between x and Δh^* [4, 84].

The activation enthalpy is defined by [4, 84]

$$\begin{aligned}\Delta h^*/R &= d \ln \tau / d(1/T) \\ &= (\Delta\mu\ \alpha_1/k_B)[1/f(T_f) + T(dT_f/dT)(1/f^2(T_f))\ df/dT_f]\end{aligned} \tag{1.15}$$

Two important results are deduced from Eq. (1.15). First, the nonlinear parameter x turns out to be [84]

$$x = F(T_g)/[F(T_g) + 1] \tag{1.16a}$$

$$= 1 - T_2/T_g \tag{1.16b}$$

[7] Fictive temperature measured from a particular property does not necessarily have to be an appropriate order parameter that specifies another property of interest [4, 84].

Substituting Eq. (1.12) in Eq. (1.16a) leads to Eq. (1.16b). Second, the enthalpy of activation at T_g is inversely proportional to the square of x [83, 84]

$$\Delta h^*/R = [\Delta\mu\ \alpha(T_g)/k_B](T_g/T_2)x^{-2} \tag{1.16c}$$

Both these predictions are in agreement with Hodge's experimental and theoretical results [4]. The crucial difference is that Hodge's arguments are based on AG configurational entropy theory [4, 14].

D. Renormalization Group Analysis

It is well known that conventional techniques for evaluating the partition function does not easily lead to the establishment of nonanalytic behavior [86, 87]. This occurs because the partition function is a finite sum of exponential quantities [86, 87]. Each such quantity is analytic in temperature. Any nonanalyticity can occur only in thermodynamic limit [86, 87]. It is therefore fruitful to recast the AG theory in differential form [13, 84]. Mohanty [13, 84] argued that the singularity in a supercooled liquid at T_2 or T_0 arises from the solutions to an appropriate set of differential equations. These differential equations must, however, be free from any singularities [13, 84, 86, 87].

To see now such differential equations arise, let us consider the variation of the transition probability with the system size in the AD model

$$dW(z^*, T)/dz^* = -A_0W(z^*, T) \qquad A_0 = \Delta\mu\ A/k_BT \tag{1.17a}$$

The cooperative region varies with the configurational entropy as

$$dz^*/dS_c = -z^*/S_c \tag{1.17b}$$

The differential equations (1.17a) and (1.17b) are reminiscent of those appearing in Wilson's theory of critical phenomena. We exploit this idea in the next two sections.

1. Size Dependence of Rearranging Region

The total volume of the system, $\mathbf{L}^3$, is partitioned into cubic cells or blocks. Each cell as z_L number of molecules or monomer segments. Few configurations are available near T_2. This observation suggests that correlations between blocks must be taken into account in formulating the temperature dependence of relaxation rate in supercooled liquids [13, 84, 86, 87]. By following critical phenomena, we introduce the notion of a correlation length ξ. The parameter ξ is a measure of the size over which such rearranging regions are correlated [86, 87]. By assumption the correlation length ξ is much larger than the block length L.

As in critical phenomena we are interested in long wavelength fluctuations [86, 87]. The short-wavelength fluctuations have been integrated out from the partition function by appropriate renormalization group transformations [86, 87]. Since the partition function is invariant to a renormalization group transformation, there is a relation between the block Gibbs free energy and the Gibbs free energy per unit volume [13, 84, 86, 87]

$$g(z, \alpha) = L^{-3} g(z_L, \alpha_L) \tag{1.18}$$

The correlation lengths $\xi(z, \alpha)$ of the system and the block $\xi(z_L, \alpha_L)$ are also related [13, 84, 86, 87]

$$\xi(z, \alpha) = L\xi(z_L, \alpha_L) \tag{1.19}$$

Here, α is a "relevant" variable in the sense of critical phenomena [12, 13, 86]. For example, one may imagine α to the number of flexed bonds in the GD theory of the glassy state [12, 13]. From Eq. (1.18) one obtains the desired relation [13] between the entropy density of the system and the entropy of a block $s(z_L, \alpha_L)$

$$s(z, \alpha) = L^{-3} s(z_L, \alpha_L) + (3/L^4)(dL/dT) g(z_L, \alpha_L) \tag{1.20}$$

There are N_A/z_L blocks in a mole of segments [13, 14, 84]. The number of blocks is also given by $\mathbf{L}^3/L^3$ [13, 14, 84]. The combination of this observation with Eq. (1.20) leads to a connection between the size and the total entropy [13, 84]

$$\begin{aligned} z_L = N_A s(z_L, \alpha_L)/S(z, \alpha) + \{3N_A/S(z, \alpha)\}(1/L) \\ \times (dL/dT) g(z_L, \alpha_L) \end{aligned} \tag{1.21}$$

Note that if z_L is approximated by the critical size z^* and if the second term in Eq. (1.21) is ignored, then one obtains the AG relation (1.2) [13]. Equation (1.21) is also in agreement with what was stated in Section I.3.1, namely, that the critical size $z^*(T)$ could be written as $\alpha_1(T)/f(T)$, where $f(T)$ is an arbitrary function of temperature (see footnote 4).

2. *Effective Couplings*

The variations of α_L and z_L with length scale is obtained by differentiating both sides of Eqs. (1.18) and (1.19) with respect to L [13, 84, 86]

$$(dz_L/dL)\,\partial g(z_L, \alpha_L)/\partial z_L + (d\alpha_L/dL)\,\partial g(z_L, \alpha_L)/\partial \alpha_L = (3/L)g(z_L, \alpha_L) \tag{1.22}$$

$$(dz_L/dL)\,\partial \xi(z_L, \alpha_L)/\partial z_L + (\partial \xi(z_L, \alpha_L)/\partial \alpha_L)\,d\alpha_L/dL = -\xi(z_L, \alpha_L)/L \tag{1.23}$$

The solutions to Eqs. (1.22) and (1.23) are of the form [13, 84, 86]

$$dz_L/dL = U(z_L, \alpha_L)/L \tag{1.24}$$

$$d\alpha_L/dL = W(z_L, \alpha_L)/L \tag{1.25}$$

The functions $U(z_L, \alpha_L)$ and $W(z_L, \alpha_L)$ are defined by Eqs. (1.24) and (1.25). An important property of these functions is that they depend on L via z_L and α_L [13, 86, 87]. If U and W are analytic at T_2, then these equations could be solved [13, 86, 87]. Equations (1.24) and (1.25) are integrated until the scale length is of the order of the correlation length [13, 86, 87]. Then, $\xi(z_L, \alpha_L)$ and $g(z_L, \alpha_L)$ are evaluated by some other techniques [13, 86, 87]. The functions $g(z, \alpha)$ and $\xi(z, \alpha)$ are constructed from Eqs. (1.18) and (1.19).

The connections and the relevance of a fixed point to T_2 are far from clear. One anticipates that at high temperatures, renormalization group trajectories flow to a disordered fixed point [13, 20]. In contrast, at low temperatures the trajectories may flow to an ordered fixed point [13, 20]. Another possibility is that a line of metastable states may end at a fixed point [13, 20]. In any case, a novel interpretation of the AG model is that there exists effective couplings z_L and α_L, which satisfy Eqs. (1.18) and (1.19) for all L [13, 84].

3. *Scale Dependence of Barrier Height*

To illustrate how renormalization group strategies are formulated to the problem at hand, consider the SW inherent structure approach to condensed phases.[8] Let us now suppose that a suitable renormalization group transformation has been carried out to eliminate irrelevant degrees of freedom within a length scale ℓ. The number of minima $\Omega(\ell)$ in the amorphous manifold increases with ℓ and number of particles N as

[8] This topic will be discussed in detail in Chapter II. The parameter Ω_p is a factor due to permutations of identical particles.

$\Omega_p \exp(N\theta(\ell))$ [10a]. The exponent $\theta(\ell)$ is a decreasing function of ℓ as coarse graining eliminates some of the minima in the potential surface [10a]. One, therefore, writes $\theta(\ell)$ as $\exp -h(\ell)$, where $h(\ell) = h_\infty \ell^q$ and q is a positive exponent. The cumulative density of relaxation rates $D_c(\lambda)$ may be approximated by [10a]

$$D_1 \exp\{-a(\tau\lambda)^{-p}\} \tag{1.26}$$

provided λ is small and positive, D_1 is larger than zero, and p and a are constants related to the Kohlrausch–Williams–Watts (KWW) exponent β_0 [10a].

Based on these considerations Stillinger has argued that if the time dependence of the correlation function is of the KWW form, then the free energy barriers $b(\ell)$ between cells increases logarithmically with ℓ [10a]

$$b(\ell) \approx (1/\beta_0)\ln(\tau/\tau_0) + (1/\beta_0 p)\ln h(\ell) \tag{1.27}$$

Observe an entropic-like contribution to $b(\ell)$. The derivation of Eq. (1.27) proceeds by evaluating the concentration of metabasins and the cumulative density of relaxation rates at fixed ℓ [see Eq. (1.26)] [10a, 13]. Then the density of localized transitions is identified with metabasins concentration for this value of ℓ [10a, 13].

The renormalization group equation for $b(\ell)$ is deduced from Eq. (1.27) [10a, 13]

$$db(\ell)/dl = q/\beta_0 p \tag{1.28}$$

A consequence of Eq. (1.28) is that if the temperature dependence of λ in Eq. (1.26) is governed by the AG relation, then $d[TS_c(\ell)]/dl = c_1[TS_c(\ell)^2]/\ell$, where c_1 is a constant [13].

E. Entropy of Supercooled Liquid

This section obtains insights into the nature of the Kauzmann paradox by obtaining a lower bound to the entropy density at low temperatures for an arbitrary one component system. The lower bound to entropy is based on a generalization of an elegant argument due to Leggett [88].

The system is described by a complete set of operators $\{A_\mu\}$.[9] The eigenvalues and the minimum energy of these operators are denoted by

[9] The operators commute with each other as well as the Hamiltonian. The parameter A_μ is an integral over a local operator: $\int A_\mu(\mathbf{r})\,d\mathbf{r}$.

$\{\alpha_\mu\}$ and $E_\mu(\alpha_\mu)$, respectively.[10] Let the states of the system satisfy the inequality $|\langle (A_\mu)^p \rangle| \leq V^{bp}$, $b < 1$ and for all p; the angular brackets $\langle\ \rangle$ denote thermal average and V is the volume of the system [88, 84]. On introducing the quantities $\mu_v = \{\partial E_{\rm m}(\alpha_v)/\partial \alpha_v\}$ and $\chi_\mu^{-1} = \{\partial^2 E_{\rm m}(\alpha_\mu)\, \partial \alpha_\mu^2\}$ both evaluated at α_μ^0 and noting that $\partial^p E_{\rm m}(\alpha_\mu)/\partial \mu_\mu^p$ scale as V^{-p+1} we obtain to $O(V^{-1+b})$, a generalization of an inequality due to Leggett that is valid at finite T [88, 84]

$$\Delta\langle E\rangle \geq \sum_v{}' \mu_v \Delta\langle \hat{A}_v\rangle + \frac{1}{2} V^{-1} \sum_\mu{}' \chi_\mu^{-1} \Delta\langle (\hat{A}_\mu)^2\rangle \tag{1.29}$$

The parameter $\{\alpha_\mu^0\}$ are eigenvalues evaluated at the Kauzmann temperature. The fluctuations of A_μ from its ground-state value is $\hat{A}_\mu$ and $\Delta\langle\ \rangle = \langle\ \rangle - \langle\ \rangle_0$ [88, 84]. The sum in Eq. (1.29) is over a subset of $\{A_\mu\}$.

Assume that the supercooled liquid is in internal equilibrium. Divide the system into n blocks each of volume V_i. Each V_i is a cube whose side is of length d. In analogy with Eq. (1.29) we postulate an inequality [84, 88]

$$\Delta\langle E\rangle \geq \sum_\nu{}' \mu_\nu \Delta\langle \hat{A}_\nu\rangle + \frac{1}{2} \sum_i \sum_\mu{}' V_i^{-1} \chi_\mu^{-1} \Delta\langle (\hat{A}_\mu^i)^2\rangle \tag{1.30}$$

The subvolume V_i has associated with it the fluctuation operators $\{\hat{A}_\mu^i\}$ [84, 88].[11] As pointed out by Leggett, one cannot obtain Eq. (1.24) by using an extensive property of the average energy [88].

The thermal average in Eq. (1.30) is now evaluated by the fluctuation-dissipation theorem [89]. This evaluation leads to a lower bound for the entropy per unit volume of a glass-forming liquid [84, 88]

$$s(T) \geq 1/(2\pi)^3 \sum{}' \int_0^T \int d\mathbf{k}\, |\phi(\mathbf{k})|^2 (dL_\mu(\mathbf{k}, T')/dT' + \frac{1}{2} d(\chi_\mu^{-1} |\Delta\langle \hat{A}_{\mathbf{k}\mu}\rangle|^2)/dT')\, dT'/T' \tag{1.31}$$

[10] The eigenvalues are extensive [88]. The $\{\alpha_\mu\}$ should not be confused with α_L in Eqs. (1.18) and (1.19).

[11] Apart from a scale factor. For solids, the analogue of Eq. (1.30) is $\Delta\langle E\rangle \geq \frac{1}{2} \sum V_i^{-1} C_{11} (\delta N_i)^2$; δN_i is the fluctuations in the particle number in V_i and C_{11} is the coefficient of elastic stiffness [88, 84].

$$L_{\mu}(\mathbf{k}, T) = h/(4\pi\chi_{\mu}) \int_0^{\infty} [\coth(h\omega/4\pi k_B T)\chi''_{\mu}(\mathbf{k}, \omega; T)$$
$$- \coth(h\omega/4\pi k_B T_k)\chi''_{\mu}(\mathbf{k}, \omega; T_k)]\, d\omega \quad (1.32)$$

The parameter $\hat{A}_{k\mu}$ are the Fourier transforms of $\hat{A}_{\mu}(\mathbf{r})$, $\phi(\mathbf{k}) = \Pi_i \sin(k_i d/2)/(k_i d/2)$ [84, 88]. The quantity $\chi''(\mathbf{k}, \omega; T)$ is the space and time Fourier transforms of the imaginary part of the retarded response function [84, 88, 89].

Assume that the product of two local operators can be expanded as an operator product expansion in the limit $T \to T_k$, the Kauzmann temperature,

$$\langle \hat{A}_{\mu}(\mathbf{r})\hat{A}_{\mu}(\mathbf{r}')\rangle = \sum D_{\mu\beta}(\mathbf{r}_{\rm rel})\langle \hat{A}_{\beta}(\mathbf{r}_{\rm cm})\rangle \quad (1.33)$$

Here, $\mathbf{r}_{\rm cm}$ and $\mathbf{r}_{\rm rel}$ are the center of mass and relative coordinates, respectively [90]. In this case, as shown by Mohanty [84], the entropy vanishes at T_k. Since μ_v vanishes, if $\{\hat{A}_v\}$ are not conserved,[12] this raises the possibility that entropy can also vanish, provided the coefficients $D_{\mu\beta}$ equal $2\delta_{\mu\beta}\alpha^0_{\mu}/n$ [84].

II. SLOW AND FAST MODES

A. Entropic Basis of the Vogel–Tammann–Fulcher Relation

Adam–Gibbs [91] and Bestul–Chang observed a correlation betwen the temperature T_0 in the Vogel–Tammann–Fulcher (VTF) relation [93–96]

$$\tau = A \exp DT_0/(T - T_0) \quad (2.1)$$

and the so-called Kauzmann temperature $T_{\rm k}$. At $T_{\rm k}$ the entropy of a supercooled liquid is equal to the entropy of the corresponding crystalline phase. Consider the configurational entropy in the AG model [91]

$$S_{\rm conf} = \int \Delta C_p d \ln T' \quad (2.2)$$

If the temperature dependence of the change in the heat capacity at constant pressure (C_p) between the liquid and the crystalline phase is given by [97–99]

$$\Delta C_p = \tilde{D}/T \quad (2.3)$$

[12] If the set $\{A_v\}$ is conserved, then $\Delta\langle A_v\rangle$ vanishes (see [88]).

where $\tilde{D}$ is a constant, then the parameters T_0 and T_k are identical.

Initially, experiments confirmed the accuracy of the approximation for ΔC_p stated in Eq. (2.3) [100–102]. However, as systematic and careful studies were carried out at higher viscosities, deviations of the VTF parameter T_0 from T_k were observed [103–106]. In fact, viscosity of some supercooled liquids showed Arrhenius-like behavior at temperatures near the glass transition temperature T_g. For viscosities η larger than 10^5 P, the relaxation time is Arrhenius for both α-phenyl-*o*-cresol and Salol [104]. The high-temperature relaxation data for α-phenyl-*o*-cresol and Salol, including *o*-terphenyl and tri-[*a*]-naphthyl benzene are well described by the VTF equation. The conclusion one may draw is that measurements of different transport properties may lead to a spread in the value of the temperature T_0 [13].

The VTF equation is also satisfied by hydrogen-bonded liquids such as 1,3-butanediol, sorbitol, propylene glycol, and ethylene glycol over 12 orders of magnitude variations in the relaxation time [103]. Thermodynamic and dielectric loss measurements for these systems revealed the VTF parameter T_0 to be identical with the Kauzmann temperature T_k [103].

For fragile liquids, the ratio T_g/T_0 approaches unity. Furthermore, relaxation measurements of these liquids indicate a non-Arrhenius region encased between a high- and a low-temperature Arrhenius behavior [107]. In contrast to fragile liquids, the change in heat capacity near T_g for strong liquids are rather small. Consequently, one finds that T_0 is usually much less than T_g [107].

Adam–Gibbs theory argued that the difference between the entropy of the glass and the crystal at $T = 0$ K

$$\Delta S_0 = S_{\text{glass}}(T=0) - S_{\text{crys}}(T=0) \tag{2.4}$$

is related to the configurational entropy of the liquid at T_g [91]. Therefore, the ratio T_g/T_0 may be estimated provided the quantities ΔC_p and ΔS_0 are known experimentally [91]. By this procedure Bestul and Chang [92] calorimetrically determined the ratio T_g/T_2 for several glass-forming liquids

$$T_g/T_2 = 1.29 \pm 10.9\% \tag{2.5}$$

From the zero-point entropy of amorphous propylene and polypropylene,

Passaglia and Kevorkian [108, 109] found the ratio T_g/T_2 to be

$$T_g/T_2 = 1.26 \tag{2.6}$$

The fragility of Ge–As–Se has recently been studied by Tatsumisago et al. [110]. By varying the composition of the various species, the temperature dependence of the viscosity can be made to mimic either strong or fragile liquids. At $\langle r \rangle = 2.4$ (the so-called percolation threshold), the jump of the heat capacity at T_g is also small and Ge–As–Se displays minimum fragility [110]. But if ΔC_p is small, the configurational entropy must be almost independent of T. Then, from the AG relation one can show that the temperature dependence of cooperative relaxation becomes Arrhenius. For Se, $T_k = 240 \pm 10$ K, while for As_2Se_3, $T_k = 236 \pm 10$ K [110–112]. The T_g values at these two compositions are 307 and 455 K, respectively [110]. Therefore, the ratio T_g/T_k is 1.23 and 1.90, respectively, for these two systems [110]. The constant D in the VTF relation is, however, found to vary between 8.5 and 33 [110].

There is another class of systems, the so-called orientationally disordered or plastic crystals. Examples of glassy crystals include appropriate phases of cyclohexanol, ethanol, and difluorotetrachloroethane. In these systems glass-like transition occurs at a temperature at or near the glass transition temperature T_g of the liquid phase of the same substance [113–119].

Cyclohexanol freezes into a face centered cubic (fcc) lattice [116]. However, its high-temperature phase shows a bump in the heat capacity. This characteristic is reminiscent of glass-forming liquids near T_g [116]. The anomaly in cyclohexanol is attributed to the freezing of orientational degrees of freedom [116]. Another example is ethanol, which can be converted into a glassy crystal by a cooling rate of $50\ \mathrm{K\ min^{-1}}$ [116]. At slower cooling rate ($\approx 2\ \mathrm{K\ min^{-1}}$) a metastable crystal II phase is obtained instead. This metastable crystal II phase transforms to a glassy crystal II via another glass transition. The glass transition temperature of the crystal II phase practically coincides with that of the corresponding supercooled liquid phase [116].

Angell et al. [119] proposed that it may be fruitful to classify the plastic crystals as fragile or strong. The data includes relaxation time measurements by brillouin absorption maximum [114], Raman line width reorientation time [114], and dielectric data [116], on such systems as $(CCl_2F)_2$, cyclohexanol, cyanoadamantane, thiophene, and $CsNO_2$. Adachi et al. [116] showed that in the case of an intermediate plastic crystal, cyclohexanol, the ratio T_g/T_k is 1.18. This value is typical of supercooled polymeric liquids. Consequently, more experiments on fragile plastic crystal are warranted to unravel the intricacies of the Kauzmann paradox [119].

B. Decoupling of Viscous and Configurational Modes

As pointed out in Section II.A, glass-forming liquids in the intermediate range satisfy the VTF equation over several decades in relaxation time provided the parameter T_0 is close to the Kauzmann temperature T_k. In contrast, the fragile liquids exhibit either a partial or in some cases a complete return to Arrhenius-like behavior as $T \to T_g$ [107, 13]. This fact has raised skepticism as to whether a supercooled liquid has a thermodynamic singularity at the Kauzmann temperature T_k [104, 107, 120].

To address this important issue, Angell and co-workers [121, 122] made a number of studies of how stress relaxes in fragile liquids. There are several important results obtained by these authors [121, 122]. First, the experimental relaxation quantity that is measured depends on the property of the state that has been perturbed. Second, stress decays with time as

$$\sigma(t) = \exp -(t/\tau)^{\beta} \tag{2.7}$$

Third, the value for the exponent β in Eq. (2.7) depends on how far the system deviates from equilibrium [123]. Finally, the relaxation of stress is not the slowest process. For example, the relaxation time towards equilibrium for the fragile liquid $3KNO_3{\cdot}2Ca(NO_3)_2$ is orders of magnitude slower than the stress relaxation time [122, 13].

DSC experiments measure enthalpy relaxation. In these experiments T_g is defined as that temperature where $\tau_H \sim 100$–200 s [124, 125, 126, 13]. Since the shear viscosity η is typically of the order of 10^{11} P, it would lead to a shear relaxation time τ_s of the order of 10^{-1} s [126, 13]. Consequently, due to the wide separation of time scales, the modes associated with structural relaxation decouple from shear stress [122, 126, 13].

If in fragile liquids the DSC experiments measure slow relaxation process [124, 125, 126, 13], then one is lead to the conclusion that it is precisely these slow modes whose temperature dependence is non-Arrhenius as $T \to T_g$ [126, 122, 13]. There is other evidence to back this conclusion. First, one usually finds larger relaxation time for enthalpy than for shear viscosity [123, 124, 126, 13]. Second, the activation energy for viscous flow in ZBLA [127] is approximately 20% less than for enthalpy relaxation [122, 126, 13]. Third, the temperature dependence of the shear viscosity for the fragile liquid *o*-terphenyl displays Arrhenius-like form for τ larger than 10^{-6} s [122, 126, 128, 129, 13]. In contrast, the temperature dependence of enthalpy relaxation is non-Arrhenius down to

T_g, which corresponds to τ approximately 100 s [122, 126, 13]. A VTF fit to the data yields $T_0 = T_k$ [128, 129]. Finally, enthalpy relaxation has been measured isothermally for glycerol by ac methods [122, 126, 128, 129]. The ac method probes the linear response of the system to a temperature perturbation. The temperature dependence of τ_H satisfies the VTF relation all the way to T_g [122, 126, 128, 129, 13]. The relaxation data agrees well with dielectric and mechanical measurements [128, 129].

In summary, the relaxation of the *entire structure* in glass-forming liquids towards an equilibrium state may be described by the VTF equation [122]. The parameter T_0 is of basic significance and is determined by the equilibrium properties of the system [91, 122, 126, 13].

III. SIGNATURES OF GLASS TRANSITION

Irrespective of whether our understanding of the dynamics of glass-forming liquids require the existence of a thermodynamic singularity [130–134] at T_0 or whether it involves kinetic arguments [135–138], there are a variety of equilibrium and dynamic characteristics of liquid–glass transitions [139]. Consider, for example, the spatial Fourier transform (FT) of density–density correlation function or the so-called structure factor [140–147]. The structure factor has been measured by neutron or X-ray scattering [145–147]. Laboratory experiments as well as computer simulations indicate no significant changes in the structure factor as a liquid is cooled through the glass transition region [138, 145b]. This finding is not surprising since the structure factor reflects only two-body correlations [146, 147]. It is well established, however, that a quantitative description of dense fluids requires back-flow effects [141–144]. Thus it is inhomogeneous distribution of molecules about a test molecule that determine dynamical motions in dense fluids [134].

A. Equilibrium Indicators

Any dynamical quantity can be expressed in terms of equilibrium quantities, that is, structural properties [134, 140]. If the first few terms in the Taylor series expansion in time of the dynamical variable under question are sufficient to describe the essential features of the liquid–glass transition, then the transition may be viewed as being thermodynamic [134].

Kivelson et al. proposed that it is fruitful to seek structural indicators defined through many-body correlations as a way to understand glass-forming liquids [134]. In Section I such an example of a structural indicator was encountered. A resolution of the Kauzmann paradox in supercooled polymeric liquids requires an analysis of the temperature dependence of configurational entropy [130–132]. Entropy is an equilibrium thermodynamic quantity. However, it does not reflect the structure at a local molecular level [134]. The reason is simply this, entropy is defined as an expected value of the Hamiltonian $\mathscr{H}$ [148, 149]

$$S = \langle \mathscr{H} \rangle / T + k_B \ln\Big(\int \cdots \int dX\, e^{-\beta\mathscr{H}} / h^N\Big) \tag{3.1}$$

h is the Planck's constant and dX is an integration over phase space with N particles. If the Hamiltonian contains short-range interactions, then it is clear from Eq. (3.1) that N-body equilibrium correlations define thermodynamic entropy [148, 149]. In other words, S depends on the total structure of the liquid [134]. In contrast, the average energy of the system, $E = \langle \mathscr{H} \rangle$, is described by pair correlations functions only [134].

A signature of the glass transition is a peak or a hump in heat capacity at constant pressure at the glass transition temperature [138, 145b]. One would thus suspect that the heat capacity may serve as a thermodynamic indicator of the glass–liquid transition. There are several reasons why this point must be viewed with caution [134]. First, extrapolations of the equilibrium heat capacity data as a function of temperature show no unusual behavior near the VTF temperature T_0. Second, experimental time scales are much larger than all relevant time scales as $T \rightarrow T_g$. Finally, since $\mathscr{H}$ is short ranged, C_v contains information about three- and four-body correlations only [134]. This follows from the identity [148, 149]

$$k_B T^2 C_v = (\langle \mathscr{H}\mathscr{H} \rangle - \langle \mathscr{H} \rangle^2) \tag{3.2}$$

Grimsditch and River [150] argued that instead, the quantity $\gamma = C_p / C_v$ may be taken as a thermodynamic indicator of the glass–liquid transition. Brillouin scattering probes fluctuations with frequencies in the gigahertz range.[13] A quantity that is measurable by Brillouin scattering and by Fabry–Perot interferometry techniques [151–159] is the dynamical structure factor. The parameter γ is evaluated from the dynamical structure factor. Measurements of such kind by Grimsditch and River [150] revealed an anomaly in γ for glycerol and $ZnCl_2$. If T_g is the temperature at which γ is a maximum, then $T_\gamma \approx 1.15 T_g$ and $1.6 T_g$ in $ZnCl_2$ and glycerol, respectively [150]. These authors proposed that the sharpness, as well as the magnitude of the so-called γ anomaly, are a measure of the fragility of the glass-forming liquid [150]. The results are not only reminiscent of current mode-coupling predictions but are similar to those found in van der Waals fluid above the critical pressure [150].

[13] The wavelengths are in the visible range.

An analysis of the specific heat data of ethanol [160–162], methanol [162, 163], and *sec*-butanol [162] by Sidebottom and Sorensen [164] leads them to identify a characteristic temperature T^* at which the heat capacity has a "hump". This temperature is close to the temperature where the power law behavior of the viscosity breaks down.[14] The authors propose that it is the temperature T^* not T_g that is basic to the glass transition and the vitreous state [164].

Consider equal-time correlation function [134]

$$\Sigma\langle A^{ij}A^{mn}\rangle \tag{3.3}$$

The parameter A could be, for example, the components of the stress tensor σ^{ij} between the ith and the jth molecules. In this case the correlation function is an infinite frequency shear modulus μ. Due to short-ranged forces, there is no significant changes in μ, near T_g [134, 139].

If the dynamic variable A is taken to be the dipole–induced–dipole tensor (DID), then correlation function (3.3) may exhibit long-range correlations [134]. This occurs because the DID tensor is proportional to r_{ij}^{-3}, where r_{ij} is the distance between the ith and the jth particles [134, 165, 166]. The DID probes the structure at a local molecular level [134]. This idea was recently explored by Kivelson, Steffan, Meier, and Patkowski (KSMP) who suggested that the DID correlation function is a nondynamic indicator of the glass–liquid transition [134].

The majority of the effects from the integrated intensity of depolarized VH light scattering from atomic liquids, I_{UV}, are due to DID interactions [134, 167]. In atomic liquids I_{VH} is dependent not only on collisional but also on short-range and multipolar interactions. For spherical molecules the DID interactions exhibit the so-called cancellation effect [134, 167–171]. This is not the case for nonspherical molecules where there are additional contributions to the VH light scattering spectrum from molecular rotations [134]. The total I_{UV} intensity is described by the equal time correlation function, and hence is a thermodynamic property. In practice, however, dynamical effects may complicate the partitioning of I_{UV} into short-range and DID components [134, 172].

Kivelson et al. [134] measured the VH depolarized Rayleigh scattering of light from *o*-terphenyl from the melting point down to T_g [134]. In addition to the usual rotational and the broader base lines, an "intermediate" line with a half-width of 50 GHz was isolated [134]. The half-width of the "intermediate" line is almost independent of tempera-

[14] As predicted by mode-coupling theories.

ture [134]. The intensity of the intermediate line $[I_{\rm UV}({\rm DID})/\rho^2]^{1/2}$, where density is denoted by g, decreases as the temperature is lowered towards the ideal glass temperature $T_0 \approx 200$ K [134]. However, the data below 200 K were obtained by extrapolations of the equilibrium data. An important result obtained by KSMP is that $I_{\rm UV}({\rm DID})/\rho^2$ vanishes near T_0 indicating substantial local ordering [134].

B. Dynamical Indicators

There are indicators of the glass transition which are dynamic in nature. As an example, consider the shear viscosity η. Assume that the frequency dependence of viscosity $\eta(\omega)$ is approximated by the Maxwell model [139].

$$\eta(\omega) = \mu\tau_{\rm m}/(1 - i\omega\tau_{\rm m}) \tag{3.4}$$

and the temperature dependence of the relaxation time $\tau_{\rm m}$ is described by the VTF relation. By the Green–Kubo formula [143], the Maxwell relaxation time $\tau_{\rm m}$ is obtained from the stress–stress correlation function. Since the experimental time scale is of the order of inverse frequency ω^{-1}, the temperature $T_{\rm g}$ satisfies [134, 139]

$$T_{\rm g}(\omega)/T_0 = 1 - D/\ln(\omega\tau_\infty) \tag{3.5}$$

The parameter τ_∞ is of the order of $10^{-12} - 10^{-13}$ s. As the temperature approaches the VTF temperature T_0 the system becomes highly sluggish. In fact, the supercooled liquid cannot equilibrate within an experimental time scale and the relaxation time diverges, that is, $\tau_{\rm m} \gg \omega^{-1}$. The supercooled liquid is locked into an amorphous state. At low frequencies the variation of $T_{\rm g}(\omega)$ is small. Consequently, the shear viscosity may be considered as a dynamic indicator of the glass liquid transition [134–140, 145].

A number of simulations studies on supercooled liquids indicate that the autocorrelations functions of the stress tensor decay slowly [173–180]. From the Green–Kubo formula for transport coefficients these correlations functions correspond to the wavevector $k = o$ Fourier components. In their simulations on a soft sphere system, Visscher and Logan [173] discovered that nonzero wavevector components of the stresses decay considerably more slowly than the corresponding $k = o$ components. This result then led to the proposal that the nonzero wavevector stress–stress correlation function is a relevant order parameter for the glass transition [178]. It is an order parameter in the sense that it is not only a function of the configurations but it also serves as an indicator of the slow relaxation towards its equilibrium state as the temperature is lowered [173].

To explain why particular elements of the stress fluctuations have slowly decaying components, Visscher and Logan introduce the concept of momentum circulation C [173]. The parameter C is a tensor quantity [173]. The discrete stress fluctuations are the curl of C [173]. It turns out that the parameter C is related to the momentum flux.[15] Visscher and Logan [173] argues that if C is a fundamental fluctuating quantity, then the results obtained for the wavevector dependence of the stress–stress correlation function could be explained [173].

Another dynamical indicator of the glass–liquid transition is obtained from the concept of a fluctuating metric defined as [180]

$$\Omega(t) = \frac{1}{N} \sum_{\alpha=1}^{N} \sum_{j\in\alpha}^{N_\alpha} [e_j(t, \alpha) - \langle e(t, \alpha)\rangle]^2 \tag{3.6}$$

where

$$\langle e(t, \alpha)\rangle = \frac{1}{N_\alpha} \sum_{j=1}^{N_\alpha} [e_j(t, \alpha)] \tag{3.7}$$

There are N_α particles that are of type α.[16] The time average of $E_j(s, \alpha)$, the kinetic plus the potential energy of the jth particle of type α, is denoted by $e_j(t, \alpha)$ [180]

$$e_j(t, \alpha) = \frac{1}{t} \int_0^t E_j(s, \alpha)\, ds \tag{3.8}$$

For ergodic systems the fluctuating metric is expressible in terms of an effective diffusion coefficient [175, 176, 180]

$$\Omega(t) \approx 1/D_\Omega t \tag{3.9}$$

The parameter D_Ω measures how fast the phase space is being explored.[17]

The fluctuation metric concept was utilized by Mountain and Thirumalai [180, 181] in a constant pressure molecular dynamics solution of soft-sphere and Lennard–Jones systems. These authors obtained the following results. (a) The temperature dependence of the self-diffusion coefficient is Arrhenius for soft-sphere molecules while it follows the VTF form for Lennard–Jones particles. (b) For Lennard–Jones particles, the

[15] This is analogous to how charge circulation is related to charge flux.
[16] $\Sigma_\alpha N_\alpha = N$.
[17] For times larger than the transient time.

diffusion coefficient decreases for temperature less than the temperature at which icosahedra clusters start to increase [180, 182]. (c) Following an analogy with fluids at the critical point,[18] one concludes that due to large values of $(T_g - T_0)/T_g$ only the cross-over region can be probed by simulations [180].

C. Diverging Correlation Length

There is growing evidence from a variety of laboratory experiments[19] for the absence of a diverging length scale in supercooled liquids near the glass transition temperature [183–185]. Several simulation studies have been carried out to probe the existence of a divergent correlation length near T_g based on appropriate order parameters to characterize the system [185b]. Despite the known weaknesses of computer experiments [186–193], these simulations have revealed important features in agreement with laboratory experiments. The properties include hysteresis effects on density as the system is cooled or heated, frequency dependent specific heat, as well as frequency independent thermal conductivity [186, 194, 195].

The so-called bond orientational order has been investigated via constant volume and pressure molecular dynamics simulations through space and time correlation function $G(\mathbf{r}, t)$. In addition, the correlations between bonds and bond angles in space and time have also been evaluated [195]. Simulations by Nagel and co-workers [185] revealed that as the system is cooled through the glass transition region, the first four peaks of the radial distribution function, at a fixed temperature, have the same relaxation rate [185b]. This finding suggests that a supercooled liquid is becoming viscous independent of the length scale used to probe the system [185a, 185b]. The characteristics of the spatial FT of $G(\mathbf{r}, t)$ does not support the idea of a diverging length scale as the temperature is lowered [185]. This is in agreement with Ullo and Yip [196]. Bond orientation freezing does occur [185]. However, this phenomena apparently does not have a length scale dependence [185]. The results are in variance with that obtained by Steinhardt et al. [197].

Dasgupta et al. investigated, via computer simulations [198], the temperature as well as the size dependence of structural relaxation in a LJ mixture. Both two-point and four-point density and certain orientational correlation functions were studied. The two- and four-point density

[18] If the system has a correlation length that diverges as $T \to T_0$, then the glass transition temperature may be considered to be far from the "critical" region [180].

[19] Via dielectric response, nonlinear as well as linear susceptibilities measurements.

correlations are defined as follows [198]:

$$C_2(t) = \langle (n(i,t)n(i,t+t_0)\rangle$$

$$C_4(\mathbf{r},t) = [\langle n(\mathbf{r}_0,t_0)n(\mathbf{r}_0,t+t_0)n(\mathbf{r}_0,t+t_0)n(\mathbf{r}_0+\mathbf{r},t+t_0))\rangle] \quad (3.10)$$

$$C_4(\mathbf{r}) = \lim_{t\to\infty} C_4(\mathbf{r},t)$$

The brackets [ ] and ⟨ ⟩ denote thermal averages over reference space-point $\mathbf{r}_0$ and time t_0, respectively. The motivations for studying $C_4(\mathbf{r})$ are twofold. First, for spin glass the analogous quantity has a diverging correlation length at the so-called spin glass transition temperature [199]. Second, mode coupling theories suggest that the quantity, $\lim_{t\to\infty} \times \langle (n(\mathbf{r},o)n(\mathbf{r},t)\rangle$, is an appropriate order parameter for glassy dynamics [135–137, 198].

The conclusions obtained by Dasgupta [198] are that (a) there was no indication of a diverging correlation length [200b]; (b) the correlation length is less than twice the hard sphere diameter [200b];[20] (b) the nonequilibrium properties could be described by the motion of few particles; (d) the four-point correlation function does not exhibit long-range spatial order; and (e) no icosahedral order was observed.

As discussed earlier, constant pressure molecular dynamics simulation of a soft-sphere and Lennard–Jones system suggest that fluctuations in supercooled liquids are due to regions of finite length [180, 200b]. This result is in agreement with independent theoretical arguments[21] by Mohanty [200]. Large values of $\{T_g - T_0\}/T_g$ obtained in simulations allow an investigation of the cross-over region only [180, 200b].

Experiments by Kiyachenko and Litvinov [201] on polystyrene spheres of various diameters show that microscopic and macroscopic viscosities differ and lend credance to the idea of an apparent length scale dependence of the glass–liquid transition [200b]. Unfortunately, Dixon et al. [202] failed to confirm this result [200b].

D. Nanoscale Inhomogeneity

As a glass-forming liquid is cooled, fluctuations in concentration usually freeze at a temperature that is usually larger than T_g [203, 200b, 204]. However, light scattering in liquids is a result of fluctuations in concentration and density [203, 200b, 204]. Thus, the scattering intensity in

[20] So long as a single dominant time scale is assumed.
[21] These arguments have been discussed in Section I.

the glass transition region is a result of the mean-square fluctuations in density ρ [203–205]

$$\langle \Delta^2 \rho \rangle / \rho^2 = k_B T \, \Delta\kappa / V \tag{3.11}$$

where $\Delta\kappa$ is the change in isothermal compressibility between the supercooled liquid and the glass at T_g [200b, 203b].

In a series of remarkable experiments, Bokov and Andreev [206] have measured the intensity of visible light scattering as a function of temperature for boric oxide (B_2O_3) [200b, 203b]. The sample is heated and then cooled through the region of the glass transition [207]. The heating curve for the scattering intensity has a maxima [200b, 203b]. In contrast, the cooling curve has a characteristic sigmoidal shape [200b, 203b].

Golubkov and Pivovarov [208, 209] observed similar anomalous behavior in small angle X-ray scattering (SAS) measurements from boric oxide [203b, 200b]. The SAS experiments indicate a maximum in Sb_2O_3–B_2O_3 glass during heating. Relative maxima in the intensity of scattered light has been observed [203b, 200b] on heating in $53ZrF_4$–$20BaF_2$–$4LaF_3$–$3AlF_3$–20NaF (ZBLAN) [203, 210, 211], sodium germanate glasses [206] and 1,3,5-tri-[*a*]-naphthyl benzene [212]. These experiments lead to the proposal that anomalous scattering of light in the glass transition region may be a universal feature of supercooled liquids [213, 200b].

Mazurin [207] proposed that the anomalous light scattering may be viewed as due to slow relaxations of regions of high density [213, 200b]. This point of view has been advocated by Robertson [204], Donth [214], and more recently by Moynihan and Schroeder [213] and Mohanty [215]. In fact, Moynihan and Schroeder [213] recently shown that the correlation volume scales as

$$V \approx [(1-x)\,\Delta h^*/RT]^2 [k_B v / (\Delta C_p \langle \Delta^2 \ln \tau \rangle)] \tag{3.12}$$

where x is a nonlinearity parameter introduced by Narayanaswamy and Gardon [216], ΔC_p is the change in specific heat between the liquid and the glass, V is the specific volume, k_B is the Boltzmann constant, $\langle \Delta^2 \ln \tau \rangle$ measures how wide is the distribution of relaxation times, and Δh^* is an Arrhenius activation energy [213, 215, 217]. The basic idea behind the derivation is that characteristics [215] of anomalous light scattering are similar to how the macroscopic specific volume v of a liquid[22] varies as it is heated and then cooled through T_g [213, 215].

[22] The volume versus temperature curve on cooling is monotonic. The corresponding heating curve is sigmoidal [213, 215].

The derivation of Eq. (3.12) proceeds by assuming the relaxation time of a local region to be governed by the AG relation [130, 213, 215]

$$\ln \tau_i = \ln \tau_0 + s^* \Delta\mu / mk_B T s_{ci} \quad (3.13)$$

Here, s_{ci} is the specific configuration entropy of the local region and m is the mass of a rearranging region [213, 215].[23] The mean-square deviation of $\ln \tau_i$, $\langle \Delta^2 \ln \tau \rangle$, follow from the relation [213]

$$\ln \tau_i - \ln \tau_i \approx -(s^* \Delta\mu / mk_B T s_c^2)(s_{ci} - s_c) \quad (3.14)$$

$$\langle \Delta^2 \ln \tau \rangle = (s^* \Delta\mu / mk_B T s_c^2)^2 (k_B v \, \Delta C_p / V) \quad (3.15)$$

To obtain Eq. (3.15) one makes use of an expression which relates the fluctuations of the specific configurational entropy [205]

$$\langle \Delta^2 s_c \rangle = k_B v \, \Delta C_p / V \quad (3.16)$$

and the difference in the specific heat between the liquid and the glass, ΔC_p [213, 215]. The correlation volume is V. Equation (3.12) follows [213, 215] from Eq. (3.15) and on assuming [218] $\Delta C_p(T) = a/T$, where a is a constant. The nonlinear parameter x is given by a relation due to Hodge [219] and extended by Mohanty [200]

$$x = 1 - T_2/T \quad (3.17)$$

The parameter x is a measure of how the relaxation time depends on structure and on temperature [213, 215, 216].

Several comments are in order regarding Eq. (3.12). First, there is a correlation between x and Δh^* [219]. Second, as $T \to T_2$, $\langle \Delta^2 \ln \tau \rangle$ diverges [195, 213, 215, 220, 221]. Finally, as the temperature increases, $\langle \Delta^2 \ln \tau \rangle$ decreases [215]. These predictions are in agreement with a variety of experiments [213, 215, 219].

The expression for the correlation volume turns out to be similar to the one given by Donth [214]. What is missing is an important nonlinear factor $(1 - x)$ [213, 215]. The Donth [214] derivation is based on relating the mean-square fluctuation in temperature to the mean-square fluctuation of $\ln \tau$ of a cooperatively rearranging region as defined by AG theory [215]. Robertson [204] relates the mean-square fluctuations in specific volume in polymeric systems to relaxation in the glassy state. Mohanty [215] showed, based neither on free-volume nor on configurational

[23] Remaining quantities in Eq. (3.13) have been defined in Section I.

entropy arguments, that the correlation volume V of such regions scale as

$$V \approx [(1-x)\,\Delta h^*/RT]^2[k_B T_g^4\,\Delta\kappa_{T_g}/\langle\Delta^2 \ln\tau\rangle] \tag{3.18}$$

where $\Delta\kappa_{T_g}$ is the change in thermal conductivity at T_g.

The correlation lengths are in the range of 0.7–3 nm for a system such as B_2O_3, glycerol, PVAC, polystyrene, As_2Se_3, crown glass, and 5-phenyl-4-ether [213–215]. The boson peak in Raman scattering inorganic glasses and polymers[24] suggest regions of size from 1 to 5 nm [213, 215, 222–225]. Some unusual features in the light scattering of OTP have been attributed by Steffen, Patkowski and co-workers [226] to density inhomogeneities of length scale 1000 A [215]. There is experimental evidence that relaxation in supercooled liquids may be due to inhomogeneous distributions of independently rearranging regions [227]. Chamberlin [228] showed that non-Arrhenius temperature dependence of relaxation rate could be explained based on a Gaussian distribution of independently relaxing regions.

IV. INHERENT STRUCTURES

A. Potential Barrier Model

Consider a system of N particles enclosed in a volume V at a temperature T. The generalized coordinates of the N particles are denoted by $\{q_i;\ i = 1, \ldots, N\}$. The potential energy $\Phi(q)$ of the particles is a function of the generalized coordinates. The Hamiltonian is denoted by $\mathcal{H}(p, q)$, where $p = \partial\mathcal{H}/\partial q$ is the generalized momentum [229]. The state of the system is depicted as a point—the so-called phase point in a $2N$-dimensional phase space [229].

The usefulness of the potential energy hypersurface to elucidate the low-temperature properties of the glassy state was realized long ago by Goldstein [230]. Goldstein hypothesized that an important characteristic of glassy systems is the existence of potential energy barriers [230]. The depth of these barriers are large compared to energies due to thermal fluctuations, that is, $k_B T$ [230]. A number of inferences follow from this assumption [230]. First, a glass-forming liquid including the glassy state has associated with it a large number of minima of various depths [230]. Second, at zero temperature the state point is trapped in one of the local

[24] For polymers the length of these inhomogeneous regions may be inferred from X-ray scattering; see [225].

minimum.[25] Third, as the temperature is raised thermal fluctuations enable the state point to jump into a nearby local minimum [230]. These motions are highly localized and involve small displacements of particles around a tagged particle [230]. Finally, at temperatures of the order of the cross-over temperature T_x or higher, thermal energies are comparable to the depth of the potential barriers. The potential barrier picture is then no longer valid [230].

The validity of Goldstein's assumptions rests on the existence of two widely separated time scales τ_0 and τ_1 [230]. The parameter τ_0 reflects vibrational motions within a typical cell, while τ_1 is the time required to jump to a nearby cell. Since the state point is assumed to be trapped at or near a potential energy minimum, we must have the inequality $\tau_1 \ll \tau_0$ [230].

The cross-over viscosity[26] may be estimated as follows [230]. The high-frequency modes in supercooled liquids are vibrational in nature. In contrast, the low-frequency modes are damped. Consequently, in a time period say, τ_s, the high-frequency modes perform several oscillations. Arguments based on the Debye density of states leads to an estimate of τ_s in the range of 10^{13} cps [230]. A modern version of these ideas are inherent in the works of Zwanzig [213], Mohanty [232], and Keyes and co-workers [233]. More insight is obtained by identifying τ_s with the time required for local rearrangements to take place after the phase point has jumped to a nearby local minimum. If ℓ is the size of the local region and κ is its thermal diffusivity, then the energy–energy correlation decays in a time scale of the order of ℓ^2/κ [230]. By identifying this time with τ_s and by taking $\ell \approx 10^{-7}$ cm and $\kappa = 10^{-3}\ \mathrm{cm}^2\ \mathrm{s}^{-1}$, which are typical values for liquids, one obtains $\tau_s \approx 10^{-11}$ s [230]. Both estimates of τ_s lead to shear viscosity in the range of 10 P [230]. This is in agreement with sophisticated mode coupling estimates [234].

B. Stillinger–Weber Inherent Structure Theory

1. *Steepest Descent Quenches*

An important generalization of the potential barrier picture was advanced by SW [235–247]. These authors proposed that in order to understand the statistical mechanics of condensed phases it may be potentially useful to separate a many-body problem into two parts [235–247]. In the first part one identifies and isolates those arrangements of particles that are mechanically stable [246, 247]. These arrangements correspond to the

[25] In discussions below the state point, Γ refers to configuration space.
[26] The value of viscosity at the cross-over temperature T_x.

local minima in the potential energy hypersurface Φ. In the second part, one takes into account vibrational motions about these stable packings [246, 247].

Stillinger and Weber [235–247] implemented these ideas by carrying out a steepest descent quench on the potential energy surface $\Phi(\mathbf{r})$. The steepest descent paths for the phase point are obtained via [246, 247]

$$m\, \partial \mathbf{r} / \partial s = -\text{grad}\, \Phi(\mathbf{r}) \tag{4.1}$$

Here m is the pass of the particle. As $s \to \infty$, the solutions $\{\mathbf{r}(s)\}$ converge to a local minimum if the starting configuration corresponds to $s = 0$. If all possible starting configurations are sampled, then the steepest descent quench would, at least in principle, identify all the local minima [246, 247].[27] The many-particle configurations could thus be mapped uniquely onto a local Φ minimum [246, 247].

2. *Cell Partition Function*

Around each minimum α in the configuration space introduce a cell variable C_α. The set $\{C_\alpha\}$ is complete and spans the configuration space. By construction, a subset of all configurations map onto each cell α. The total number of minima for an N particle system is expected to scale as [246, 247]

$$\Omega \approx \Omega_\text{p} e^{\omega N} \qquad \omega > 0 \tag{4.2}$$

where Ω_p is the number of equivalent minima for each Φ minima. It is useful to classify the minima on the potential surface by an order parameter ϕ [246, 247]. The parameter ϕ measures the depth of each local minima. Then, the number of minima between ϕ and $\phi + d\phi$ is proportional to [246, 247]

$$e^{N\sigma(\phi)}\, d\phi \tag{4.3}$$

From Eqs. (4.2) and (4.3) we obtain

$$e^{\omega N} \approx \int e^{N\sigma(\phi)}\, d\phi \tag{4.4}$$

Using Eq. (4.4) the canonical ensemble partition function is written as

[27] Except for a set of measure zero.

quadratures over ϕ [246, 247]

$$Q(T, V, N) = (\lambda^{-3N}) \int e^{N[\sigma(\phi) - \beta\phi - \beta f(\beta,\phi)]}\, d\phi \tag{4.5}$$

Here $f(\beta, \phi)$ is a mean vibrational free energy per particle[28] and λ is the thermal wavelength. As $N \to \infty$, the integrand may have a sharp maximum as a function of ϕ. If so, the partition function could be evaluated by the maximum term method. The maximum term occurs at $\phi = \phi_{\mathrm{m}}(\beta)$, which is a solution to [246, 247]

$$\partial\sigma(\phi)/\partial\phi - \beta\, \partial f/\partial\phi = \beta \tag{4.6}$$

3. *Structural Transitions*

Stillinger and Weber [236–239] carried out extensive constant energy molecular dynamics to study the transition between inherent structures at various temperatures and densities. Newton equations of motion were solved by standard fifth-order Gear algorithm at fixed density. A typical run lasted $10^4\, \Delta t$, where Δt is the time step length in the Gear algorithm [248]. The temperature of the system was calculated from a knowledge of the average kinetic energy, while the total energy was varied by scaling the velocities of each particle.

At certain time intervals during a molecular dynamics run the configurations of the system were subjected to quenching operations [236–239]. The steepest descent quenches locate the various local minima by conjugate gradient methods [236–239]. There are several important results obtained by Stillinger and Weber [236–247]. (a) The inherent structures include crystalline and amorphous packings. The rate of transition between different inherent structures increases with the number of particles [236–247]. (b) The transitions between different inherent structures are due to thermal fluctuations. These fluctuations lead to a large number of independently rearranging regions [236–247]. (c) Transitions due to permutations of the particles rarely occur. (d) In supercooled and amorphous states, the quench-potential energy versus time had intermittent-like characteristics [236, 237]. (e) As the system jumps between neighboring cells, the rearranging process involves a small subset of all the particles [236, 237]. (f) If the system is in an amorphous state, transitions between two quench values occur at low temperatures. In other words, transition is between two localized levels [236, 237]. (g) In

[28] Note that the minimum lies at $\Phi = N\phi$.

the supercooled state, the system evolves through regions of the configuration space whose topology is complex [236, 237, 247]. For example, the surface may contain long narrow channels that wind around one another [236, 237, 247].

4. *Supercooled States*

A supercooled liquid is dynamically metastable with respect to both the crystalline and the glassy states. Consequently, it remains in that part of the configuration space that is free of crystalline packings [247]. The distribution of the various minima in a supercooled liquid, $\sigma_s(\phi)$, is obtained from $\sigma(\phi)$, in particle at least, by a suitable projection [247]

$$\sigma_s(\phi) = \hat{A}\sigma(\phi) \tag{4.7}$$

where the projection operator $\hat{A}$ projects onto the manifold of amorphous states.

The partition function for the supercooled liquid is obtained from a knowledge of the vibrational free energy and the distribution of the various minima in the amorphous manifold [247]

$$Q_s(T, V, N) = (1/\lambda^{-3N}) \int e^{N[\sigma_s(\phi) - \beta\phi - \beta f_s(\beta, \phi)]} \, d\phi \tag{4.8}$$

As before $Q_s(T, V, N)$ is evaluated in the thermodynamic limit by a maximum term method [247]

$$\partial\sigma_s(\phi)/\partial\phi - \beta \, \partial f_s/\partial\phi = \beta \tag{4.9}$$

Laboratory as well as computer experiments indicate that various properties of glass-forming liquids depend on the cooling rate. The steepest descent quenches by Stillinger [246, 247] indicate that transitions between cells involve localized motions of few particles. These localized motions change the potential energy by a small amount.[29] These facts suggest, in agreement with Goldstein's proposal, that the configuration point in a glass-forming liquid gets trapped in minima of various depths and that these depths are of order N [230, 246, 247]. In other words, the topography of the Φ surface is rough over a wide range of scale lengths [230, 246, 247].

To investigate how the roughness of the potential energy landscape is related to properties of glass-forming liquids, Stillinger [247] coarse

[29] It is of order 1.

grained small-scale roughness in Φ_s

$$\Phi(\mathbf{r}, \ell) = \int K(\mathbf{r}, \mathbf{r}'; 1)\Phi_s(\mathbf{r}')\, d\mathbf{r}' \tag{4.10}$$

where $K(\mathbf{r}, \mathbf{r}'; 1)$ is a function that smoothes the repulsive part of the intermolecular potential over a length scale ℓ.[30] The integration regions in Eq. (4.10) are over the accessible parts of the configuration space [247].

The number of minima of $\Phi(\mathbf{r}, \ell)$ scales as [246, 247]

$$\Omega \approx \Omega_p e^{\theta(\ell)N} \tag{4.11}$$

As ℓ increases, some of the minima of $\Phi(\mathbf{r}, \ell)$ are eliminated. Hence, $\theta(\ell)$ decreases as ℓ increases [247].

An analysis similar to that done in Section IV.2.2 could be carried out. This analysis would lead to a complete set of cells $\{C_\alpha(\ell)\}$ whose average number decreases with increasing ℓ [246, 247]

$$\langle C_\alpha(\ell)\rangle \approx e^{-\theta(\ell)N} \tag{4.12}$$

In analogy with Eq. (4.4) one has a relation between $\theta(\ell)$ and the distribution of minima $\sigma(\psi, 1)$ on the order $\Phi(\mathbf{r}, \ell)$ surface [246, 247]

$$e^{\theta(\ell)N} \approx \int e^{N\sigma(\psi, \ell)}\, d\psi \tag{4.13}$$

Stillinger [246, 247] argued that $\theta(\ell)$ scales as

$$\theta(\ell) \approx \exp - (h_\infty \ell^q) \tag{4.14a}$$

where q is a constant greater than zero. His argument is summarized as follows. Let M be the number of distinct packings that are enclosed in volume $V(M, \ell)$. By construction, $V(M, \ell)$ is related to M, the volume of the system V, and $\theta(\ell)$ [246, 247]

$$V(M, \ell) = V \ln M/(N\theta(\ell)) \tag{4.14b}$$

Introduce a function $h(\ell)$ via $\theta(\ell) = \exp -h(\ell)$. As ℓ increases, the

[30] It is assumed that the intermolecular potential is separated into a short-ranged repulsive part $\Phi_s(\mathbf{r})$ and a long-ranged attractive part. Configurations that do not overlap are represented by hard-core repulsions.

bistable degrees of freedom decrease.[31] If this decrease is exponential, then Eq. (4.14a) is obtained.[32]

C. Applications

As applications of SW inherent structure approach let us now study a number of topics that include AG and GD configurational entropy models, the Stokes–Einstein relation, stress relaxation in stabilized glass and interbasin dynamics.

1. *Generalization of the Adam–Gibbs Model*

This section obtains insights into the AG theory based on the SW concept of inherent structures and thermodynamics of small systems.[33]

Following AG theory we consider a cooperative rearranging region V^* [249]. The macroscopic system consists of a number of equivalent subvolumes V^* [250, 249]. The subvolume V^* rearranges itself into a configuration independent of its environment [250, 249]. The parameter V^* is not of macroscopic dimensions [250, 251].

Let us consider an ensemble of M equivalent but independent subsystems [249, 250]. Each subsystem has identical pressure P, number of monomer segments N, and temperature T. If the time average of a small system is equal to its ensemble average, then its thermodynamics could be expressed in terms of mean values of fluctuating extensive quantities [251, 252]:

$$dE_t = T\,dS - P\,dV_t + \mu M\,dN + G\,dM \tag{4.15}$$

where S_t, E_t, and μ are, respectively, the entropy, the energy, and the chemical potential of the macroscopic system.[34] G could be identified as the Gibbs free energy of the small system.[35]

[31] This occurs because these degrees of freedom are converted into monostability. One would thus have larger cells. These degrees of freedom, however, involve localized motion; see [247].

[32] The parameter $h(\ell)$ would scale as ℓ if coarse graining leads to attrition of the cells; see [247].

[33] Details of the AG model are given in Section I.

[34] The subscript t denotes macroscopic system.

[35] If P, N, and T are held fixed, E_t is a homogeneous function of S_t, N_t, and M. This, in conjunction with Eq. (4.15) leads to the relation, $\bar{E} = TS - P\bar{V} + G$. The time average is denoted by a bar; see [250, 252].

The subsystems are equivalent and independent; we therefore have[36]

$$S_t = MS \tag{4.16}$$

The stable packing configurations have been isolated from the vibrational motions about the packings by using SW construction. Consequently, $S_{tc} = MS_c$, where S_c is the configurational entropy [249, 250]. If the macroscopic system is taken to be a mole of molecules, then the configurational entropy is given by $S_{tc} = (N_A/N^*)S_c^*$, where N_A is the Avogadro number [249, 250].

The probability $P(T)$ of a fluctuation, small or large, is given by [253, 250]

$$\exp(-N\,\Delta G/Tk_B) \tag{4.17}$$

where ΔG is the maximum reversible work. But, the partition function of a rearranging region is $\exp(-N\mu/k_B T)$.[37] The transition probability, which is proportional to Eq. (4.17), is now averaged over all possible values of N, from a critical value N^* to infinity; this leads to the expression [249, 250]

$$W(T) = A\exp(-N^*\,\Delta\mu/k_B T) \tag{4.18}$$

where A is a constant that is weakly temperature dependent. Substituting the expression for S_{tc} in Eq. (4.18) yields the AG relation [249, 250].

2. *Generalization of the Gibbs–DiMarzio Model*

In the GD model the system consists of n_x polymers each having x monomers [254–257]. The configurations of each polymer are assumed to fit a lattice. The state of the system depends on two quantities f and n_0, where f is related to the chain stiffness, while n_0 is the number of vacant sites on the lattice [254–257]. Based on the Flory–Huggins approximation for the number of accessible conformational states for a given value of f and n_0, Gibbs and DiMarzio predicted a second-order Ehrenfert-like transition for the SPT equation of state [254, 255, 257]. This resolved the Kauzmann paradox since the configurational entropy is zero below a temperature T_2 [254, 255, 257].

[36] The parameter S does not fluctuate and depends on overall probability distribution; see [250, 252].

[37] The partition function of the system is $Z_t(P_t, N_t, T) = \{Z(P, N, T)\}^M$, where $Z(P, N, T)$ is the partition function of a subsystem; see [250, 252].

A generalization[38] of the GD model was recently proposed by Montero, Mohanty and Brey [257]. The starting point is an isothermal-isobaric partition function $\gamma(T, p, n_x)$, which is obtained via SW-like construction. The parameter $\gamma(T, p, n_x)$ is expressed as a sum over all possible states of the product of a vibrational partition function $Z_v(T, n_0)$ and the number of configurational states $P(f, n_0)$ [257]

$$\gamma(T, p, n_x) = \sum_{f,n_0}{}' P(f, n_0) Z_v(T, n_0) \exp -(\beta(E + pV) \quad (4.19)$$

The prime over the sum denotes that values of $P(f, n_0)$ less than 1 have been omitted.[39] The factorization in Eq. (4.19) is possible because the time scale for vibrational motions are much faster than configurational changes [255, 257]. The total energy E is a function of n_0 and f. The total volume V varies due to configurational as well as vibrational changes [255, 257]

$$V = (xn_x + n_0)C(p) \quad (4.20)$$

where the volume of the lattice site $C(p)$ is a function of pressure only.

The vibrational partition is of the form [255, 257]

$$Z_v(T, n_0) = D(\lambda)^{3xn_x} \quad (4.21)$$

where $\lambda = \hbar\omega / k_B T$, the vibrational frequency is ω, and D is an arbitrary function of λ.

The partition function is evaluated by a maximum term method [257]

$$\ln \gamma(T, p, n_x) = \ln \gamma(T, p, n_x; \tilde{f}, \tilde{n}_0) \quad (4.22)$$

where $\tilde{f}$ and $\tilde{n}_0$ are those values that maximize the partition function with the constraint $P(f, n_0) \geq P_0$ and where

$$\ln \gamma(T, p, n_x; f, n_0) = P(f, n_0) D(\lambda)^{3xn_x} \exp -\beta(E + PV) \quad (4.23)$$

The parameter T_2 is determined by the constraint [257]

$$P(\tilde{f}(T, p), \tilde{n}_0(T, p)) = P_0 \quad (4.24)$$

[38] The approach bypasses the use of the mean-field Flory–Huggins approximation.

[39] The parameter $P(f, n_0)$ increases with n_0 and f. Consequently, the constraint implies that for each n_0 there is a minimum value for f and vice versa and f should not be confused with the free energy defined in Eq. (4.5); see [257].

Montero et al. [257] argued that the maximum term of $\ln \gamma(T, p, n_x)$ is a valid approximation for temperatures $T \geq T_2$.

The maximum term approach is not valid for temperatures less than or equal to T_2 [255, 257]. To proceed further, assume that $\gamma(T, p, n_x; f, n_0)$ has a single maximum at $(\tilde{f}, \tilde{n}_0)$ and that this maximum decreases with temperature towards smaller values of the parameters [257]. Then for temperatures less than or equal to T_2, the contribution that is a maximum is obtained from terms that correspond to minimum values of the parameters [257]. For $T \geq T_2$ the isothermal-isobaric partition function[40] in the thermodynamic limit is [257]

$$\ln_{>}\gamma(T, p, n_x) = \ln\{P(\tilde{f}, \tilde{n}_0)D(\tilde{\lambda})^{3xn}x \exp -\beta(\tilde{E} + P\tilde{V})\} \quad (4.25)$$

while for $T \leq T_2$ it is obtained from [257]

$$\ln_{<}\gamma(T, p, n_x) = \ln\{P_0[D(\hbar\omega_2/k_B T)]^{3xn}x \exp -\beta(\tilde{E}_2 + P\tilde{V}_2)\} \quad (4.26)$$

The various quantities in Eqs. (4.25) and (4.26) are defined by

$$\tilde{\omega} = \omega(\tilde{n}_0), \quad \tilde{E} = E(\tilde{f}, \tilde{n}_0), \quad \tilde{V} = C(p)[xn_x + \tilde{n}_0] \quad (4.27)$$

$$\tilde{\omega}_2 = \omega(\tilde{n}_{02}), \quad \tilde{E}_2 = E(\tilde{f}_2, \tilde{n}_{02}), \quad \tilde{V}_2 = C(p)[xn_x + \tilde{n}_{02}] \quad (4.28)$$

The equilibrium properties of glass-forming polymeric liquids are readily deduced from the partition function. Montero et al. showed that at a temperature T_2,[41] there is a second-order transition in the Ehrenfest sense. For example, the entropy of the system for $T \geq T_2$ is [257]

$$S_{>}(T, p) = k_B \ln P(\tilde{f}, \tilde{n}_0) + 3k_B xn_x[\ln D(\tilde{\lambda}) - \hbar\tilde{\omega}/(k_B T)d \ln D(\tilde{\lambda})/d\tilde{\lambda}] \quad (4.29)$$

while for $T \leq T_2$ it is

$$S_{<}(T, p) = S_c(T_2, p) + S_v(T, \tilde{\omega}_2) \quad (4.30)$$

The parameters $S_c(T, p)$ and $S_v(T, \omega)$ are, respectively, the configurational and the vibrational contributions to the total entropy. Note that the

[40] For a detailed justification of the conditions under which the maximum term analysis is a good approximation to the partition function, see [255, 257].

[41] The parameter T_2 is a solution to Eq. (4.24). In Eq. (4.28), $\tilde{f}_2$, $\tilde{n}_{02}$ are the minimum values of parameters.

entropy of the liquid is continuous at T_2. On the other hand, second derivatives of the Gibbs free energy have a jump at T_2 [255, 257].

Observe that below T_2 the total entropy is not zero. Assuming configuration contributions only, let $T_{2,c}$ be the temperature at which an Ehrenfest transition occurs. This temperature is nonzero if the slope of the density of states at the minimum is finite [257]. In addition, one finds that if vibrational motions are taken into account, then the temperature T_2 is always less than $T_{2,c}$ [257]. The derivation of these results depend on stability conditions [257]. The predictions are significant because they are not based on mean-field Flory–Huggins approximations. Systematic improvements to mean-field Flory–Huggins theory is a formidable problem that has been recently attempted by Bawendi, Freed and Mohanty [256].

3. *Stokes–Einstein Relation: Zwanzig-Mohanty Model*

The shear viscosity η of a fluid and the self-diffusion constant D are related by the Stokes–Einstein (SE) relation

$$D = Ck_B T/\eta\sigma \tag{4.31}$$

where σ is a molecular diameter and the constant C depends on hydrodynamic boundary condition [232, 258]. The derivation of the SE relation is usually based on hydrodynamic or kinetic theory-like arguments [259].

Zwanzig [231] and Mohanty [232] proposed an analysis of the SE relation based on a synthesis of ideas inherent in the theories of SW [235–247], Goldstein [230], and AG [249]. The model makes a number of plausible assumptions about the dynamics of dense fluids that are supported by computer simulations [231, 232]. The states of the system are described by "fixed" configurations $\Gamma^{(\alpha)}$ [232, 260]. The system performs harmonic oscillations near $\Gamma^{(\alpha)}$ [232, 260]. The oscillations about each α are Debye like. There is a waiting-time distribution $\psi(t)$ for jumps between α and another configuration, say ν [232, 260]. The time dependence of the waiting time distribution is $e^{-t/\tau}$, where τ is the lifetime for a cell jump [231]. The mean time for a cell jump is much larger than a Debye period. As a result, the system performs several oscillations before it jumps to a nearby local minimum. Another assumption is that motions after a jump are uncorrelated with cell motions before a jump [231, 232].

Any transport coefficient J, such as diffusion and viscosity, can be expressed in terms of a time correlation function [231, 232, 260]

$$J = (Vk_B T)^{-1} \lim_{t\to\infty}(2t)^{-1}[A(t) - A(0)]^2 \tag{4.32}$$

This is the so-called Green–Kubo formula. If J is the shear viscosity, then $A(t)$ is [231, 232]

$$\sum_j y_j P_{jx}$$

where P_{jx} is the x components of the momentum relative to the center of mass. We replace the $\lim_{t\to\infty}(\,)$ in Eq. (4.32) with [231, 232, 260]

$$(2t)^{-1}\int_0^\infty dt\,\psi(t) < [A(t) - A(0)]^2\rangle \tag{4.33}$$

The angular brackets denote an equilibrium average over initial states.[42] Since the states have fixed configurations, $A(t)$ is approximated by $\Sigma_j P_{jx} Y_j$, where Y_j denotes lattice configurations [231, 232, 260]. By evoking momentum conservation the integrand in Eq. (4.33) is rewritten as

$$\langle [A(t) - A(0)]^2\rangle = \sum_j \sum_k (Y_j - Y_k)^2 \langle P_{jx}(t) P_{jx}(0)\rangle \tag{4.34}$$

Using spatial homogeneity and Eq. (4.34) the expression for the shear viscosity reduces to [231, 232, 260]

$$\eta = (N/Vk_B T)\int_0^\infty dt\,\psi(t)\sum_j Y_{j0}^2 C_{jo}(t) \tag{4.35}$$

where $C_{jk}(t) = \langle P_{jx}(t) P_{jx}(0)\rangle$.

The momentum autocorrelation function in Eq. (4.35) is evaluated by a normal-mode analysis [231, 232, 260]. From the Debye wavevector cutoff q_D and the volume of the system, one obtains the total number of modes, $[V/(2\pi)^3](4\pi/3)q_D$ [232, 260]. The integral over time in Eq. (4.35) is readily evaluated [231, 232, 260]

$$\eta = \rho c^2 t \tag{4.36}$$

Here ρ is the mass density and c is an average velocity of sound.[43]

In the normal mode picture, the self-diffusion coefficient is found to be

[42] The replacement of the limit by Eq. (4.33) is feasible because the mean lifetime in a state is τ.

[43] The average velocity is expressed in terms of longitudinal and transverse sound velocities.

[231, 232, 260]

$$D = (k_B T/3Nm) \int_0^\infty dt \sum_\omega \cos(\omega t) e^{-t/\tau} \tag{4.37}$$

where m is the mass of the particle. The sum over ω is converted to an integral over the Debye density of states. The transverse and longitudinal oscillations are treated independently [231, 232]. Performing the integral over time leads to an explicit expression for self-diffusion [231, 232]

$$D = k_B T/6\pi\rho[1/\tau^2 c_1^3 + 2/\tau^2 c_t^3] \tag{4.38}$$

On combining the expression for shear viscosity with the expression Eq. (4.38) for self-diffusion coefficient, one finds the SE relation is satisfied [231, 232]. Further analysis shows that self-diffusion coefficient is, however, more sensitively dependent on the form of cell-to-cell jump distribution than shear viscosity η [232]. The temperature and the density dependence of the parameter τ, which is a measure of the average lifetime in a cell, may be estimated from a sum rule due to de Gennes [261, 232].

4. *Stress Relaxation*

Experimental data compiled by Douglass, Kurkjian, Angell, and co-workers, and others have shown that the time dependence of stress relaxation in a glass is of the KWW form [262–277]. In particular, the KWW exponent β is approximately $\frac{1}{2}$ for $t \approx \tau$. For $t \gg \tau$, β is approximately 1, while for $t \ll \tau$, β is approximately $\frac{1}{3}$ [273–277].

These results have been explained based on a dynamical generalization of the inherent structure approach [260, 277, 278]. Let us denote the cells $C_\alpha(\Gamma)$ in configuration space by a characteristic function [260, 277, 278]

$$C_\alpha(\Gamma) = \begin{cases} 1, & \Gamma \text{ is in cell } \alpha \\ 0, & \text{otherwise} \end{cases} \tag{4.39}$$

The cells are complete and span the configuration space [260, 277, 278]

$$\begin{aligned} \sum_\alpha C_\alpha(\Gamma) &= 1 \qquad \text{for all } \Gamma \\ C_\alpha(\Gamma) C_\beta(\Gamma) &= \delta_{\alpha\beta} C_\alpha(\Gamma) \end{aligned} \tag{4.40}$$

The time dependence of the probability of a state to be in a cell α at time

t is [260, 277, 278]

$$P_\alpha(t) = \int dX\, C_\alpha(X) f(X, t) \tag{4.41}$$

where $f(X, t)$ is the phase space distribution function at a point X at time t.

The initial condition [260, 277, 278]

$$f(X, 0) = f_{\text{equil}}(X) T(C) \tag{4.42}$$

and the use of Mori–Zwanzig algorithm leads to the time dependence of probabilities in a cell α

$$\dot{P}_\alpha(t) = -\sum_\gamma \int_0^t dt'\, K_{\alpha\gamma}(t - t') P_\gamma(t') \tag{4.43}$$

where $T(C)$ is an arbitrary function of C. If the memory of the initial cell is lost, then the time dependence of residence probabilities and fluctuations in stress $\sigma(t)$ are [277, 278]

$$\dot{P}_\alpha(t) = -\sum_{\gamma \neq \alpha} [K_{\alpha\gamma} P_\gamma(t) - K_{\gamma\alpha} P_\alpha(t)] \tag{4.44}$$

$$\sigma(t) = \sum_\alpha \sigma_\alpha P_\alpha(t)$$
$$= \sum_\alpha A_n \phi_n \exp(-\lambda_n t) \tag{4.45}$$

where λ_n are the eigenvalues of the transition matrix $K_{\alpha\gamma}$.[44]

At times much less than $\tau_0 \approx \varphi^2/D$, where D is the self-diffusion coefficient and φ is of the order of short-range order in glass, the probability that thermal fluctuations would excite a mode of wavelength ϑ is approximately $\exp(-\vartheta/\varphi)$ [277, 279]. On introducing a continuum approximation to the sum in Eq. (4.45) with a lower and an upper cutoff and performing an asymptotic analysis of the integral yields [277, 279]

$$\sigma(t) \approx [(k^*)^2/A] \phi_{k^*} (\pi/3Dt)^{1/2} \exp -(t/\tau_s)^{1/3} \tag{4.46}$$

where $k^* = (2Dt\varphi)^{-1/3}$, $A = 2/\pi\vartheta_{\max}$, and $\tau_s = (4/27)\varphi^2/D$ [277, 279]. In deriving Eq. (4.46) an important physical assumption [277] has been

[44] The quantity ϕ_n may be expanded in terms of the eigenvectors $B_\alpha^{(n)}$ of the matrix $K_{\alpha\gamma}$ and $\phi_n = \Sigma_\alpha B_\alpha^{(n)} \sigma_\alpha$.

made, namely, that structural rearrangements in viscous fluids occur on molecular length scale.[45]

For times of the order of τ_0, the probability of exciting a mode is $\exp(-\pi E\vartheta^2/k_B T)$, where E is an activation energy per unit area for slipping motion [277, 279]. In this case, the time dependence of stress relaxation is found to be [277, 279]

$$\sigma(t) \approx [(k^*)^2/A]\phi_{k^*}(\pi/4Dt)^{1/2} \exp -(t/\tau_{1s})^{1/2} \tag{4.47}$$

where $\tau_{1s} = k_B T/4\pi^2 E^2 D$. As $t \to \infty$, a single mode only contributes and the stress decays exponentially [277, 279]

$$\sigma(t) \approx [4\pi/A\vartheta^2_{\max}]\phi_{k^*} \exp -(t/\tau_\infty) \tag{4.48}$$

Observe that the various times τ_s, τ_{1s}, τ_∞ can be expressed in terms of the fundamental time τ_0 [277].

5. *Interbasin Dynamics*

To study the dynamics of transitions between inherent structures it is useful to classify the various minima by a set of intensive order parameters $\{\xi_\alpha\}$ [246]. The probability that the system is in a cell α satisfies the master equation [see Eq. (4.44)]. The transition probabilities $K_{\alpha\gamma}$ satisfy detailed balance and can be written in the form

$$K_{\alpha\to\gamma} = \{\langle M_\alpha(E)\rangle / \langle M_\gamma(E)\rangle\}^{1/2} A_{\alpha\to\gamma}(E) \tag{4.49}$$

where $\langle M_\alpha(E)\rangle$ is the volume in phase space of cell α and $A_{\alpha\to\gamma}(E)$ is a symmetric matrix. Consider a change of variables to an order parameter basis [246]

$$P(\xi_\alpha, t) = \exp(N\sigma(\xi_\alpha))P_\alpha(t) \tag{4.50}$$

In Eq. (4.50), $\sigma(\xi_\alpha)$ is proportional to the configurational entropy [246][46]

$$\exp[S(\xi_\alpha, E)/k_B] = \langle M_\alpha(E)\rangle \exp N\sigma(\xi_\alpha) \tag{4.51}$$

Here, E is the energy of the system. Then, from Eqs. (4.44) and (4.49)–(4.51) one obtains the master equation in the continuum limit

[45] Consequently, $(\tau_k)^{-1} = Dk^2$. Lower and upper cutoffs are denoted, respectively, by $\vartheta_{\min}$ and $\vartheta_{\max}$. The parameter τ_∞ in Eq. (4.48) is defined as $\vartheta_{\max}/4\pi^2 D$. See [277, 279].

[46] The parameter $\sigma(\xi_\alpha)$ should not be confused with stress $\sigma(t)$ introduced in Eq. (4.45). Instead, see Eq. (4.3).

[246]

$$dP(\xi,t)/dt = \int d\xi'\, A(\xi',\xi;E)[\exp(S(\xi,E)-S(\xi',E))/2k_B P(\xi',t) - \exp(S(\xi',E)-S(\xi,E))/2k_B P(\xi,t)] \quad (4.52)$$

where $A(\xi', \xi; E)$ is an average transition matrix that is weighted in an appropriate way [246]. The transition between cells is due to localized rearrangements of particles [246]. Consequently, changes of the order parameters are of $O(1/N)$ and one can carry out a Van Kampen system size expansion in Eq. (4.52) [246]. This finding leads to the desired Fokker–Planck equation [246]

$$dP(\xi,t)/dt = -\mathrm{div}[(\zeta\cdot\mathbf{F})P(\xi,t)] + 1/\beta\ \mathrm{div}\cdot[(\zeta\cdot\mathrm{grad}\,P)] \quad (4.53)$$

where the thermodynamic force $\mathbf{F}$ and the mobility tensor ζ are

$$\mathbf{F}(\xi,E) = (k_B\beta)^{-1}\,\mathrm{grad}\,S(\xi,E) \quad (4.54)$$

$$\zeta(\xi,E) = \tfrac{1}{2}\beta\int d\xi'\, A(\xi,\xi-\xi';E)(\Delta\xi')^2 \quad (4.55)$$

Several comments are in order. First, observe that the thermodynamic force as given by Eq. (4.54) is due to entropic-like barriers [280]. Second, one can define a diffusion coefficient (tensor) in the order parameter space $D(\xi, E) = k_B T\zeta(\xi, E)$. Third, on the extremum surface, the force vanishes [280]. Mohanty, Oppenheim, and Taubes [280] recently showed that a study of the dynamics of the supercooled liquids on this extremum surface leads to temperature dependence of relaxation rate, which is of the AG form. The approach is remarkably similar to Cooper's formulation of the Cooper pair gap equation in the Bardeen–Cooper–Schrieffer theory of superconductivity [281, 282].

V. MODELS FOR DIFFUSION, VISCOSITY, AND DIELECTRIC RELAXATION

A. Diffusion in Rough Potentials

The potential surface $\Phi(x)$ of a glass-forming liquid has minima of various depths. This section explores the dynamical consequences of hierarchical-like structure in $\Phi(x)$ on the self-diffusion coefficient D. Suppose the potential is partitioned into a random or rough part $\Phi_1(x)$ superimposed on a smooth background $\Phi_0(x)$ [283]. The potential $\Phi_1(x)$ has an amplitude A_μ and varies over a length scale, Δx.

Let us assume that the motion of the system on the potential surface is described by the Smoluchowski equation. The Smoluchowski current density can be expressed in terms of the probability distribution $\rho(t)$ [283]

$$J(x) = -D \exp\{-\beta\Phi(x)\}\, \partial/\partial x \exp\{\beta\Phi(x)\}\rho(t) \tag{5.1}$$

where $\beta = 1/k_B T$. The mean first-passage time $\langle t; x\rangle$ for the system to reach the point x if it starts at x_0 is given by [283][47]

$$\langle t; x\rangle = \int_{x_0}^{x} dy \exp\{\beta\Phi(y)\} 1/D \int_{a}^{y} dz \exp\{-\beta\Phi(z)\} \tag{5.2}$$

Zwanzig [283] evaluated Eq. (5.2) based on the following idea. Integrations over y or z are equivalent to spatial averages. The integral over dx is, therefore, approximated as

$$\int \mathrm{dz} \exp\{\beta\Phi(z)\} \approx \int dz \exp\{-\beta\Phi_0(z)\} \exp \Psi_-(z) \tag{5.3}$$

where $\exp \Psi_-(z) = \langle \exp -\beta\Phi_1(z)\rangle$. Similarly, integration over y results in a term of the form $\exp \Psi_+(z) = \langle \exp \beta\Phi_1(z)\rangle$.

The mean first-passage time turns out to be given by an expression identical with Eq. (5.2). The only difference is that one has an effective potential $\Phi^*(x)$ and an effective diffusion coefficient $D^*(x)$ [283]

$$\Phi^*(x) = \Phi_0(x) - \Psi_-(x)/\beta \tag{5.4a}$$

$$1/D^*(x) = \exp \Psi_+(x)(1/D) \exp \Psi_-(x) \tag{5.4b}$$

If the roughness is independent of x, then the diffusion coefficient is [283]

$$D^* = D \exp(-\Psi_+) \exp(-\Psi_-) \tag{5.5}$$

Zwanzig [283] applied Eq. (5.5) to two cases. The first deals with a periodic perturbation of the form $A_\mu \cos kx$, while the second case deals with a perturbation in which the amplitude is a Gaussian random variable with a probability distribution proportional to $\exp -(\Phi_1^2/2\varepsilon^2)$. The essential results obtained by Zwanzig [283] is that at low temperatures diffusion is strongly temperature dependent while at temperatures large in comparison with the amplitude, diffusion is not effected by the

[47] A reflective boundary condition is assumed at the barrier located at $x = a$.

multitude of small barriers.

$$D^* \approx \exp(-2A_\mu/k_B T) \qquad \text{periodic} \tag{5.6}$$

$$D^* = D \exp[-(\varepsilon/k_B T)^2] \qquad \text{random} \tag{5.7}$$

B. Diffusion Past Entropic Barriers

Recently, Mohanty, Oppenheim, and Taubes [284] formulated the dynamics of configuration point past entropic barriers in an appropriate hypersurface of the $3N$-dimensional configurational space. The results lead naturally to temperature dependence of cooperative relaxation that are a nonequilibrium generalization of the AG theory. To gain insight into why entropic surfaces are important in glassy dynamics, we will study diffusion of molecules in one and two dimensions past simple entropic barriers.

Consider a molecule performing a random walk through a tube. The tube has varying cross-section $A(x)^3$ with the x axis through the center of the tube. Clearly, the molecule slows down at the bottleneck or constriction [285]. This is the simplest example of an entropic barrier. To describe dynamics in such systems, Jacobs and Fick (JF) introduced local concentration $c(x, y, t)$ as the appropriate dynamical variable. Let $F(x, t)$ be the amount of molecules that passes through a differential cross-section dx of the tube in a time t. The parameter $F(x, t)$ satisfies the JF equation [285–287]

$$\partial F(x, t)/\partial t = D\, \partial/\partial x\, A(x)\, \partial/\partial x[F(x, t)/A(x)] \tag{5.8}$$

The similarity of the JF equation with the Smoluchowski equation leads to the correspondence

$$A(x) \leftrightarrow \exp -\beta\Phi(x) \tag{5.9}$$

where $\Phi(x)$ is an arbitrary potential. As pointed out by Zwanzig [285], it is natural to relate the cross-section $A(x)$ to entropy

$$A(x) \leftrightarrow \exp S(x)/k_B \tag{5.10}$$

Equations (5.8) and (5.10) then describe the dynamics of the system past entropic barriers.

A derivation of Eq. (5.8) proceeds by starting with the Smoluchowski equation in two dimensions for the local concentration $c(x, y, t)$ and rewriting it in terms of $F(x, t)$

$$F(x, t) = \int dy\, c(x, y, t) \tag{5.11}$$

where $F(x, t)$ satisfies the differential equation [285]

$$\partial F(x, y)/\partial t = D\, \partial/\partial x \int dy \exp -\beta\Phi(x, y)\, \partial/\partial x \times \exp \beta\Phi(x, y) c(x, y, t) \tag{5.12}$$

If the concentration is at local equilibrium, then $c(x, y, t) \approx F(x, t) \exp -\beta\Phi(x, y)/\exp -\beta G(x)$. The JF equation then follows from Eq. (5.12) [285]

$$\partial F(x, t)/\partial t \approx D\, \partial/\partial x \exp -\beta G(x)\, \partial/\partial x \exp \beta G(x) F(x, t) \tag{5.13}$$

where $G(x)$ is a free energy that depends on x [285]

$$\exp -\beta G(x) = \int dy \exp -\beta\Phi(x, y) \tag{5.14}$$

Zwanzig [285] has also taken into account deviations of the concentration from local equilibrium. The principal result is that the JF equation has an entropic barrier modified by a position-dependent diffusion coefficient $D(x)$ [285].

1. *Diffusion in Periodic Potentials*

To study motions through channels one could consider one-dimensional periodic potentials. The effective diffusion coefficient in a periodic potential may be obtained either by a mean first-passage time method or by an eigenvalue analysis [288a]

$$1/D^* = \langle 1/[D^*(x) A(x)] \rangle \langle A(x) \rangle \tag{5.15}$$

where $\langle\ \rangle$ is an average over one period of the potential.

In two dimensions, diffusion through periodic channel is readily handled by conformal transformations [285, 288b]. Suppose the channels are defined by orthogonal coordinates

$$\begin{aligned} w &= u + iv \\ z &= f(w) \\ z &= x + iy \end{aligned} \tag{5.16}$$

As an example, consider the channel studied by Zwanzig [285, 288b]

$$f(w) = w + a \sin w$$
$$x = u + a \cosh v \sin u \qquad (5.17)$$
$$y = v + a \sinh v \cos u$$

where $0 < u < 2\pi$ and $-V < v < V$. For this case, an exact expression is obtained for the diffusion coefficient D^* in the long time limit [285]

$$(D^*/D)_{\text{exact}} = [1 + (\alpha^2/2V) \tanh V]^{-1} \qquad (5.18)$$

where $\alpha = a \cosh V$. The exact JF equation and its modified form[48] lead to the following results [285]

$$(D^*/D)_{\text{exact}} = [1 + \alpha^2/(\tanh V/V)^2 + O(\alpha^4)]^{-1}$$
$$(D^*/D)_{\text{mod}} = [1 + \alpha^2/(\tanh V/V)^2 + \alpha^2/6(\tanh^2 V) + O(\alpha^4)]^{-1} \qquad (5.19)$$

The difference between the exact and the modified JF resides in the α^4 term.

2. *Diffusion in a Hyperboloidal Tube*

Another exactly solvable model is diffusion through a hyperboloidal tube. The steady-state diffusion in spheroidal coordinates (ζ, η, ϕ) is given by

$$\partial/\partial\zeta(1 - \zeta^2)\, \partial c/\partial\zeta + \partial/\partial\eta(1 - \eta^2)\, \partial c/\partial\eta = 0 \qquad (5.20)$$

Particles that reach the small end of the hyperboloidal tube are removed. The appropriate boundary conditions are, therefore $c \rightarrow c_0$ as $\zeta \rightarrow \infty$, $c = 0$ for $\zeta = 0$, and $(\partial c/\partial\eta)_{\eta_0} = 0$, that is, with a reflecting wall at $\eta = \eta_0$ [285]. The solution is straightforward [285]

$$c(\zeta, \eta, \phi) = c_0(1 - 2/\pi \tan^{-1} 1/\zeta) \qquad (5.21)$$

The steady-state flux, $-k_{\text{ss}}$, is obtained from the flux normal through the hole and from the area of the hole of radius a, $\pi a^2(1 - \eta_0^2)$,

$$k_{\text{ss}} = 4Da(1 - \eta_0)c_0 \qquad (5.22a)$$

The solution to the JF equation yields [285]

$$k_{\text{FJ}} = k_{\text{exact}}(1 + \eta_0)/2\eta_0 \qquad (5.22b)$$

[48] Defined as the JF equation with fluctuations taken into account; see [285].

Thus, the essential result is that as long as the crosssection of the tube is not varying rapidly, the JF equation is a good approximation [285]; it fails when $|A'(x)|$ exceeds unity [285].

C. Normal Mode Analysis

The basis of normal mode expansion is the observation that the potential energy $\Phi(\mathbf{R})$ can be expanded as [289–294]

$$\Phi(\mathbf{R}) = \Phi(\mathbf{R}_0) + \mathbf{F}(\mathbf{R}_0) \cdot \mathbf{x} + \tfrac{1}{2}\mathbf{K}(\mathbf{R}_0) \colon \mathbf{xx} \tag{5.23}$$

where $\mathbf{R}_0$ is the coordinate of the N particles, $\mathbf{K}$ is a matrix of the second derivative of the potential, $\mathbf{x} = \mathbf{R} - \mathbf{R}_0$ and $\mathbf{F}(\mathbf{R}_0)$ are the forces at $\mathbf{R}_0$. The matrix $\mathbf{K}$ could be diagonalized to yield the eigenvalues and the eigenfrequencies of the system.

For a solid, $\mathbf{R}_0$ is at the minimum and the force vanishes. The eigenvalues of $\mathbf{K}$ are therefore positive. This is not so in a liquid. In fact, $\mathbf{K}$ would have negative eigenvalues. Since the frequencies are obtained by taking the square roots of the eigenvalues, one would thus have imaginary frequencies. One is further faced with an ambiguity regarding the choice of $\mathbf{R}_0$. Since there is no unique choice for $\mathbf{R}_0$, a configuration averaging is needed for properties of interest.

Keyes and co-workers [289–292] carried out a normal mode analysis by simulations of a Lennard–Jones system. They analyzed the density of states averaged over many configurations, $\langle \rho(\omega) \rangle$, at various densities and temperatures, which are relevant to glassy and supercooled states. A quantity f_u was introduced, which measures the fluidity of the system [289–292]. The parameter f_{u} is the number of unstable modes in the system. It is an indicator of the number of directions away from a barrier region [289–292]. The density of states $\langle \rho(\omega) \rangle$ as a function of $\omega\tau$ displays[49] the following characteristics [289–292]. (a) As temperature declines the number of unstable modes decreases. (b) The unstable mode is located at a characteristic frequency ω_{u}. The parameter ω_{u} shifts towards zero with decreasing temperature [289–292]. (c) The position of the stable mode peak ω_{s} is of the order of Einstein frequency. It increases with decreasing temperature [289–292]. Overall, the structure of the stable mode is complex. For example, the high frequency but stable part of $\langle \rho(\omega) \rangle$ decays slowly [289–292].

Keyes and co-workers argued that the quantity f_{u} determines the time spent on the peaks while ω_{u} is a reflection of the curvature of the peaks [289–292]. The ratio of the time spent in a valley τ_{v} to the time spent in

[49] Here τ is 2.15 ps; see [289]. In Eq. (5.25) $\omega_{\mathrm{v}} \sim 1/\tau_{\mathrm{v}}$ and m is the mass of the particle.

crossing a barrier τ_b is therefore proportional to $(1-f_u)/f_u$. The self-diffusion coefficient D is then approximately given by [289–292]

$$D \approx T^{1/2} f_u/(1-f_u) \tag{5.24}$$

Since the stable mode part of $\langle \rho(\omega) \rangle$ describes phonons, another expression for self-diffusion is obtained from the Zwanzig–Mohanty model [295–298] by approximating the lifetime in a cell by τ_v [289–292]

$$D = (k_B T/m)\omega_v(1-f_u)^{-1} \int d\omega \, \langle \rho(\omega) \rangle_s (\omega^2 + \omega_s^2)^{-1} \tag{5.25}$$

The temperature and density dependence of D is calculated from a knowledge of the normal modes, f_u, ω_s, and $\langle \rho(\omega) \rangle_s$; these quantities are, however, evaluated via simulations [289–292]. By this procedure Keyes and co-workers [289–292] obtained insights into the Zwanzig–Mohanty model and showed that it is an appropriate starting point to understanding dynamics in glassy and supercooled liquids.

The connection between self-diffusion coefficient and f_u was anticipated several years ago by Cotterill and Masden [293] who studied the normal modes of a Lennard–Jones system for particular cross sections of the configuration space. An important result obtained by these authors are that these sections are single wells for the glassy state while they are double wells for supercooled fluids above T_g [293].

D. Models for Self-Diffusion

1. *Damped Harmonic Picture*

The self-diffusion coefficient of a cold dense fluid was studied long ago by Rahman et al. [299] using a normal mode analysis. The basic assumptions in this model are that rapidly varying degrees of freedom are similar to that in a solid, while the slowly varying degrees of freedom perform Brownian motions and satisfy a Langevin equation.

If the displacement of the atoms are expressed in terms of normal modes ξ_s, then the Langevin equation for these modes are [299]

$$d^2\xi_s/dt^2 + \lambda_s \, d\xi_s/dt + \omega_s^2 \xi_s = f_s(t) \tag{5.26}$$

where f_s is a random force, λ_s is a friction constant, and ω_s is the frequency of oscillation of the sth mode. Since harmonic restoring forces are not present at low frequencies, the diffusion modes instead obey [299]

$$d^2\xi_s/dt^2 + \lambda_s \, d\xi_s/dt = f_s(t) \qquad \omega_s < \omega' \tag{5.27}$$

To account for damping of the high-frequency modes, Rahman et al. [299] assume that the friction is proportional to frequency

$$\lambda_s = 2\gamma\omega_s \qquad \omega' \leq \omega_s \leq \omega_D \tag{5.28}$$

where ω_D is the Debye frequency. For frequencies less than ω', the damping term is set equal to λ.

The velocity autocorrelation function is readily evaluated [299] by approximating the density of states by the Debye spectrum with an upper frequency cutoff ω_D,

$$\langle \mathbf{v}(0) \cdot \mathbf{v}(t) \rangle = 3k_B T/m\{(\omega'/\omega_D)^3 \exp(-\lambda t) + (3/\omega_D)^3 \int_{\omega'}^{\omega_0} d\omega_s \\ \times \exp -\Gamma\omega_1 t(\cos \omega_1 t - \Gamma \sin \omega_1 t) \tag{5.29}$$

where $\omega_1 = [\omega_s - \lambda_s^2/4]^{1/2}$ and $\Gamma^2 + 1 = (1 - \gamma^2)^{-1}$. The diffusion constant is obtained by integrating Eq. (5.29) over time [299]

$$D = (k_B T/m\lambda)(\omega'/\omega_D)^3 \tag{5.30}$$

We can compare Eq. (5.30) with the expression for D obtained by Keyes and co-workers [289–292]. The term $(\omega'/\omega_D)^3$ represents the fraction of modes that participate in diffusion; hence, this quantity is related to the fraction of unstable modes.[50]

Another model for self-diffusion along the lines described above was independently formulated by Tobolsky [300]. Tobolsky was also exploiting the concept of a damped harmonic lattice. The equations of motion of the normal modes satisfy Eq. (5.26). A relaxation time was associated with each mode of frequency ω_α

$$\tau_\alpha = \lambda/\omega_\alpha^2 \tag{5.31}$$

For a cubic lattice the maximum frequency ω_{max} is approximately $18.7v^{1/3}G_\infty/\text{MW}$, where G_∞ is the infinite frequency shear modulus, v is the volume per particle, and MW is the molecular mass [300]. Associated with ω_{max} is a minimum relaxation time τ_{min}. The shear viscosity is found to be [300]

$$\eta \approx 18.7 G_\infty \tau_{min}/3\pi \tag{5.32}$$

Tobolsky [300] evaluates the shear modulus by using the SE relation

[50] U. Mohanty, unpublished work.

and a Debye approximation,

$$G(t) \approx G_\infty [1 + t/\tau_{\min}]^{-3/2} \tag{5.33}$$

Note that Eq. (5.33) leads to an expression for shear viscosity $\eta = 2G_\infty \tau_{\min}$ in quantitative agreement with Eq. (5.32).

2. *Disorder Models*

The concept of disorder has been used by various authors to describe viscous flow in undercooled melts [301–303]. Avramov and Mitchev (AM) [301, 302] proposed that viscosity and diffusion coefficients of undercooled melts and even crystals may be obtained by relating structural disorder to a distribution of energy barriers. The basic idea is that the barriers may be regarded as random variables with a width w_σ around some average value of energy E [301, 302]. The average transition rate for the system to hop over a barrier is [301]

$$\langle \bar{\omega} \rangle = \int_0^{E_{\max}} W(E) \bar{\omega}(E)\, dE \tag{5.34}$$

Here, $W(E)$ is the probability of jumping over a barrier of height E and $\bar{\omega}(E)$ is the transition rate. If there are z escape channels and the distribution of such channels is $P(E)$, then [301]

$$W(E)\, dE = zP(E) \left[\int_E^{E_{\max}} P(\varepsilon)\, d\varepsilon \right]^{z-1} dE \tag{5.35}$$

where $E_{\max}$ is the maximum value of E. By using uniform Poisson and Gaussian distributions for $P(E)$, Avramov and Milchev [301] showed that the entropy is given in terms of the dispersion w_σ

$$S - S_0 \approx (Rz/2) \ln(w_\sigma / w_{\sigma 0}) \tag{5.36}$$

The parameter S_0 is a reference entropy.[51] From Eqs. (5.34–5.36) and assuming a Poisson distribution,[52] an expression is obtained for the

[51] The parameter $w_{\sigma 0}$ is the dispersion of the reference system.
[52] The Poisson distribution for $P(E)$ holds for $0 \leq E \leq E_c$.

viscosity in terms of three parameters a, b, and η_0 [301]

$$\ln(\eta/\eta_0) = b/(T/T_{\mathrm{m}})^a \tag{5.37}$$

$$a = 2C_p/Rz \tag{5.38}$$

$$b = (E_{\mathrm{c}}/w_{\sigma 0})(T_{\mathrm{m}}/T_0)^a \tag{5.39}$$

Here T_{m} is the melting temperature, R is the gas constant, and C_p is the heat capacity under constant pressure.[53] Taking the heat capacity of the reference ordered state to be $3R$ and the entropy of the disordered state over $3R$ to be the experimental excess heat capacity, AM found the temperature dependence of viscosity for a variety of glass forming melts is well described by relations (5.37–5.39) [301]. The substances include organic liquids (glycerin and glucose), phosphates, chalcogenide substances (As_2Se_2), transition metals, and silicates [302]. Their results also provide an explanation of the empirical Bimen–Kauzmann rule, namely, that an upper bound to T_{g} is $0.7T_{\mathrm{m}}$ [302].

Another example of how the concept of disorder may be utilized to elucidate the glassy state is the work of Bassler [303]. Bassler approximates the multidimensional potential surface of the supercooled liquid by a Gaussian density of states [303]. This leads to several plausible assumptions. First, as the temperature is lowered, the molecules tend to be localized among the lowest states accessible [303]. Second, the evolution of the density of states is slow in comparison to all relevant time scales of interest [303]. At large times the particles equilibrate with a mean energy $\langle e_\infty \rangle = -w_\sigma^2/k_{\mathrm{B}}T$, where as before w_σ is the width of the density of states [303]. The mean transition to a nearby minima is found to be proportional to $\exp[-(T_0/T)^2]$ [303]. Bassler then obtains the viscosity $\eta(T)$ by assuming that it is proportional to the mean relaxation time, that is, $\eta(T) = \eta_\infty \exp[-(T_0/T)^2]$ [303]. An analysis of the experimental viscosity data indicates that the relation is obeyed over several order of magnitude decades in η. This includes the region near T_{g}, where Arrhenius behavior is supposed to occur [303]. A correlation is found between T_0 and T_{g}, as well as between η_∞ and T_{g} [303]. However, computer simulations are apparently in variance with the approximation that the width of the Gaussian density of states is independent of supercooling [304].

[53] The parameter C_p is taken to be a constant [301].

E. Validity of Debye–Stokes–Einstein Relations

The experimental validity of Debye–Stokes–Einstein (DSE) relations in supercooled liquid has received renewed interest both experimentally and theoretically. In a remarkable paper that appeared over three decades ago, McCall et al. [305] showed, using pulsed nuclear magnetic resonance (NMR) techniques, that for *o*-terphenyl the SE relation is valid within a factor of 2 from 0 to 180°C [305, 295, 298].

Using high-pressure NMR techniques Walker et al. [306] measured both the viscosity and the self-diffusion coefficients of 2-ethylhexylbenzoate[54] over a broad range of temperatures (−20–100 °C) and pressures from 1 to 4500 bar. They conclude that the SE relation is valid over a range that covers five orders of magnitude of data for both viscosity and self-diffusion [306]

$$D = k_B T / C\pi a \eta \tag{5.40}$$

The constant C is 4.2 at 1 bar while it approaches the value 5.6, that is, the stick boundary condition at 4500 bar. The coupling between translation and rotation increases as the pressure increases, thereby changing the boundary conditions from slip to stick [306].

Artaki–Jones [307–309] tested the validity of the Debye relation

$$\tau_\theta = \kappa V_H \eta / k_B T + \tau_h \tag{5.41}$$

for several supercooled liquids, such as isopropylbenzene-d_5, *cis*-decalin-d_{10}, and *sec*-butylbenzene-d_5, by using high-pressure effects to separate kinetic and volume effects. In Eq. (5.41) η is the shear viscosity, V_H is a hydrodynamic volume, τ_θ is a reorientation correlation time, τ_h is a relaxation time when viscosity vanishes, and κ reflects the coupling between rotational and translational motions of the molecule. Wang et al. [307] also studied the validity of the Debye relation in α-phenyl *o*-cresol molecules by light-scattering techniques. The results obtained by Artaki and Jonas [308, 309] are that (a) at fixed T, the parameter κV_H increases with density if the molecules lack an axis of symmetry. Thus, κV_H is sensitive to volume effects [308, 309]; (b) at high pressure the linearity between τ_θ and η / T is absent [308, 309]; (c) volume plays the crucial role in the coupling of rotational and translational motions [308, 309]; (d) under isochoric conditions, κV_H is insensitive to changes in kinetic energy [308, 309]; and (e) in the viscosity range of up to 10 P, the DSE relations adequately represent the dynamics of isopropylbenzene in the slip limit [308, 309].

[54] The chemical formula is [$C_6H_5COOCH_2CH(CH_2CH_3)(CH_2)_3CH_3$].

Various modification of the SE relation have been proposed in the literature such as [310]

$$D = \text{const}/\eta^{p} \qquad p \text{ is a constant} \tag{5.42}$$

Zwanzig and Harrison [311] argued that such modifications of the SE relation should not be viewed as if the law is incorrect. They offer two reasons to support their assertion. The first reason is based on hydrodynamics. The relation between D and the friction coefficient is based on linear response. This relation does not depend on the mechanism that leads to frictional dissipation [311, 297]. Consequently, if deviations from SE are observed experimentally, then it may be fruitful to look for additional nonviscous mechanisms for dissipation [311, 297]. The second reason is based on the experiments of Artaki and Jones [309], which measured the pressure dependence of η and rotational relaxation time τ at fixed temperature. By introducing activation volumes ΔV_τ and ΔV_η, which depend on temperature, η and τ are fitted to the functional form [309, 311]

$$\tau \approx \exp P\,\Delta V_\tau/RT \tag{5.43a}$$

$$\eta \approx \exp P\,\Delta V_\eta/RT \tag{5.43b}$$

These results imply [311]

$$\tau = \eta^{p} \tag{5.44}$$

where $p = \Delta V_\tau/\Delta V_\eta$ is not a universal quantity. In fact, the values for p lie between 0.66 and 1.24 [311].

At constant volume, the SE relation satisfies Eq. (5.44) with $p = 1$. But the effective hydrodynamic volume κV_{H} is [311]

$$\kappa V_{\mathrm{H}} = \exp P(\Delta V_\tau - \Delta V_\eta)/RT \tag{5.45}$$

This result indicates that pressure and temperature are independent variables. However, what is relevant is the equation of state, connecting the two independent variables [311].

Below the mode coupling temperature T_{x}, the temperature dependence of rotational as well as translational diffusion times in several undercooled melts deviate from predictions of hydrodynamic models [312]. Fujara et al. [313] argued that in OTP, deviations should occur only in translational diffusion coefficient. The differences have been attributed by these authors to correlated dynamics near T_{g} [313].

Cicerone and Ediger [314a] used photobleaching techniques to carefully measure the rotational diffusion of probe molecules in glass-forming liquids both above and below T_g. For tetracene in OTP these authors measured the rotational correlation times that span the range 10^{-2}–$10^{+4.5}$ s [314a]. On combining these results with data from nanosecond experiments of anthracene rotations, Cicerone and Ediger obtained the following results [314a]. (a) The rotational diffusion of tetracene is in agreement with that predicted from the DSE relation below and near T_g [314a]; (b) the high-temperature rotational data for anthracene indicates that the DSE relation is satisfied over 14 decades in viscosity and time [314a]. In contrast, the SE relation for translation diffusion is off by a factor of 2 or so at T_g [314b]. An intriguing result obtained by Blackburn et al. [314b] is that as $T \to T_g$, translational and rotational motions decouple in OTP and polystyrene.

Recently, Zeng et al. [315] derived a generalized SE relation or Walden's rule

$$\tau_A T/\eta = \omega_A^2 T/\omega_a^2 \rho c_{\text{sh}}^2 \tag{5.46}$$

Here A is an arbitrary dynamical variable, ρ is an average density, c_{sh} is the infinite frequency shear speed, and the amplitudes ω_A^2 and ω_a^2 are defined by [315]

$$\omega_a^2 = \langle \dot{A}(0)\dot{A}(0)^* \rangle / \langle A(0)A(0)^* \rangle \tag{5.47a}$$

$$\omega_A^2 = \langle \ddot{A}(0)\ddot{A}(0)^* \rangle / \langle \dot{A}(0)\dot{A}(0)^* \rangle + \omega_a^2 \tag{5.47b}$$

where τ_A is the auocorrelation time of $A(t)$. The derivation of Eq. (5.46) is based on two assumptions about the second memory function $\phi(t)$ [315]. First, $\phi(t)$ has a bifurcation [315]. The second assumption is that $\phi(t)$ is independent of the long-time behavior of $A(t)$, and hence is universal [315].

F. Dielectric Relaxation

The frequency dependence of dielectric relaxation ($\varepsilon(\omega) - 1$) and retardational compliance $J_r(\omega)$ in supercooled liquids are remarkably similar [316, 317]. Both quantities are expressible in terms of the Cole–Davidson function, $(1 - i\omega\tau_{\text{cd}})^{-\beta}$, where β is a spectral width parameter [316]. The Cole–Davidson relaxation time τ_{cd} diverges as $T \to T_g$; at high T, τ_{cd} is small and the exponent β is approximately unity.[55] At low temperatures

[55] In this section the exponent β is defined by the Cole–Davidson function and should not be confused with $1/k_B T$.

not only is $\mu J_r(0)$ approximately constant[56] but so also is the quantity [316, 317]

$$\mu J_r(0)[\tau_m/\tau_{cd}]^{\beta} \approx 2 \tag{5.48}$$

where τ_m is the Maxwell relaxation time. The empirical relation (5.48) is satisfied for a variety for glass-forming liquids [316].

The molecular expressions for dielectric relaxation $(\varepsilon(\omega)-1)$ and retardational compliance $J_r(\omega)$ are given, respectively, by [317]

$$(\varepsilon(\omega)-1)/(\varepsilon(0)-1) = \int_0^{\infty} dt \exp(i\omega t)\langle Y_1(t)\dot{Y}_1^*(0)\rangle/\langle|Y_1(0)|^2\rangle \tag{5.49}$$

$$\mu J_r(\omega) = \int_0^{\infty} dt \exp(i\omega t)\langle \sigma(t_\sigma) - \sigma(t_\sigma \to \infty)]\sigma(0)^*\rangle/\langle|\sigma(0)|^2\rangle \tag{5.50}$$

The angled brackets denotes equilibrium average, Y_1 is a collective reorientational variable, and μ is the shear modulus,

$$\mu = \langle|\sigma(0)|^2\rangle/k_B TV \tag{5.51}$$

where σ is the stress tensor that includes only off-diagonal components [317]. The time evolution t_σ is governed by a propagator whose components along σ have been eliminated [317, 318]. The relation (5.50) for the retardational compliance is due to MacPhail and Kivelson [317].

Consider a dynamical variable A. Let $\hat{G}_A(\omega)$ be the Laplace transform of an unprojected correlation function of A [317–319]

$$\dot{G}_A(t) = -\langle A(t)\dot{A}(0)^*\rangle/\langle|A(0)|^2\rangle \tag{5.52}$$

Similarly, let $\hat{\Gamma}(\omega)$ be the Laplace transform of the projected correlation function of A [317–319]

$$\dot{\Gamma}_A(t) = -\langle[A(t_A) - A(t_A \to \infty)]\dot{A}(0)^*\rangle/\langle|A(0)|^2\rangle \tag{5.53}$$

The relation between $\hat{G}_A(\omega)$ and $\hat{\Gamma}_A(\omega)$ is [317–319]

$$\hat{G}_A(\omega) = [-i\omega\hat{\Gamma}_A(\omega) - \tau_A^{-1}]\{-i\omega[-\hat{\Gamma}_A(\omega)+1] + \tau_A^{-1}\}^{-1} \tag{5.54}$$

[56] The value for $\mu J_r(0)$ is between 2 and 80 [262].

where

$$\tau_A^{-1} = \int_0^\infty dt\, \langle \dot{A}(t_A)\dot{A}(0)^* \rangle / \langle |A(0)|^2 \rangle \tag{5.55}$$

To explain the observed correlation (5.48), MacPhail and Kivelson [317] proposed that the projected and the unprojected correlation functions have similar functional forms for Y_1 and σ. If $\hat{\Gamma}_A(\omega)$ is expressed in terms of a function $F(\omega)$ [317]

$$\hat{\Gamma}_A(\omega) = BF(\omega) \tag{5.56}$$

then as shown by MacPhail and Kivelson [317], this leads to the requirement

$$\hat{\Gamma}_A(\omega) = B\hat{G}_A(\omega) \tag{5.57}$$

In Eq. (5.56), B does not depend on frequency and $F(0)$ is unity. For the problem at hand, $\hat{\Gamma}_A(\omega) = \mu J_r(\omega)$. The important results [317] obtained by these authors are as follows: (a) there is a solution for all ω, if $B < 0 \leq 1$; (b) the Cole–Cole plots of $\hat{\Gamma}_A(\omega)$ and $\hat{G}_A(\omega)$ are similar provided $B[\tau_A/\tau_{cd}]^\beta \approx 0.7$ [317]; (c) if τ_A is the Maxwell relaxation time and $B = \mu J_r(0)$, then from (a) and (b) one obtains Eq. (5.48) [317].

In summary, Eq. (5.48) is obtained by requiring the projected and the unprojected correlations functions to be similar [317]. To put if differently, the characteristics of the Cole–Davidson function may be attributed to properties of the projected and the unprojected correlation functions [317].

ACKNOWLEDGMENTS

I am indebted to I. Oppenheim, S. A. Rice, R. Marcus, and M. Fixman for their encouragement. Over the years the author has benefited from discussions on supercooled liquids with C. A. Angell, R. Zwanzig, E. DiMarzio, H. C. Anderson, S. Yip, I. Oppenheim, S. A. Rice, P. G. Wolynes, F. Stillinger, C. Yu, M. Fayer, K. Freed, C. H. Taubes, S. Nagel, J. Jonas, M. H. Cohen, A. Sjolander, D. Fisher, D. Nelson, F. Spaepen, D. Stein, S. Das, T. Keyes, J. J. Brey, M. R. Montero, T. V. Ramakrishnan and K. Kundu. I thank W. Kob for proofreading the manuscript. I thank T. Keyes and S. Yip for collaboration which led to a revised version of the Introduction (I.A). This work was supported by the National Science Foundation and the Petroleum Research Funds (ACS).

REFERENCES

1. (a) M. I. Klinger, *Phys. Rep.*, **165**, 257 (1988); P. K. Gupta, *Rev. Solid State Sci.*, **3**, 221 (1989); *J. Jackle, Rep. Prog. Phys.*, **49**, 171 (1986); R. Zwanzig, *Kinam*, **3A**, 5 (1981). (b) G. Harrison, *The Dynamical Properties of Supercooled Liquids*, Academic, New York, 1976. See also the various articles and references in *Ann. NY Acad. Sci.*, **484** (1986), special issue. J. Wong and C. A. Angell, *Glass Structure by Spectroscopy*, Dekker, New York, 1976. C. A. Angell, J. H. R. Clark, and L. V. Woodcock, *Adv. Chem. Phys.*, **48**, 397 (1981).
2. M. Williams, R. Landel, and J. Ferry, J. Ferry, *J. Am. Chem. Soc.*, **77**, 3701 (1955); H. Vogel, *Phys. Z.*, **22**, 645 (1921); G. Tammann and G. Hesse, *Z. Anorg. Allgem. Chem.*, **156**, 245 (1926); D. Davidson and R. H. Cole, *J. Chem. Phys.*, **19**, 1484 (1951).
3. A. Doolittle, *J. Appl. Phys.*, **22**, 147 (1951).
4. I. M. Hodge, *Macromolecules*, **20**, 2897 (1987), and references cited therein; O. S. Narayanaswamy, *J. Am. Ceram. Soc.*, **54**, 491 (1971); I. Avramov, Bulgarian Academy of Science preprint, 1992; C. Moynihan, S. Opalka, and S. Crichton, *J. Non-Cryst. Solids*, **131**, 420 (1991).
5. W. Kauzmann, *Chem. Rev.*, **43**, 219 (1948).
6. J. H. Gibbs and E. DiMarzio, *J. Chem. Phys.*, **28**, 373 (1958); E. A. DiMarzio, *Ann. N.Y. Acad. Sci.*, **371**, 1 (1981), and references cited therein.
7. M. Bawendi, K. Freed, and U. Mohanty, *J. Chem. Phys.*, **87**, 807 (1987).
8. F. Stillinger, *J. Chem. Phys.*, **88**, 7818 (1988), and references cited therein.
9. M. R. Montero, U. Mohanty, and J. Brey, *J. Chem. Phys.* **99**, 9979 (1993).
10. (a) F. Stillinger and T. A. Weber, *Phys. Rev. A*, **28**, 2408 (1983); F. Stillinger, *Phys. Rev. B*, **32**, 3134 (1985); F. H. Stillinger and R. A. LaViolette, *J. Chem. Phys.*, **85**, 6027 (1986). (b) M. Goldstein, *J. Chem. Phys.*, **51**, 3728 (1969).
11. R. Zwanzig, *J. Chem. Phys.*, **79**, 4507 (1983).
12. U. Mohanty, *J. Chem. Phys.*, **89**, 3778 (1988).
13. U. Mohanty, *Physica A*, **162**, 362 (1990); U. Mohanty, *Physica A*, **188**, 692 (1992).
14. G. Adams and J. H. Gibbs, *J. Chem. Phys.*, **43**, 139 (1965).
15. B. Madan, T. Keyes, and G. Seeley, *J. Chem. Phys.* **94**, 6762 (1991); G. Seeley, T. Keyes, and B. Madan, *J. Phys. Chem.*, **96**, 4074 (1992), and references cited therein. See also M. Buchner, B. Ladanyi, and R. M. Stratt, *J. Chem. Phys.*, **97**, 8522 (1992).
16. G. Fredrickson, *Ann. Rev. Phys. Chem.*, **39**, 149 (1988); G. H. Fredrickson and S. Brawer, *J. Chem. Phys.*, **84** (1986) 3351; G. H. Fredrickson and H. C. Andersen, *J. Chem. Phys.*, **83**, 5822 (1985).
17. S. Butler and P. Harrowell, *J. Chem. Phys.*, **95**, 4454 (1991); M. Foley and P. Harrowell, *J. Chem. Phys.*, **98**, 5069 (1992).
18. T. Kirkpatrick and P. G. Wolynes, *Phys. Rev. B*, **36**, 8552 (1987); T. Kirkpatrick, D. Thirumalai, and P. G. Wolynes, *Phys. Rev. A*, **40**, 1045 (1989).
19. R. W. Hall and P. G. Wolynes, University of Illinois at Urbana preprint, 1987.
20. D. L. Stein and R. G. Palmer, **B38**, 12035 (1988).
21. N. Walker, D. M. Lamb, S. Adamy, J. Jonas, and M. P. Dare-Edwards, University of Illinois at Urbana preprint, 1989.

22. E. Rossler, *Phys. Rev. Lett.*, **65**, 1595 (1990); E. Rossler, *Phys. Rev. Lett.*, **69**, 1620 (1990).
23. M. T. Cicerone and M. D. Ediger, University of Wisconsin-Madison preprint, 1993; F. R. Blackburn, M. T. Cicerone, G. Hietpas, P. A. Wagner, and M. D. Ediger, University of Wisconsin-Madison preprint, 1993.
24. N. O. Birge and S. R. Nagel, *Phys. Rev. Lett.*, **54**, 2674 (1985); For a theoretical interpretation of these results see [26].
25. C. A. Angel, *J. Non-Cryst. Solids*, **102**, 205 (1988); C. A. Angel, *J. Phys. Chem. Solids*, **49**, 863 (1988). See also R. Bohmer and C. A. Angell, *Phys. Rev. B*, **45**, 10091 (1992).
26. R. Zwanzig, *J. Chem. Phys.*, **88**, 583 (1988) and D. Oxtoby, *J. Chem. Phys.*, **85**, 1549 (1986).
27. E. Leutheusser, *Phys. Rev. A*, **29**, 2765 (1984).
28. U. Bengtzelius, W. Gotze, and A. Sjolander, *J. Phys. C*, **17**, 8915 (1984).
29. W. Gotze, in *Liquids. Freezing and the Glass Transition*, J. P. Hansen, D. Levesque, and J. Zinn Justin, Eds., Elsevier, New York, 1991.
30. S. P. Das and G. F. Mazenko, *Phys. Rev. A*, **34**, 2265 (1986); S. P. Das, *Phys. Rev. A*, **36**, 211 (1987); B. Kim and G. Mazenko, *Phys. Rev. A*, **45**, 2393 (1992); University of Chicago preprint, 1991.
31. W. Gotze and L. Sjogren, *Z. Phys. B*, **65**, 415 (1987).
32. J. R. Fox and H. C. Andersen, *J. Phys. Chem.*, **88**, 4019 (1984); H. Jonsson and H. C. Andersen, *Phys. Rev. Lett.*, **60**, 2295 (1988).
33. J. J. Ullo and S. Yip, *Phys. Rev. A*, **39**, 5877 (1989), and references cited therein.
34. R. Taborek, R. N. Kleinman, and D. J. Bishop, *Phys. Rev. B*, **34**, 1835 (1986); S. Sridhar and R. Taborek, *J. Chem. Phys.*, **88**, 1170 (1988).
35. R. D. Mountain and D. Thirumalai, *Phys. Rev. A*, **36**, 3300 (1987); J. N. Roux, J. L. Barrat, and J. P. Hansen, *J. Phys. Cond. Matter*, **1**, 7171 (1989).
36. U. Bengtzelius, *Phys. Rev. A*, **34**, 5059 (1986).
37. S. P. Das, private communication, 1990.
38. F. Mezei, in *Dynamics of Disordered Material*, D. Richter et al., Eds., Springer Verlag, 1989.
39. S. H. Chen and J. S. Huang, *Phys. Rev. Lett.*, **55**, 1888 (1985); P. N. Pusey and W. van Megan, *Phys. Rev. Lett.*, **59**, 2089 (1987); Y. X. Yan, L. T. Cheng, and K. A. Nelson, *J. Chem. Phys.*, **88**, 6477 (1988); S. M. Silence, S. R. Goates, and K. A. Nelson, *J. Non-Cryst. Solids*, **131**, 37 (1991); S. N. Gomperts, J. E. Variyar, and D. Kivelson, *J. Chem. Phys.*, **98**, 31 (1993), and references cited therein.
40. See the various articles in *Book of Abstracts, 2nd International Discussion Meeting on Relaxations in Complex Systems*, Alicante, Spain (June 28–8 July).
41. See the contributions in *Relaxation in Complex Systems*, Proceedings of the Workshop on Relaxation Processes, Blacksburg, VA, July 1983, K. Ngai and G. B. Smith, Commerce, Washington, DC, 1985.
42. G. P. Johari and M. Goldstein, *J. Chem. Phys.*, **53**, 2372 (1970).
43. G. P. Johari, *J. Chem. Phys.*, **58**, 1776 (1973).
44. G. P. Johari, in *Plastic Deformation of Amorphous and Semicrystalline Materials*, 1982 Les Houches Lectures, Les Editions de Physique, France, 1982, pp. 109–141; G. P. Johari, Lecture Notes in Physics, Vol. 277, Springer, Berlin, 1987.

45. J. Perez, *Rev. Phys. Appl.*, **21**, 93 (1986).
46. M. H. Cohen and G. S. Grest, *Phys. Rev. B*, **20**, 1077 (1979); *B*, **21**, 4113 (1980).
47. M. H. Cohen and G. S. Grest, *Solid State Commun.*, **39**, 1077 (1979); M. H. Cohen and G. S. Grest, **B21**, 143 (1981).
48. S. A. Brawer, *J. Chem. Phys.*, **81**, 954 (1984).
49. J. C. Dyre, *Phys. Rev. Lett.*, **58**, 792 (1987).
50. (a) S. S. N. Murthy, *J. Chem. Soc., Faraday Trans. 2*, **85**, 581 (1989). (b) P. D. Hyde, T. D. Evert, and M. D. Ediger, **131**, 42 (1991); F. Fujara, B. Geil, H. Sillescu, and G. Fleischer, *Z. Phys. B Cond. Matter*, **88**, 195 (1992); J. A. Hodgdon and F. H. Stillinger, *Phys. Rev. E*, **48**, 207 (1993). (c) X. C. Zeng, D. Kivelson, and G. Tarjus, University of California at Los Angeles preprint, 1993; D. Kivelson and D. Miles, *J. Chem. Phys.*, **88**, 1925 (1988).
51. J. Y. Cavaille, J. Perez, and G. P. Johari, *Phys. Rev. B*, **39**, 2411 (1989).
52. R. G. Palmer, D. L. Stein, E. Abrahams, and P. W. Anderson, *Phys. Rev. Lett.*, **53**, 985 (1984); R. W. Zwanzig, *Phys. Rev. Lett.*, **54**, 364 (1985).
53. P. D. Condo, I. C. Sanchez, C. G. Panayiotou, and K. P. Johnson, *Macromolecules*, **25**, 6119 (1992).
54. R. V. Chamberlin, Arizona State University preprint, 1993; R. V. Chamberlin, in *On Clusters and Clustering, from Atoms to Fractals*, P. J. Reynolds, Ed., Elsevier Science Publication, 1993.
55. W. Kob and R. Schilling, *Phys. Rev. A*, **42**, 2191 (1990); University of Basel preprints, 1990.
56. R. Zwanzig, *J. Chem. Phys.*, **94**, 6147 (1991).
57. R. Zwanzig, *J. Phys. Chem.*, **96**, 3926 (1992).
58. U. Mohanty, I. Oppenheim, and C. Taubes, unpublished results.
59. U. Buchenau, *Europhys. News*, **24**, 77 (1993); U. Buchenau, Forschungszentrum Julich preprint, 1993.
60. U. Buchenau, M. Prager, N. Nucker, A. J. Dianoux, N. Ahmad, and W. A. Phillips, *Phys. Rev. B*, **34**, 5665 (1986).
61. H. R. Schober and B. B. Laird, *Phys. Rev. B*, **44**, 6746 (1991).
62. V. G. Karpov, M. I. Klinger, and F. N. Ignat'ev, *Sov. Phys. JETP*, **57**, 439 (1983).
63. A. Ansari, J. Berendzen, S. Browne, H. Frauenfelder, I. E. T. Iben, T. B. Sauke, E. Shyamsunder, and R. D. Young, *Proc. Natl. Acad. Sci. USA*, **82**, 5000 (1985).
64. I. E. T. Iben, D. Braunstein, W. Doster, H. Frauenfelder, M. K. Hong, J. B. Johnson, S. Luck, P. Ormos, A. Schulte, P. J. Steinbach, A. H. Xie, and R. D. Young, *Phys. Rev. Lett.*, **62**, 1916 (1989).
65. P. G. Wolynes, in *Biologically Inspired Physics*, L. Peliti, Ed., Plenum, New York, 1991, and references cited therein.
66. H. Frauenfelder, S. Sligar, and P. G. Wolynes, *Science*, **254**, 1598 (1991), and references cited therein.
67. R. A. Goldstein, Z. A. Luthey-Schulten, and P. G. Wolynes, *Proc. Natl. Acad. Sci. USA*, **89**, 4918 (1992).
68. D. L. Stein, *Proc. Natl. Acad. Sci. USA*, **82**, 3671 (1985).
69. P. L. Privalov, *Adv. Protein Chem.*, **33**, 167 (1979).
70. See also J. M. Sturtevant, *Annu. Rev. Phys. Chem.*, **38**, 463 (1987).

71. R. Baldwin, *Proc. Natl. Acad. Sci. USA*, **83**, 8069 (1986).
72. J. F. Brandts, *J. Am. Chem. Soc.*, **86**, 4291 (1964).
73. J. F. Brandts and L. Hunt, *J. Am. Chem. Soc.*, **89**, 4826 (1967).
74. F. Franks, R. H. M. Hatley, and H. L. Friedman, *Biophys. Chem.*, **31**, 307 (1988).
75. W. J. Becktel and J. A. Schellman, *Biopolymers*, **26**, 1859 (1987).
76. K. P. Murphy, P. L. Privalov, and S. J. Gill, *Science*, **247**, 559 (1990).
77. P. L. Privalov and S. J. Gill, *Adv. Protein Chem.*, **39**, 191 (1988), and references cited therein.
78. J. M. Sturtevant, *Proc. Natl. Acad. Sci. USA*, **74**, 2236 (1977).
79. U. Mohanty, unpublished results, 1991.
80. A. B. Bestul and S. S. Chang, *J. Chem. Phys.*, **40**, 731 (1964).
81. E. Passaglia and H. K. Kevorkian, *J. Appl. Phys.*, **34**, 90 (1963).
82. C. A. Angell and D. L. Smith, *J. Phys. Chem.*, **86**, 3845 (1982); C. A. Angel, A. Dwarkin, P. Figuiere, A. Fuchs, and H. Szwarc, *J. Chim. Phys.*, **82**, 773 (1985).
83. U. Mohanty, *J. Chem. Phys.*, **93**, 8399 (1990); **97**, 4575 (1992).
84. U. Mohanty, *Physica*, **177**, 345 (1991); **A183**, 579 (1992).
85. The situation is more complex, however. See C. M. Roland and K. L. Ngai, *Macromolecules*, **25**, 5765 (1992).
86. K. G. Wilson, *Phys. Rev. B*, **4**, 3174 (1971); **4**, 3184 (1971).
87. K. G. Wilson and J. Kogut, *Phys. Rep.*, **12**, 75 (1974).
88. A. Leggett, *Ann. Phys.*, **72**, 80 (1972).
89. D. Foster, *Hydrodynamics Fluctuations, Broken Symmetry, and Correlation Functions*, Benjamin, New York, 1975.
90. K. G. Wilson, *Phys. Rev.*, **179**, 1499 (1969).
91. G. Adam and J. H. Gibbs, *J. Chem. Phys.*, **43**, 139 (1965).
92. A. B. Bestul and S. S. Chang, *J. Chem. Phys.*, **40**, 731 (1964).
93. H. Vogel, *Phys. Z.*, **22**, 645 (1921).
94. G. Tammann and G. Hesse, *Z. Anorg. Allg. Chem.*, **156**, 245 (1926).
95. G. S. Fulcher, *J. Am. Chem. Soc.*, **8**, 339 (1925).
96. J. Sethna, *Europhys. Lett.*, **6**, 529 (1988).
97. C. A. Angell and R. D. Bressel, *J. Phys. Chem.*, **76**, 3244 (1972).
98. C. A. Angell and W. Sichina, *Ann. N.Y. Acad. Sci.*, **279**, 53 (1976).
99. V. P. Privalko, *J. Phys. Chem.*, **84**, 3307 (1980).
100. A. A. Miller, *J. Poly. Sci., Part A-2*, **6**, 249 (1968).
101. C. A. Angell and C. T. Moynihan, Ed. G. Mamantov, Marcel Dekker, New York, 1969.
102. C. A. Angell and J. C. Tucker, *J. Phys. Chem.*, **78**, 278 (1974).
103. C. A. Angell and D. L. Smith, *J. Phys. Chem.*, **86**, 3845 (1982).
104. W. T. Laughlin and D. R. Uhlmann, *J. Phys. Chem.*, **76**, 2317 (1972).
105. J. H. Ambrus, C. T. Moynihan, and P. B. Macedo, *J. Electrochem. Soc.*, **119**, 192 (1972).
106. (a) H. Tweer, J. H. Simmons, and P. B. Macedo, *J. Chem. Phys.*, **54**, 1952 (1971). (b) N. Menon, K. P. O'Brien, P. K. Dixon, L. Wu, S. R. Nagel, B. D. Williams, and J. P. Carni, *J. Non-Cryst. Solids*, **141**, 61 (1992).

107. C. A. Angell, *J. Non-Crystalline Solids.*, **73**, 1 (1985).
108. E. Passagli and H. K. Kevorkian, *J. Appl. Polymer Sci.*, **7**, 119 (1963).
109. E. Passagli and H. K. Kevorkian, *J. Appl. Phys.*, **34**, 90 (1963).
110. M. Tatsumisago, B. L. Halfpap, J. L. Green, S. M. Lindsay, and C. A. Angell, *Phys. Rev. Lett.*, **64**, 1549 (1990).
111. S. S. Chang and A. B. Bestul, *J. Chem. Thermodyn.*, **6**, 325 (1974).
112. M. B. Meyers and E. J. Felty, *J. Electrochem. Soc.*, **117**, 818 (1970); A. J. Easteal, J. Wilder, R. K. Mohr, and C. T. Moynihan, *J. Am. Ceram. Soc.*, **60**, 134 (1977).
113. G. P. Gohari, in *Relaxations in Complex Systems*, K. L. Ngai and G. B. Wright, Eds., National Technical Information Service, U.S. Department of Commerce, Springfield, VA 22161, 3, 1985; G. P. Gohari, in *Plastic Deformation of Amorphous and Semicrystalline Materials*, 1982 Les Houches Lectures, Les Editions de Physique, France, 1982; G. P. Gohari, in *Phase Transitions*, Vol. 5, Gordon and Breach, Science Publishers, and OPA Ltd., 1985, p. 277.
114. C. H. Wang and S. K. Satija, *Chem. Phys. Lett.*, **97**, 330 (1982); S. K. Satija and C. H. Wang, *J. Chem. Phys.*, **69**, 1101 (1978).
115. J. Pelous, R. Vacher, J. P. Bonnet, and J. L. Ribet, *Phys. Lett. A.*, **78**, 195 (1980).
116. K. Adachi, H. Suga, and H. Seki, *Bull. Chem. Soc. Jpn.*, **41**, 1073 (1968); K. Adachi, H. Suga, S. Seki, S. Kubota, S. Yamaguchi, O. Yano, and Y. Wada, *Mol. Cryst. Liq. Cryst.*, **18**, 345 (1972); H. Suga, in *Dynamic Aspects of Structural Change in Liquids and Glasses*, Vol. 484, Annals of the New York Academy of Sciences, C. A. Austin and M. Goldstein, Eds., The New York Academy of Sciences, New York, 1986.
117. (a) K. Moriya, T. Matsuo, and H. Suga, *J. Phys. Chem. Solids*, **41**, 1103 (1983). (b) J. P. Amoureux, G. Noyel, M. Foulon, M. Bee, and L. Jorat, *Mol. Phys.*, **52**, 161 (1984).
118. C. A. Angell, L. E. Busse, E. I. Copper, R. K. Cooper, R. K. Kadiyala, A. Dworkin, M. Ghelfenstein, H. Szwarc, and A. Vassai, *J. Chim. Phys.*, **82**, 267 (1985).
119. C. A. Angell, A. Aworkin, P. Figuiere, A. Fuchs, and H. Szwarc, *J. Chim. Phys.*, **82**, 773 (1985).
120. H. Tweer, N. Laberge, and P. B. Macedo, *J. Am. Cer. Soc.*, **54**(2), 121 (1971).
121. H. G. K. Sundar and C. A. Angell, XIVth International Congress on Glass, Collected Papers, Indian Ceramic Society Pub. II, 161 (1986).
122. C. A. Angell, *J. Non-Crystalline Solids*, **102**, 205 (1988).
123. C. T. Moynihan, H. Sasabe, and J. C. Tucker, in *Molten Salts*, Proceeding of the International Conference on Molten Salts, J. P. Pemsler, Ed., The Electrochemical Society, 188, 1976.
124. C. T. Moynihan, P. B. Macedo, C. J. Montrose, P. K. Gupta, M. A. DeBolt, J. F. Dill, D. E. Dom, P. W. Drake, A. J. Easteal, P. B. Elterman, R. P. Moeller, H. Sasabe, and J. A. Wilder, *Ann. N.Y. Acad. Sci.*, **279**, 15 (1976).
125. A. Barkatt and C. A. Angell, *J. Chem. Phys.*, **70**, 901 (1979).
126. C. A. Angell, *J. Phys. Chem. Solids*, **49**, 863 (1988).
127. C. T. Moynihan, S. M. Opalka, R. Mossadegh, S. N. Crichton, and A. J. Bruce, *Lecture Notes in Physics*, Springer-Verlag, **277**, 16 (1987).
128. N. O. Birge and S. R. Nagel, *Phys. Rev. Lett.*, **54**, 2674 (1985).
129. N. O. Birge, Y. H. Jeong, and S. R. Nagel, *Ann. N.Y. Acad. Sci.*, **484**, 101 (1986).

130. G. Adam and J. Gibbs, *J. Chem. Phys.*, **43**, 139 (1965); E. A. DiMarzio and J. H. Gibbs, *J. Chem. Phys.*, **28**, 373 (1958).
131. U. Mohanty, *J. Chem. Phys.*, **89**, 3778 (1988).
132. U. Mohanty, *J. Chem. Phys.*, **93**, 8439 (1990).
133. J. Sethna, *Europhys. Lett.*, **6**, 529 (1988).
134. D. Kivelson, W. Steffan, G. Meier, and A. Patkowski, *J. Chem. Phys.*, **95**, 1943 (1991).
135. E. Leutheusser, *Phys. Rev. A*, **29**, 2765 (1984).
136. U. Bengtzelius, W. Gotze, and A. Sjolander, *J. Phys. C*, **17**, 5915 (1984); W. Gotze, in *Liquids, Freezing and the Glass Transition*, D. Levesque, J. P. Hansen, and Z. Zinn-Justin, Eds., Elsevier, New York, 1991.
137. S. P. Das and G. F. Mazenko, *Phys. Rev. A*, **34**, 2265 (1986); S. P. Das, *Phys. Rev. A*, **36**, 211 (1987); B. Kim and G. Mazenko, *Phys. Rev. A*, University of Chicago preprint, 1991.
138. G. H. Fredrickson, *Annu. Rev. Phys. Chem.*, **39**, 149 (1988).
139. G. Harrison, *The Dynamical Properties of Supercooled Liquids*, Academic, London, 1976.
140. B. Berne and R. Pecora, *Dynamic Light Scattering*, Wiley, New York, 1986.
141. D. Forster, *Hydrodynamics Fluctuations, Broken Symmetry, and Correlation Functions*, Benjamin, New York, 1975.
142. T. Keyes, in *Statistical Mechanics*, Part B: Time-Dependent Process, B. J. Berne, Ed., Plenum, New York, 1977.
143. J. P. Boon and S. Yip, *Molecular Hydrodynamics*, McGraw-Hill, New York, 1980.
144. N. H. March and M. P. Tosi, *Atomic Dynamics in Liquids*, MacMillan, London, 1977.
145. (a) S. W. Lovesey, in *Dynamics of Solids and Liquids by Neutrons Scattering*, S. W. Lovesey and T. Springer, Eds., Springer-Verlag, Heidelberg, 1977. (b) D. Frenkel and J. P. MacTauge, *Annu. Rev. Phys. Chem.*, **31**, 491 (1980).
146. J. A. Barker and D. Henderson, *Rev. Mod. Phys.*, **48**, 587 (1976).
147. J. P. Hansen and I. R. McDonald, *Theory of Simple Liquids*, 2nd ed., Academic, London, 1986; H. E. Stanley, *Introduction to Phase Transition and Critical Phenomena*, Oxford University Press, 1971.
148. D. A. McQuarrie, *Statistical Mechanics*, Harper & Row, Publishers, 1976.
149. T. L. Hill, *Statistical Mechanics*, McGraw-Hill, New York, 1956; R. C. Tolman, *Statistical Mechanics*, Oxford University Press, 1938.
150. M. Grimsditch and N. River, preprint, 1990.
151. J. R. Sandercock, *Light Scattering in Solids III*, M. Cardoa and G. Guntherodt, Eds., Springer, New York, 1982, p. 173.
152. L. M. Torell, *J. Chem. Phys.*, **76**, 3467 (1987).
153. P. B. Bezot and C. Hesse-Bezot, *Can. J. Phys.*, **60**, 1709 (1982).
154. L. M. Torell and R. Aronsson, *J. Chem. Phys.*, **78**, 1121 (1983).
155. P. B. Bezot, C. Hesse-Bezot, and Ph. Pruzan, *Can. J. Phys.*, 61, 1291 (1983).
156. H. E. Gunilla Knape, *J. Chem. Phys.*, **80**, 4788 (1984).
157. J. Schroeder, L. G. Hwa, G. Kendall, C. S. Dumais, M. C. Shyong, and D. A. Thompson, *J. Non-Cryst. Solid*, **102**, 240 (1988).

158. M. Stoltwisch, J. Sukmanowski, D. Quitmann, and K. Kuenzler, *Mater. Sci. Forum*, **6**, 571 (1985).
159. M. Stoltwisch, J. Sukomanowski, and D. Quitmann, *J. Chem. Phys.*, **86**, 3207 (1987).
160. K. K. Kelly, *J. Am. Chem. Soc.*, **51**, 779 (1929).
161. O. Haida, H. Suga, and S. Seki, *J. Chem. Thermodyn.* **9**, 1133 (1977).
162. N. B. Vargaftik, Ed., *Handbook of Physical Properties of Liquids and Gases*, Hemisphere Publishing, New York, 1983.
163. M. Sugisaki, H. Suga, and S. Seki, *Chem. Soc. Jpn.*, **41**, 2586 (1968).
164. D. L. Sidebottom and C. M. Sorensen, Kansas State University preprint, 1990.
165. G. Tarjus, V. Friedrich, and D. Kivelson, *J. Mol. Struct.*, **223**, 253 (1990).
166. V. Friedrich, G. Tarjus, and D. Kivelson, *J. Chem. Phys.*, **93**, 2246 (1990).
167. J. P. McTague and G. Birnbaum, *Phys. Rev. Lett.*, **21**, 661 (1968).
168. H. Levine and G. Birnbaum, *Phys. Rev. Lett.*, **20**, 439 (1968).
169. G. Birnbaum, Ed., *Phenomena Induced by Intermolecular Interactions*, Vol. 127 in Series B, Plenum, New York, 1985.
170. B. Guillot, S. Bratos, and G. Birnbaum, *Phys. Rev. A*, **22**, 2230 (1980).
171. P. A. Fleury, J. M. Worlock, and H. L. Carter, *Phys. Rev. Lett.*, **27**, 1493 (1971); P. A. Fleury, J. M. Worlock, and H. L. Carter, *Phys. Rev. Lett.*, **30**, 591 (1971).
172. T. Keyes, J. McTague, and D. Kivelson, *J. Chem. Phys.*, **55**, 4096 (1971).
173. P. B. Visscher and W. T. Logan, preprint (1990); W. T. Logan and P. B. Visscher, *Phys. Rev. A*, **39**, 1298 (1989).
174. R. D. Mountain and D. Thirumalai, *Phys. Rev. A*, **36**, 3300 (1987).
175. J. Erpenbeck, *Phys. Rev. Lett.*, **52**, 1333 (1984).
176. J. G. Amar and R. D. Mountain, *J. Chem. Phys.*, **86**, 2236 (1987).
177. W. G. Hoover, M. Ross, K. Johnson, D. Henderson, J. Barker, and B. Brown, *J. Chem. Phys.*, **52**, 4931 (170).
178. J. Kiefer and P. B. Visscher, *J. Stat. Phys.*, **27**, 389 (1982).
179. S. Begum and P. B. Visscher, *J. Stat. Phys.*, **20**, 641 (1978).
180. R. D. Mountain and D. Thirumalai, University of Maryland preprint, 1991.
181. R. D. Mountain and D. Thirumalai, *J. Phys. Chem.*, **93**, 6975 (1989); R. D. Mountain and D. Thirumalai, *Phys. Rev. A*, **42**, 4574 (1990); R. D. Mountain and D. Thirumalai, *J. Chem. Phys.*, **92**, 6616 (1990).
182. H. Jonsson and H. C. Andersen, *Phys. Rev. Lett.*, **60**, 2295 (1988).
183. P. K. Dixon, L. Wu, S. R. Nagel, B. D. Williams, and J. P. Carini, *Phys. Rev. Lett.*, **66**, 960 (1991).
184. Y. H. Jeong and S. R. Nagel, unpublished results; L. Wu, preprint; see Ref. 3 cited in [184].
185a. R. M. Ernst, S. R. Nagel, and G. S. Grest, *Phys. Rev. B*, **43**, 8070 (1991); University of Chicago preprint, 1991.
185b. See [200b].
186. J. Fox and H. C. Andersen, *J. Phys. Chem.*, **88**, 4019 (1984); J. Fox and H. C. Andersen, *Ann. N.Y. Acad. Sci.*, **371**, 123 (1981).
187. G. S. Grest and S. R. Nagel, *J. Phys. Chem.*, **91**, 4916 (1987).
188. See [174, 180, and 181].
189. S. Nose and F. Yonezawa, *J. Chem. Phys.*, **84**, 1803 (1986).

190. A. Rahman, M. J. Mandell, and J. P. McTahue, *J. Chem. Phys.*, **64**, 1564 (1976).
191. H. R. Wendt and F. F. Abraham, *Phys. Rev. Lett.*, **41**, 1244 (1978).
192. B. Bernu, J. P. Hansen, Y. Hiwatari, and G. Pastore, *Phys. Rev. A*, **36**, 4891 (1987).
193. S. T. Chui, G. O. Williams, and H. C. Frisch, *Phys. Rev. B*, **26**, 171 (1982).
194. N. O. Birge and S. R. Nagel, *Phys. Rev. Lett.*, **54**, 2674 (1985); N. O. Birge, *Phys. Rev. B*, **34**, 1631 (1986).
195. P. K. Dixon and S. R. Nagel, *Phys. Rev. Lett.*, **61**, 341 (1988).
196. J. J. Ullo and S. Yip, *Phys. Rev. Lett.*, **54**, 1509 (1985); J. J. Ullo and S. Yip, *Phys. Rev. A*, **39**, 5877 (1989); J. J. Ullo and S. Yip, *Chem. Phys.*, **149**, 221 (1989); J. L. Barrat, J. L. Roux, and J. P. Hansen, *Chem. Phys.*, **149**, 197 (1990).
197. P. J. Steinhardt, D. R. Nelson, and M. Ronchetti, *Phys. Rev. Lett.*, **B28**, 784 (1983); P. J. Steinhardt, D. R. Nelson, and M. Ronchetti, *Phys. Rev. Lett.*, **47**, 1297 (1981).
198. C. Dasgupta, A. V. Indrani, S. Ramaswamy, and M. K. Phani, *Europhys. Lett.*, **15**, 307 (1991); University of Minnesota Supercomputer Institute Research report, preprint, 1991.
199. K. Binder and A. P. Young, *Rev. Mod. Phys.*, **58**, 801 (1986).
200a. U. Mohanty, *Physica A*, **177**, 345 (1991); see also [132].
200b. U. Mohanty, *J. Chem. Phys.* **100**, 5905 (1994).
201. Y. F. Kiyachenko and Y. I. Litvinov, *JETP Lett.*, **4**, 2669 (1985).
202. P. K. Dixon, S. R. Nagel, and D. Weitz, University of Chicago preprint; see Ref. 5 cited in [185].
203a. J. Schroeder, M. Whitmore, M. R. Silvestri, and C. T. Moynihan, *J. Non-Cryst. Solids*, in press.
203b. See also [213].
204. R. E. Robertson, *J. Polym. Sci. Polym. Sym.*, **68**, 173 (1978); **17**, 597 (1979).
205. L. D. Landau and E. M. Lifshitz, *Statistical Physics*, Addison-Wesley, Reading, MA, 1969.
206. N. A. Bokov and N. S. Andreev, *Sov. J. Glass Phys. Chem.*, **15**, 243 (1989).
207. O. V. Mazurin and E. A. Porai-Koshits, *The Physics of Non-Crystalline Solids*, L. D. Pye, W. C. LaCourse, and H. J. Stevens, Eds., Taylor and Francis, Washington, DC, 1992).
208. V. V. Golubkov, *Sov. J. Glass Phys. Chem.*, **15**, 280 (1989).
209. V. V. Golubkov and M. Pivavarov, *Sov. J. Glass Phys. Chem.*, **17**, 135 (1991).
210. J. Schroeder et al., *Mater. Sci. Forum*, **67**, **68**, 471 (1991).
211. J. Schroeder, *Treatise on Materials Science and Technology*, Vol. 12, M. Tomozawa and R. H. Doremus, Eds., Academic, New York, 1977.
212. R. J. Ma, T. J. He, and C. H. Wang, *J. Chem. Phys.*, **88**, 1497 (1988).
213. C. Moynihan and J. Schroeder, *J. Non-Cryst. Solid*, in press; preprint, 1993.
214. E. Donth, *J. Non-Cryst. Solids*, **53**, 325 (1982); E. W. Fischer, E. Donth, and W. Steffen, *Phys. Rev. Lett.*, **68**, 2344 (1992).
215. U. Mohanty, *J. Chem. Phys.*,**100**, 5905 (1994).
216. R. Gardon and O. S. Narayanaswamy, *J. Am. Ceram. Soc.*, **53**, 148 (1970); O. S. Narayanaswamy, *J. Am. Ceram. Soc.*, **51**, 691 (1971).
217. C. T. Moynihan, L. P. Boesch, and N. L. Laberge, *Phys. Chem. Glasses*, **14**, 122 (1973).

218. G. W. Scherer, *J. Am. Ceram. Soc.*, **75**, 1060 (1992).
219. I. M. Hodge, *Macromolecules*, **20**, 2897 (1987); *J. Non-Cryst. Solids*, **131–133**, 435 (1991).
220. S. R. Nagel and P. K. Dixon, *J. Chem. Phys.*, **90**, 3885 (1989).
221. P. K. Dixon, *Phys. Rev. B*, **42**, 8179 (1990).
222. E. Duval, A. Boukenter, and T. Achibat, *J. Phys: Condens. Matter*, **2**, 10227 (1990).
223. T. Achibat, A. Boukenter, E. Duval, G. Lorrentz, and S. Etienne, *J. Chem. Phys.*, **95**, 2949 (1991).
224. E. Duval, A. Boukenter, T. Achibat, and B. Champagnon, in *The Physics of Non-Crystalline Solids*, L. D. Pye, W. C. LaCourse, and H. J. Stevens, Eds., Taylor and Francis, Washington, DC, 1992.
225. V. K. Malinovsky and A. P. Sokolov, *Solid State Comun.*, **57**, 757 (1986).
226. W. Steffen, A. Patkowski, G. Meier, and E. W. Fischer, *J. Chem. Phys.*, **96**, 4171 (1991).
227. K. Schmidt-Rohr, J. Clauss, and H. W. Spiess, *Macromolecules*, **25**, 3273 (1992); K. Schmidt-Rohr and H. W. Spiess, *Phys. Rev. Lett.*, **66**, 3020 (1991).
228. R. V. Chamberlin, Arizona State University preprint, 1993; R, V. Chamberlin, in *On Clusters and Clustering, from Atoms to Fractals*, P. J. Reynolds, Ed., Elsevier Science Publishers, 1993.
229. K. Huang, *Statistical Mechanics*, Wiley, New York, 1963.
230. M. Goldstein, *J. Chem. Phys.*, **51**, 3728 (1969).
231. R. Zwanzig, *J. Chem. Phys.*, **79**, 4507 (1983).
232. U. Mohanty, *Phys. Rev. A*, **32**, 3055 (1985).
233. G. Seeley, T. Keyes, and B. Madan, *J. Phys. Chem.*, **96**, 4074 (1992).
234. For a critical review with emphasis on experimental relevance see A. Angell, *J. Phys. Chem. Solid.*, **49**, 863 (1988).
235. F. H. Stillinger and T. A. Weber, *Phys. Rev. A*, **25**, 978 (1982).
236. F. H. Stillinger and T. A. Weber, *Phys. Rev. A*, **28**, 2408 (1983).
237. T. A. Weber and F. H. Stillinger, *Phys. Rev. B*, **32**, 5402 (1985); T. A. Weber and F. H. Stillinger, *J. Chem. Phys.*, **80**, 2742 (1984); F. H. Stillinger and T. A. Weber, *J. Chem. Phys.*, **80**, 4434 (1984); T. A. Weber and F. H. Stillinger, *J. Chem. Phys.*, **81**, 5089 (1984).
238. L. E. Root and F. H. Stillinger, *J. Chem. Phys.*, **90**, 1200 (1989).
239. F. H. Stillinger and T. A. Weber, *Phys. Rev. B*, **31**, 5262 (1985).
240. F. H. Stillinger, *J. Chem. Phys.*, **88**, 380 (1988).
241. F. H. Stillinger, *J. Chem. Phys.*, **89**, 4180 (1988).
242. F. H. Stillinger, *J. Chem. Phys.*, **88**, 7818 (1988).
243. F. H. Stillinger, *J. Chem. Phys.*, **89**, 6461 (1988).
244. F. H. Stillinger, *J. Chem. Phys.*, **88**, 6494 (1984).
245. F. H. Stillinger and T. A. Weber, *J. Chem. Phys.*, **83**, 4767 (1985).
246. F. H. Stillinger, AT & T Bell Laboratories preprint, 1990.
247. F. H. Stillinger, *Phys. Rev. B*, **32**, 3134 (1985).
248. See, for example, M. P. Allen and D. J. Tidesley, *Computer Simulations in Liquids*, Claredon Press, Oxford, 1987.
249. G. Adam and J. H. Gibbs, *J. Chem. Phys.*, **31**, 1164 (1965).
250. U. Mohanty, *J. Chem. Phys.*, **89**, 3778 (1988).

251. B. Wunderlich, *J. Phys. Chem.*, **64**, 1052 (1960).
252. T. Hill, *J. Chem. Phys.*, **36**, 3182 (1962).
253. L. D. Landau and E. M. Lifshitz, *Statistical Physics*, Oxford, Pergamon, 1958.
254. J. H. Gibbs and E. A. DiMarzio, *J. Chem. Phys.*, **28**, 373 (1958).
255. E. A. DiMarzio and J. H. Gibbs, *J. Chem. Phys.*, **28**, 807 (1958); E. A. DiMarzio and F. Dowell, *J. Appl. Phys.*, **50**, 6061 (1979).
256. M. Bawendi, K. Freed, and U. Mohanty, *J. Chem. Phys.*, **84**, 7036 (1986).
257. M. R. Montero, U. Mohanty, and J. Brey, *J. Chem. Phys.* **99**, 9979 (1993).
258. H. J. Parkhurst, Jr., and J. Jonas, *J. Chem. Phys.*, **63**, 2698 (1975); see also J. T. Hynes, *Annu. Rev. Phys. Chem.*, **28**, 301 (1977).
259. T. Keyes and I. Oppenheim, *Phys. Rev. A*, **8**, 937 (1973); A. J. Masters and P. A. Madden, *J. Chem. Phys.*, **74**, 2450 (1981); J. Mehaffey and R. I. Cukier, *Phys. Rev. A*, **17**, 1181 (1978); R. Peralta and R. Zwanzig, *J. Chem. Phys.*, **70**, 504 (1979); J. R. Dorfman, H. VanBeijeren, and C. F. McClure, *Arch. Mech.*, **28**, 333 (1976); J. T. Hynes, R. Kapral, and M. Weinberg, *J. Chem. Phys.*, **70**, 1456 (1979).
260. R. Zwanzig, unpublished notes.
261. P. G. de Gennes, *Physica (Utrecht)*, **25**, 825 (1959).
262. G. Harrison, *The Dynamical Properties of Supercooled Liquids*, Academic, New York, 1976.
263. R. Kohlrausch, *Ann. Phys. (Leipzig)*, **12**, 393 (1847).
264. G. Williams and D. C. Watts, *Trans. Faraday Soc.*, **66**, 80 (1970).
265. A. J. Barlow and A. Erginsv, *Proc. R. Soc. London, Ser. A*, **175** (1972).
266. R. Piccirelli and T. A. Litovitz, *J. Acoust. Soc. Am.*, **29**, 1009 (1957).
267. R. Meister, C. J. Marheffer, R. Sciamanda, L. Cotter, and T. Litovitz, *J. Appl. Phys.*, **31**, 854 (1960).
268. G. Williams and D. C. Watts, *Trans. Faraday Soc.*, **66**, 80 (1970).
269. H. S. Chen and M. Goldstein, *J. Appl. Phys.*, **43**, 1642 (1972).
270. G. D. Patterson, C. P. Lindseay, and J. R. Stevens, *J. Chem. Phys.*, **70**, 643 (1979).
271. S. Matsuoka, G. Williams, G. Johnson, E. Anderson, and T. Furukawa, *Macromolecules* **18**, 2652 (1985).
272. R. W. Douglas, in *Proceedings of the Fourth International Congress in Rheology*, E. H. Lee and A. L. Copley, Eds., Wiley-Interscience, New York, 1965.
273. C. R. Kurkjian, *Phys. Chem. Glass*, **4**, 1289 (1963).
274. C. R. Kurkjian and R. W. Douglass, *Phys. Chem. Glass*, **1**, 19 (1960); C. R. Kurkjian and R. W. Douglass, *Phys. Chem. Glasses*, **4**, 128 (1963).
275. J. de Bast and P. Gilard, *Phys. Chem. Glasses*, **4**, 117 (1963).
276. A. Kulkarni, H. Serapati, C. Liu, and C. A. Angell, Purdue University preprint, 1986.
277. U. Mohanty, *Phys. Rev. A*, **34**, 4993 (1986), and references cited therein.
278. R. Zwanzig, *J. Stat. Phys.*, **30**, 255 (1983).
279. C. Mazumdar, *Solid State Commun.*, **9**, 1087 (1971).
280. U. Mohanty, I. Oppenheim, and C. Taubes, unpublished results.
281. L. Cooper, *Phys. Rev.*, **104**, 1189 (1956).
282. J. Bardeen, L. Cooper, and J. Schrieffer, *Phys. Rev.*, **106**, 162 (1957); J. Bardeen, L. Cooper, and J. Schrieffer, *Phys. Rev.*, **108**, 1175 (1957).
283. R. Zwanzig, *Proc. Natl. Acad. Sci. USA*, **85**, 2029 (1988).
284. U. Mohanty, I. Oppenheim, and C. Taubes, unpublished results.

285. R. Zwanzig, *J. Phys. Chem.*, **96**, 3926 (1992).
286. M. H. Jacobs, *Diffusion Process*, New York, 1967; p. 68.
287. A. Ficks, *Poggendorfs. Ann.*, **94**, 59 (1985).
288. (a) S. Lifson and J. L. Jackson, *J. Chem. Phys.*, **36**, 2410 (1962); R. Festa and E. G. d'Agliano, *Physica A*, **90**, 229 (1978). (b) R. Zwanzig, *Physica A*, **117**, 277 (1983).
289. G. Seeley and T. Keyes, *J. Chem. Phys.*, **91**, 5581 (1989).
290. G. Seeley and T. Keyes, in *Spectral Line Shapes*, Vol. 5, J. Szudy, Ed., Ussolineum, Wroclaw, Poland, 1989, p. 649.
291. G. Seeley, T. Keyes, and B. Madan, *J. Chem. Phys.*, **95**, 3847 (1991).
292. B. Madan, T. Keyes, and G. Seeley, *J. Chem. Phys.*, **92**, 7565 (1990); **94**, 6762 (1991).
293. R. Cotterill and J. Masden, *Phys. Rev. B*, **33**, 262 (1988).
294. R. LaViolette and F. Stillinger, *J. Chem. Phys.*, **83**, 4079 (1985).
295. R. Zwanzig, *J. Chem. Phys.*, **79**, 4507 (1983).
296. U. Mohanty, *J. Chem. Phys.*, **89**, 3778 (1988).
297. U. Mohanty, *Phys. Rev. A*, **34**, 4993 (1986).
298. U. Mohanty, *Phys. Rev. A*, **32**, 3055 (1985).
299. A. Rahman, K. S. Singwi, and A. Sjolander, *Phys. Rev.*, **126**, 997 (1962).
300. A. V. Tobolsky, *J. Chem. Phys.*, **58**, 1223 (1973).
301. I. Avramov and A. Milchev, *J. Non-Cryst. Solids*, **104**, 253 (1988), and references cited therein.
302. I. Avramov, *J. Chem. Phys.*, **95**, 4439 (1991); I. Avramov, Bulgarian Academy of Science preprint, 1992.
303. H. Bassler, *Phys. Rev. Lett.*, **58**, 767 (1987).
304. R. Mountain and D. Thirumalai, University of Maryland preprint, 1991.
305. D. W. McCall, D. C. Douglass, and D. R. Falcone, *J. Chem. Phys.*, **50**, 3839 (1969).
306. N. A. Walker, D. M. Lamb, S. T. Adamy, J. Jonas, and M. P. Dare-Edwards, University of Illinois preprint, 1989.
307. C. H. Wang, R. J. Ma, G. Fytas, and Th. Dorfmuller, *J. Chem. Phys.*, **78**, 5863 (1983), and references cited therein.
308. I. Artaki and J. Jones, *Mol. Phys.*, **55**, 867 (1985).
309. I. Artaki and J. Jones, *J. Chem. Phys.*, **55**, 867 (1985).
310. G. L. Pollack and J. J. Enyeart, *Phys. Rev. A*, **31**, 980 (1985).
311. R. Zwanzig and A. K. Harrison, University of Maryland preprint, 1988.
312. E. Rossler, *Phys. Rev. Lett.*, **65**, 1595 (1990).
313. F. Fujara, B. Geil, H. Sillescu, and G. Fleischer, *Z. Phys. B. Cond. Matter*, **88**, 195 (1992), and references cited therein.
314. (a) M. T. Cicerone and M. D. Ediger, University of Wisconsin preprint, June 1993. (b) F. R. Blackburn, M. T. Cicerone, G. Hietpas, P. A. Wagner, and M. D. Ediger, University of Wisconsin preprint, 1993, and references cited therein.
315. X. C. Zeng, D. Kivelson, and G. Tarjus, University of California at Los Angeles preprint, 1993, and references cited therein.
316. G. Harrison, *Dynamical Properties of Supercooled Liquids*, Academic, London, 1976.
317. R. A. MacPhail and D. Kivelson, *J. Chem. Phys.*, **90**, 649 (1989).
318. B. Berne, J. P. Boon, and S. A. Rice, *J. Chem. Phys.*, **45**, 1086 (1966).
319. B. Berne and R. Pecora, *Dynamic Light Scattering*, Wiley, New York, 1976.

TERNARY SYSTEMS CONTAINING SURFACTANTS

MOHAMED LARADJI, HONG GUO, MARTIN GRANT, AND MARTIN J. ZUCKERMANN

Centre for the Physics of Materials and Physics Department, Ernest Rutherford Building, McGill University, Montréal, Québec, Canada H3A 2T8

CONTENTS

Advances in Chemical Physics, Volume LXXXIX, Edited by I. Prigogine and Stuart A. Rice.
ISBN 0-471-05157-8

1. INTRODUCTION

Over the past 25 years, fluid mixtures containing surfactant molecules have attracted considerable attention from both pure science and industry. From the industrial point of view, these systems have important applications in oil recovery, cosmetics, skin care, detergents, and food preparation. Consumer products such as floor waxes, lotions, beverage concentrates, pesticide preparations, and creams are common examples of the use of surfactant mixtures. Langmuir monolayers composed of amphiphilic lipids adsorbed at an air–water interface are used to fabricate thin films of inorganic materials such as silicon dioxide. Biological research involving these systems stems from attempts to understand the phase behavior of lipid bilayers and the rhythmic movements of red blood cells [1, 2]. Furthermore, researchers are intrigued by the remarkable properties of these mixtures. Indeed, mixtures containing surfactants exhibit many distinct fluid phases with interesting properties, such as microemulsions, micelles, vesicles, monolayers, bilayers, and liquid crystalline structures. These phases have complex structures on a mesoscopic scale, which are different from those of simple or classical fluids, and they are therefore examples of complex or structured fluids. Their structures often depend on the chemical nature of the surfactant molecules, the relative concentrations of the components forming the mixture, the details of sample preparation, and external conditions such as the temperature. This makes it difficult to formulate a global theory. Fortunately, experiments have shown that these systems share several generic features. This finding leads to the possibility of developing models for complex fluids that are able to capture the generic behavior, and can therefore be used to make qualitative and even quantitative predictions.

The most common examples of this type of system are ternary mixtures of water, oil, and amphiphilic surfactant molecules. We also consider the limiting case of a binary mixture of water and surfactants. The purpose of this chapter is to examine the phase behavior and related properties of such mixtures based on theoretical considerations. In addition, our focus will be on the universal features of the mixtures, rather than the particular chemical nature of the surfactants. There are several excellent reviews in the literature [3] on similar topics, and we will therefore present our own point of view in this chapter. A typical phase diagram for the ternary mixtures is displayed in Fig. 1.1. The richness of the phase behavior is a direct consequence of the chemical and physical properties of the amphiphilic surfactants. Binary mixtures also exhibit a rich-phase behavior that is quite similar in many aspects to the ternary case.

Amphiphilic surfactants are characterized by the presence of two distinct groups in the same molecule that differ greatly in their solubility. Such molecules are termed amphiphiles to denote the presence of a *lyophilic* group, which has an affinity for the solvent, and a second group referred to as *lyophobic*, since it is antipathetic to the solvent. When the solvent is water, the two groups are termed the *hydrophilic* and the *hydrophobic* groups for the sympathetic and the antipathetic parts, respectively [2]. If amphiphiles are mixed with water and oil, the hydrophilic tail prefers to be dissolved in oil. Other types of surfactant include polymeric surfactants, such as A–B diblock copolymers that

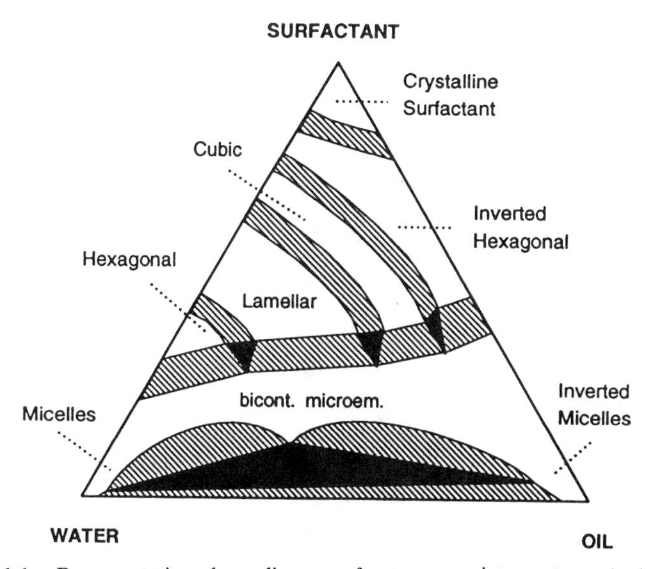

Figure 1.1. Representative phase diagram of a ternary mixture at constant temperature. White regions indicate a single phase, gray regions indicate a two-phase coexistence, while black regions indicate a three-phase coexistence.

behave similarly to amphiphiles when mixed with a binary mixture of A and B homopolymers [4–7].

The effect of amphiphiles on surface and interfacial tensions can be measured using a Langmuir trough. In this apparatus, amphiphilic molecules can be spread as monolayers on either air–water or oil–water interfaces. The amphiphiles then arrange themselves in such a way that the hydrophobic group moves away from the water surface or water–oil interface while the hydrophilic group is in contact with water due to its polar nature. For oil–water interfaces, the hydrophobic tail dissolves in bulk oil [2]. This arrangement of amphiphiles leads to a considerable reduction of the interfacial tension between water and oil, due to a screening of the interaction between water and oil molecules. Hence, amphiphiles are termed surface-active agents or simply *surfactants*.

As seen in the phase diagram of Fig. 1.1, ternary mixtures containing two immiscible components plus surfactants are in general structured into segregated mescopic domains. These domains are composed of one or other of the immiscible components and they are separated by monomolecular surfactant monolayers as in a Langmuir trough. This occurs because the mutual interaction between surfactants is usually weaker than that between a surfactant and the other two components, and hence the surfactants preferentially adsorb at the interfaces between the two immiscible components. Furthermore, the average size of the domains is quite large compared to the size of a surfactant molecule. The aggregation of these domains can be ordered or disordered depending on the strength and concentration of the surfactant and on external conditions.

We now describe several phases and other properties of both ternary water–oil–surfactant mixtures and the limiting case of binary mixtures of water and surfactants.

A. Microemulsions

One of the most interesting problems in the description of thermodynamic phases is an understanding of the microemulsion. The term "microemulsion" was coined by Schulman and co-workers in 1943 [8, 9]. *Macroscopically*, a microemulsion is a homogeneous, isotropic, and transparent solution, which has the appearance of a simple fluid. *Microscopically*, it consists of small but well-segregated domains of water and oil, where the surfactants form thin monolayers at the interfaces separating the water from the oil regions. The size of these domains is usually mesoscopic, that is, of the order of a few hundred angstroms. The structure of a microemulsion depends on the relative concentration of water and oil plus the concentration of surfactants. The domains are usually in the form of globules of water in oil for low concentrations of water (reverse inflated micelles), or globules of oil in water for low concentrations of oil (inflated micelles). These systems are usually referred to as globular microemulsions. When the concentrations of water and oil are comparable, the microemulsion becomes an interconnected network of water and oil domains, known as a bicontinuous microemulsion (see Fig. 1.1). The microemulsion "phase" can be found as a single phase, or in coexistence with water- or oil-rich phases. When the microemulsion coexists with both water and oil, it resides at the interface between the water and oil phases.

In the case of globular microemulsions where the compositions of the two immiscible components are very different, the observed water–water scattering intensity exhibits a broad peak at wavevector $q = 0$. This finding implies that the droplets are not correlated but are characterized by an average domain size given by the width of the scattering intensity. However, a peak at $q \neq 0$ is generally observed [10, 11] when the concentrations of water and oil are comparable. Furthermore, the position of this peak increases while its height decreases as the surfactant concentration is increased [11–14]. This occurs because a higher concentration of surfactant creates more interfaces, thereby leading to a smaller average size for the domains. In real space, the presence of a diffuse peak in the scattering intensity implies that the water–water correlation function has long wavelength decaying oscillations. Water and oil domains are therefore correlated over short distances but become uncorrelated at longer distances. The width of the peak is an indication of the disordered character of the microemulsion.

A direct way of studying the structural properties of microemulsions is by electrical conductivity or fluid viscosity measurements. For example, if the conductivity is more "oil-like," it implies that the microemulsion is formed of droplets of water in oil. Another method is freeze fracture electron microscopy, which is a good tool in the demonstration of micellar and bicontinuous real space structures found for microemulsions [2]. Neutron spin-echo and nuclear magnetic resonance (NMR) techniques have been used to study shape and size fluctuations of microemulsion droplets [15–17]. Some of the most useful tools in the study of microemulsion structure are small angle neutron scattering (SANS) and X-ray scattering [10–14, 18–25]. For polymeric surfactants, light scattering is usually used since in this case the size of the microdomains is comparable to the wavelength of light [22]. Experiments have also been performed using transmission electron microscopy to observe various morphologies of the solution [7], and in certain cases direct observation is made by simply taking photographs [26] of the real space structure of critical microemulsions.

B. Spontaneous Curvature

The shape of a surfactant molecule is important in determining microemulsion structure. When the polar head is less bulky than the hydrophobic tail, the surfactant monolayer tends to bend towards the oil regions, thereby promoting the formation of water droplets in oil even when the concentrations of water and oil are comparable. On the other hand, if the polar head is larger than the hydrophobic tail, the surfactant monolayer tends to bend towards the water regions thereby creating an oil-in-water microemulsion. For nonionic surfactants the tail becomes bulkier than the polar head due to the presence of *gauche* defects as the temperature increases. This lead to the creation of oil-in-water microemulsions for low temperatures, a bicontinuous structure at intermediate temperatures, and water-in-oil microemulsions at high temperatures [25]. Furthermore, the repulsive electrostatic interaction between the hydrophilic groups plays a major role for ionic surfactants. In this case, the spontaneous curvature of the surfactant film can be varied by adding a fourth component to the mixture, which is usually a salt. Salt ions screen the repulsive interaction between surfactants, and therefore oil-in-water microemulsions are observed for low-salt concentrations. The structure becomes bicontinuous as the salt concentration is increased and water-in-oil microemulsions are observed for even higher salt concentrations [10]. The salt concentration is usually quite low ($<1\%$) under experimental conditions.

C. Liquid Crystal Phases

When the concentration of surfactants in the mixture is relatively high, their interfacial density becomes very high making the interfaces rigid. This gives rise to phases with long-range order having periodic structures characterized by Bragg peaks at low-angle scattering. For macroscopic length scales, these periodic structures behave like weak solids with a high but finite viscosity. Consequently, they are often termed liquid crystal phases. The most common liquid crystal phase is the lamellar phase [2], which is shown schematically in Fig. 1.2. In this phase, water and oil are organized into quasi-one-dimensional (1-D) sheets characterized by a mesoscopic length scale and separated by surfactant monolayers. This phase usually occurs for comparable concentrations of water and oil when the surfactant film has zero spontaneous curvature. It can also occur for different concentrations of water and oil with nonzero spontaneous curvature of the surfactant film. Other liquid crystal phases are also found. An example is the hexagonal phase in which water (or oil) domains are organized in cylinders, which are arranged in a quasi-two-

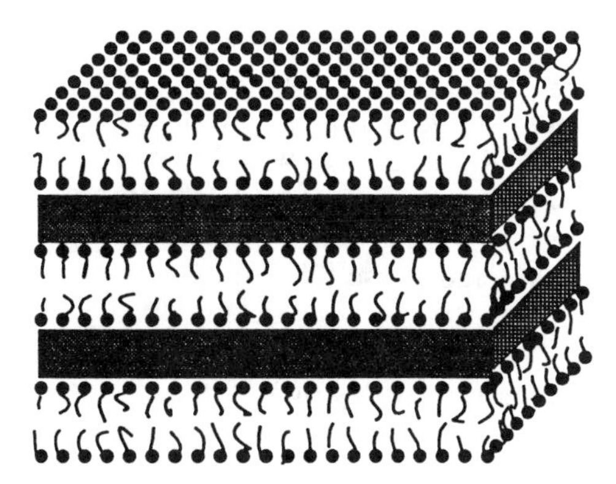

Figure 1.2. Schematic picture of lamellar phase. Gray represents water, white represents oil, and surfactants are represented by dots with tails.

dimensional (2-D) triangular lattice, as schematically shown in Fig. 1.3. Another example is the cubic phase in which water (or oil) domains have a spherical shape and are organized in a quasi-three-dimensional (3-D) body centered cubic (bcc) lattice (see Fig. 1.4). Liquid crystal phases can coexist with each other or with the microemulsion.

D. Interfacial Behavior

The phase behavior of a microemulsion is strongly related to the magnitude of the various interfacial tensions. In fact, it is always observed that when a microemulsion coexists with water, oil, or both, the interfacial tension between the microemulsion and water or oil, and that between water and oil, can be as low as 10^{-2} dyn sm^{-1}, that is, a few

Figure 1.3. Schematic picture of the hexagonal phase.

Figure 1.4. Schematic picture of the cubic phase.

percent of the water–oil interfacial tension [10] when surfactants are absent. It has been observed that simple disordered ternary mixtures containing surfactants wet the water–oil interface completely. In contrast, bicontinuous microemulsions do not wet the water–oil interface [27, 25] but rather form a drop at the interface. Furthermore, it is believed that the nature of the wetting transition is strongly related to the mesoscale structure of the disordered phase [28, 29].

E. Binary Mixtures of Water and Surfactants

As stated above, binary mixtures of water and surfactants have a rich phase behavior. In fact, many experiments have shown that these mixtures exhibit phases such as lamellar, hexagonal, and cubic phases and also a disordered phase characterized by a finite peak in the structure factor whose position also increases with increasing surfactant concentration [30, 31]. This disordered phase is known as the "sponge phase" and it is the analogue of the microemulsion phase in the case of comparable densities of water and oil. Similar behavior is observed in binary mixtures of A–B diblock copolymers and A homopolymers [7].

From a theoretical point of view, it should be possible to construct models for ternary mixtures of water, oil, and surfactants that are able to account for some or all of the above mentioned generic features. Such models have indeed been constructed, using a lattice gas approach, a fully microscopic description, or a continuum Ginzberg–Landau formalism. Lattice gas models are microscopic interaction models in which the interactions are based on the physical behavior of the components. Once the details of the interactions are established, the equilibrium phase behavior of the system can be determined. Since all the molecules are restricted to move on a lattice, the model is a considerable simplification of reality. However, lattice gas models are relatively easy to treat both analytically and using computer simulations. Many important results have been obtained through the study of lattice gas models in the case when the characteristic length scales of the phases are greater than the lattice constant.

Continuum Ginzberg-Landau models are formulated in terms of coarse grained variables, such as the local densities of the various molecular species. Physically meaningful terms in the density expansion based on symmetry considerations must be included in order to obtain the correct behavior of the ternary mixture. Methods such as mean-field theory or the renormalization group analysis can then be applied to study various phase properties. Numerical procedures are relatively easy to implement for this class of model. Furthermore, such models can also be coupled with hydrodynamic equations to study the flow of the mixture. We note that proper coarse graining of a microscopic model will lead to the corresponding Ginzberg–Landau model, but this has not been carried out so far.

Most of the theoretical and experimental studies on systems containing surfactants are concerned with equilibrium properties, though there are some studies on the dynamic properties of microemulsions. These are usually related to near-equilibrium shape fluctuations in globular microemulsions [32]. The far from equilibrium behavior of the ternary mixtures has not been thoroughly explored. For instance, it would be of interest to understand the dynamical process for the formation of a microemulsion or a liquid crystal phase. This situation is, however, more complicated than the well-studied case of phase ordering in simple binary mixtures. Since interface motion plays a major role in phase-ordering dynamics and surfactants change interface properties, we expect that the ordering dynamics is sufficiently different in the ternary mixtures as compared to the more familiar case of binary mixtures [33–38].

The purpose of this chapter is not to give an exhaustive review but rather to show how simple models can capture the fundamental equilibrium and dynamic properties of ternary mixtures containing surfactants. This is based on our own research, which makes considerable use of Monte Carlo simulations and the Ginzburg–Landau approach. The chapter is organized as follows. In Section II we review our work on a microscopic lattice gas model after briefly describing several other related lattice models. In Section III we present a continuum Ginzberg–Landau model and compare the results with those found using the lattice model of Section II. Our work on the ordering dynamics of ternary mixtures containing surfactants is described in Section IV. This work is based on a reduced version of the continuum model of Section III. A summary and the outlook for the future are presented in Section V.

II. LATTICE MODELS FOR BINARY AND TERNARY MIXTURES CONTAINING SURFACTANTS

This section studies the phase equilibria of binary and ternary mixtures containing amphiphilic surfactants using a version of the model proposed by Ciah et al. [39, 40]. This vector model has been analyzed in two dimensions using both Monte Carlo numerical simulations and a local mean-field analysis. A classification of existing lattice models is given in Section II.A. According to this classification the model used in this section belongs to Class B3 (see below). The model and the numerical methods are discussed in Section II.B. The model was studied for ternary mixtures on a square lattice and for both ternary and binary mixtures on a triangular lattice. The Monte Carlo results are described in Section II.C. Section II.D gives a mean-field theory for ternary mixtures on a square lattice, thereby allowing a comparison with the simulation results of Section II.C. In Section II.E we survey recent developments and discuss future work.

One of our main interests in this section is to introduce the reader to the use of modern techniques in the characterization of phase transitions and their application to the determination of the phase diagrams of binary and ternary mixtures containing surfactants. These methods are discussed in Section II.B and their application to phase behavior is presented in Section II.C.

A. Introduction

A description of existing lattice models for binary and ternary mixtures containing amphiphiles is given in the recent review article of Gompper and Schick [3] together with a discussion of the phase behavior and physical properties found from these models. We therefore restrict ourselves in this brief introduction to a short description of the models themselves and then discuss our results for a vector model.

The existing lattice models may be divided into three major classes. For ease of description, we will refer to the two immiscible fluids as "oil" and "water", respectively. The surfactant will be assumed to be an amphiphilic molecule. The three types of models can then be described as follows:

Class A.

In these models all three types of molecule are identified with lattice *bonds*. The basic models in this class are the Widom–Wheeler and the Widom models [41–44]. The concentration of all molecular species is controlled by chemical potentials and there are attractive interactions between the water molecules and between the oil molecules to ensure that water and oil are immiscible below a critical temperature. Each amphiphile can take two directions along the bond with its head and tail interacting attractively with water and oil, respectively. In contrast, the interactions between the water and the tail and between the oil and the head are made infinitely repulsive. The total Hamiltonian can then be written in terms of Ising spin $\frac{1}{2}$ operators with two-body interactions between nearest and next-nearest neighbors and a three-body interaction between all distinct contiguous spin triplets. Interfacial curvature is included in this description and the coupling constant of the last interaction represents the surfactant strength of the amphiphile. Three phase coexistence requires the addition of a four-body interaction as demonstrated by Hansen and Stauffer [45]. One advantage of this model is that multispin coding can be used for numerical simulations as this is essentially a two-component model as exemplified by the numerical simulations of Stauffer and co-workers [46–48].

Class B.

This class is composed of all models in which the molecules occupy the *sites* of a crystalline lattice. There are at present four subcategories in this class: (B1) models with three-body interactions, (B2) "bead" models, (B3) vector models without hydrogen bonding, and (B4) vector models with hydrogen bonding.

Class B1. As noted by Gompper and Schick, the phase behavior of a weak amphiphilic system can be expected for an ordinary ternary mixture. Such mixtures are described by the spin 1 lattice model due to Blume, Emery, and Griffiths (BEG) [49]. The typical effect of the amphiphile in this case is initially to lower the critical temperature of the immiscible water–oil mixture. This step is followed at higher concentrations by the appearance of a tricritical point. Schick and Shih [50] extended the BEG model and included the surfactant nature of the amphiphile in their Hamiltonian by introducing a three-body interaction in which it is energetically favorable for the amphiphile to sit in the same row between a water and an oil molecule. In order to extend the Hamiltonian to binary systems of water and amphiphile, these authors proposed an additional four-particle interaction that allows bilayer formation in oil-rich and water-rich phases. The results for this model are reported in a series of articles [31, 51–55].

Class B2. "Bead" models are originally due to Larson [56], who represented the water and oil molecules by "beads" of spin ± 1 as in the BEG model. He then constructed the amphiphilic molecules by connecting water and oil molecules to form a small polymer on the lattice. For example, an E_nC_m amphiphile would then be composed of n connected water beads followed by m connected oil beads. The amphiphilic polymers were then allowed to move on the lattice via standard polymer moves, such as reptation. There is only one interaction parameter in the model and numerical simulations are employed to infer the phase diagram.

Class B3. In the vector models [29, 31, 39, 40, 57–60], the oil and water molecules are again treated as in the BEG model but the amphiphile is represented by a vector at a lattice site. The vectors are usually made to point along lattice directions. The head of the vector has an attractive interaction with neighboring water molecules while the tail interacts favorably with neighboring oil molecules. Two-body interactions between the amphiphiles can also be included. These are, for example, aligning and skew vector–vector interactions and interactions favoring head-to-head configurations between neighboring amphiphiles. A model in which the vector describing the amphiphile points in all directions has been proposed by Gunn and Dawson [61]. However, Gompper and Schick [3] point out that the authors in their analysis do not allow the amphiphile to distinguish between heads and tails of surfactants, thereby restricting its use somewhat.

Class B4. In this class the hydrogen bonding between water and amphiphiles is included in the model. The recent model of Matsen et al. [62] extends the vector model by including orientations of the water molecule. This enables the authors to include hydrogen bonding between the head of the water molecule and that of the amphiphile. The mean field results for this model are in good agreement with experiment for binary and ternary systems.

Class C.

The third and final class of model places the oil and water molecules on the sites of the lattice as in Class B models and amphiphiles on the bonds as for Class A models. The Alexander model [63] is the basic model in this class, which has also been examined by Chen et al. [64] and Stockfisch and Wheeler [65]. The interactions between oil and water are the same as in the BEG model and amphiphiles on neighboring sites that make an angle of π or $\pi/2$ with one another interact favorably. As in Class B1 models, the coupling between the oil and water molecules and the amphiphile is provided by a three-body interaction between amphiphiles on a given lattice bond and the molecules on the sites joined by this bond. Gompper and Schick [3] comment that the model is restricted to describing ternary mixtures with considerable amounts of oil and water.

These models succeed in varying degrees in describing the generic phase diagram, the structure, and the interfacial properties of ternary and binary mixtures containing surfactants. These models have been studied by a variety of analytical methods and numerical simulation techniques, particularly in the case of vector models. These include the mean-field approximation, the Bethe approximation, transfer matrix methods, Monte Carlo simulations, and the molecular dynamics calculations of Smit [66] for bead models. We refer the reader to the review article by Gompper and Schick [3] for further details of models of types A to C and the related bibliography.

B. Model and Numerical Methods

1. Model

In this subsection we describe in detail the model used in the remainder of this section. As in the BEG model, the three species are represented by a spin 1 scalar field σ_i, which takes on values $+1, 0$, and -1 for water, amphiphile, and oil molecules, respectively. Furthermore, an orientation $\mathbf{m}_i$ is assigned to amphiphilic molecules. An amphiphile can have four possible orientations on a square lattice, and six possible orientations on a triangular or a simple cubic lattice. The Hamiltonian can be written as follows:

$$\mathcal{H}_1 = -\sum_{i,j} [J_1\sigma_i\sigma_j + A_{\text{lat}}J_2(\sigma_j\mathbf{m}_i \cdot \mathbf{r}_{ij} + \sigma_i\mathbf{m}_j \cdot \mathbf{r}_{ji})] - \mu_s N_s - \mu_w N_w - \mu_o N_o \tag{2.1}$$

where the first term in the first sum corresponds to the BEG model. The second term in the first sum is important when the third component corresponds to amphiphiles, since it mimics two-body interactions between an amphiphile and water or oil molecules. This term reflects the fact that the tail of an amphiphile prefers a hydrophobic environment such as oil, and the head interacts preferentially with water molecules. $\mathbf{r}_{ij}$ is a unit vector pointing from the site i to a nearest neighbor site j.

In this section we examine the model for two 2-D lattices, the square lattice in which the amphiphiles orient along the lattice bonds and a triangular model in which the amphiphiles orient along the bisector between two lattice bonds. On the square lattice each amphiphile only interacts with two neighboring molecules, whereas an amphiphile interacts with four neighboring molecules on the triangular lattice. The factor $A_{\text{lat}} = 1$ for the square lattice and $2/\sqrt{3}$ for the triangular lattice.

The last three terms in Eq. (2) represent the chemical potentials of the three species where N_s, N_w and N_o correspond to the total number of amphiphiles, water, and oil molecules, respectively. The first sum in Eq. (2.1) is over first nearest neighbors only. Noting that $N_s = \Sigma_i (1 - \sigma_i^2)$ and that $N_w - N_o = \Sigma_i \sigma_i$, and defining

$$\Delta = \mu_s - \frac{1}{2}(\mu_w - \mu_o) \tag{2.2}$$

the Hamiltonian in Eq. (2.1) can be rewritten as

$$\mathcal{H}_1 = -\sum_{i,j} [J_1\sigma_i\sigma_j + A_{\rm lat}J_2(\sigma_j\mathbf{m}_i \cdot \mathbf{r}_{ij} + \sigma_i\mathbf{m}_j \cdot \mathbf{r}_{ji})] + \Delta \sum_i \sigma_i^2$$
$$-\frac{1}{2}(\mu_w - \mu_o)\sum_i \sigma_i \tag{2.3}$$

While remaining relatively uncomplicated, the model of Eq. (2.3) contains the essential ingredients of three-component systems composed of water, oil, and amphiphiles. Interactions between amphiphiles could become important, especially in the case of an aqueous solution of surfactants. The influence of such interactions on the phase diagram has been studied in detail using the mean-field theory by Matsen and Sullivan [58, 59]. They are not included here since we are interested in examining the simplest possible models for surfactant mixtures using numerical simulation. In this section we will be interested in ternary water–oil–amphiphile mixtures where the average concentrations of water and oil are equal and in binary water–amphiphile mixtures. The following reduced parameters will be used here

$$j_2 = \frac{J_2}{J_1} : \delta = \frac{\Delta}{J_1} : t = \frac{T}{J_1} \tag{2.4}$$

where T is the temperature in units of Boltzmann's constant.

2. *Numerical Methods*

Monte Carlo simulation is a useful tool for the study of both equilibrium and nonequilibrium properties of thermodynamic systems. The Metropolis Monte Carlo method [67] used here includes thermal fluctuations in a natural manner and it can therefore be used to solve problems in statistical physics that are too complex to be resolved by an exact analytic treatment. There are other numerical methods that are often used to determine phase equilibria. These include, for example, transfer-matrix calculations [68]. This method is, however, usually limited to relatively small systems due to the computational effort required.

At equilibrium, a problem often encountered in calculating phase diagrams is the accurate determination of the phase boundaries and the order of the related phase transitions. For example, a weak first-order transition can easily be confused with a second-order phase transition. This occurs because, for a finite system, the discontinuities in the thermodynamic quantities are smeared out. A finite size scaling analysis must then be performed to resolve such ambiguities often encountered in Monte Carlo simulations. This requires a study of many system sizes together with a well-resolved scan over temperatures and fields around the phase transition.

Ferrenberg and Swendsen [69] recently introduced a numerical extrapolation technique that helps to overcome some of these technical problems. Their method is not difficult to apply, since it is based on the Metropolis Monte Carlo method, and continuous thermodynamic quantities can be generated over a considerable range of temperatures and fields. This method is therefore useful in the transition regions of the phase diagram where sharp changes occur. While standard Monte Carlo simulations require high resolution in temperatures and fields in order to locate the transition, the use of the method of Ferrenberg and Swendsen [69] allows us to determine the location of the transition using a single simulation at a point close to the transition. The accuracy of this method has been proven in various models such as the spin $\frac{1}{2}$ Ising model [69, 71], the ϕ^4 model [72], and the 10-state Pink model for lipid bilayers [73, 74].

The combination of the extrapolation method for finite size scaling analysis allows us to determine the order of the phase transition. At a second-order phase transition, finite size scaling is well understood, and the dependence of the thermodynamic quantities on the system size is governed by universality classes [75]. For example, the specific heat and the susceptibility diverge as $L^{\alpha/\nu}$ and as $L^{\gamma/\nu}$, respectively, where α, ν, and γ are the critical exponents for the heat capacity, correlation length, and susceptibility. At a first-order transition, the specific heat and the susceptibility diverge as L^d, that is, at a faster rate. Another useful quantity for determining the nature of a phase transition is the fourth cumulant of the energy, defined by Binder [76].

For weak first-order transitions where the size of the correlation length is very large, very large systems, and therefore a huge amount of computer time, are required to examine the nature of the phase transition. In order to overcome such difficulties, Lee and Kosterlitz [70] proposed a numerical method that consists of calculating the free energy at the transition point. In our case, the free energy can be calculated as a function of the amphiphile density. It can also be calculated as a function of the difference in the water and oil concentrations. The reader is referred to the original articles for details of these methods.

The structure of the disordered phase and the comparison with experimental results are studied by calculating the structure functions that are directly proportional to the scattering intensities obtained from small angle neutron scattering on microemulsions. In particular, the water–water structure factor allows us to distinguish the structured disordered states (or microemulsions) from the ordinary (or classical) disordered fluid. The water–water and the amphiphile–amphiphile structure factors are calculated as follows:

$$S_{\alpha\alpha}(\mathbf{q}) = \frac{1}{L^d}\left\langle \left| \sum_{\mathbf{x}_i} \sigma(\mathbf{x}_i) \exp(i\mathbf{q}_i \cdot \mathbf{x}_i) \right|^2 \right\rangle \tag{2.5}$$

where $\mathbf{q} = 2\pi/L(n\mathbf{i} + m\mathbf{j})$ and $m, n = 0, 1, \ldots, L$ and α refers to the molecular species. The brackets in Eq. (2.5) denote an ensemble average. Here, the variable σ can be either the spin variable if we calculate the water–water structure factor, or $1 - \sigma_i^2$ if we calculate the amphiphile–amphiphile structure factor. Microemulsions are isotropic structures since they are disordered. It is therefore useful to calculate the circularly averaged structure factor defined as $\mathcal{S}(q) = \Sigma'(\mathbf{q})/\Sigma' 1$, where Σ' denotes a sum over circular shells (in 2D) defined by $n - \frac{1}{2} \leq L|\mathbf{q}|/2\pi < n + \frac{1}{2}$.[1]

C. Monte Carlo Phase Diagrams

1. Phase Diagram for Ternary Mixtures

In this subsection, we present the phase diagrams for the model of Eq. (2.3) as calculated from Monte Carlo simulations for the square lattice and for equal average densities of water and oil, that is, when $\mu_w = \mu_o$ for $j_2 = 3$. An approximate determination of the phase boundaries was obtained from standard Monte Carlo simulations at various chemical potentials ranging from $\mu = -8$ to $\mu = 12$, and at several temperatures. This determination gave an estimate of the transition points but not their order. In these simulations, the grand canonical ensemble was used with the amphiphilic chemical potential held fixed. The amphiphile, water, and oil densities were therefore subject to fluctuations.

Once the transition points were estimated, very long runs ranging from 6×10^5 to 2×10^6 MCSPS (Monte Carlo steps per site) were performed extremely close to the transition points, in order to use the Ferrenberg–Swendsen extrapolation technique [69] discussed in Section II.C. These simulations were performed on the following system sizes: $L = 4, 8, 16, 20, 24, 32$, and 40.

[1] We note that the underlying lattice might create some anisotropy in the structure factor even in the disordered phase.

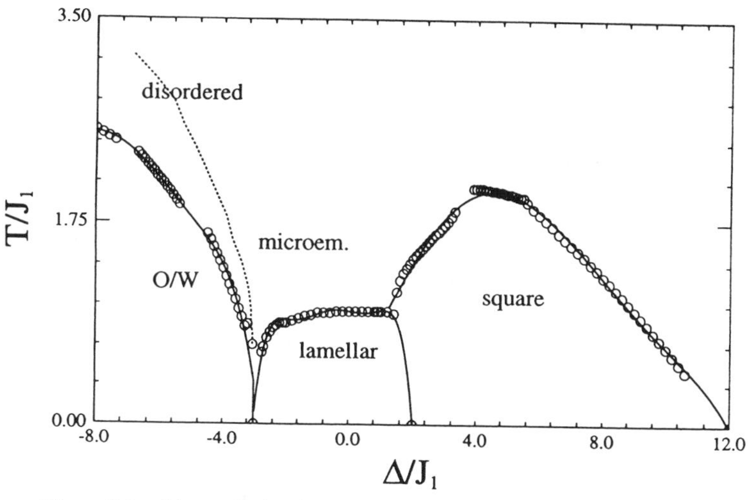

Figure 2.1. Monte Carlo phase diagram of the square lattice model.

The phase diagram of the model obtained from the extrapolation technique for $j_2 = 3$ is shown in Fig. 2.1, a wide water–oil coexistence region extending to $\delta \rightarrow -\infty$, with a lamellar phase, a square phase, and a broad disordered region. The transition from the water–oil coexistence to the disordered phase is mostly second order. However, the transition becomes first order for low temperatures close to the lamellar phase. There is no phase transition between the two-phase region and the lamellar phase, that is, at any finite temperature, there is no three-phase coexistence between the water, the oil, and the lamellar phase in two dimensions. This implies that the disordered phase extends to vanishingly small temperatures. A similar behavior was observed in a different model for 2-D systems of water, oil, and amphiphiles by Gompper and Schick [52–54].

The lamellar phase consists of layers of water and oil in either the (01) or the (10) directions separated by amphiphilic monolayers. The period of the lamellae can be relatively large as seen in Fig. 2.2. However the width of the water or oil domains is usually only one lattice spacing. If the average width is larger than one lattice spacing, the interfaces become rough, and only quasi long-range order is present. This occurs because the interfaces are 1-D, and are therefore rough at any finite temperature. The same behavior is observed when the Ginzburg–Landau model is used (see Section III).

The lamellar phase is separated from the disordered phase by a first-order transition, as will be shown later by finite size scaling analysis.

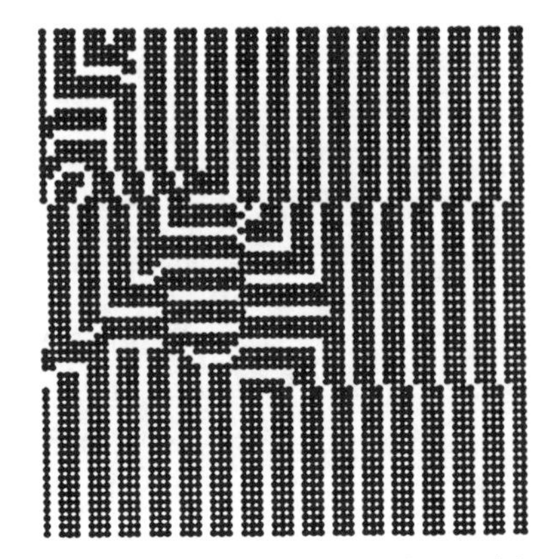

Figure 2.2. Two typical configurations in the lamellar phase of the square lattice model. The left configuration corresponds to $\delta = -2.4$ and $T = 0.4$, and the right configuration corresponds to $\delta = -0.5$ and $T = 0.55$. Open circles represent amphiphiles and closed circles represent water molecules. Blank spaces represent oil molecules.

As δ is further increased, the lamellar phase evolves towards a square phase via a first-order transition. A configuration obtained in the square phase from Monte Carlo simulations is displayed in Fig. 2.3. This phase is analogous to the liquid crystalline cubic phase observed in experiments [2]. The transition line from the square phase to the disordered phase is also first order (see below).

The disordered phase does not exhibit the same behavior at all points in the phase diagram. In particular, this phase can be divided into two broad regions by a Lifshitz line shown as a dotted line in Fig. 2.1. To the left of this line, the water–water structure factor shows the usual monotonic decay with a peak at $q = 0$. We therefore interpret the disordered phase in this region as a classical disordered fluid that has no structure except for length scales comparable to the size of the molecules. However, a diffuse peak in the structure factor at $q > 0$ is observed to the right of the Lifshitz line. The disordered phase in this region therefore exhibits short-range rather than long-range order. This region of the disordered phase can thus be identified with a microemulsion. No singularities are encountered at the Lifshitz line in the derivatives of the free energy implying that this line is not a transition line in the thermodynamic sense. Four configurations in the microemulsion regime of the disordered phase are displayed in Fig. 2.4. for $T = 0.9$ and $-2.9 \leq \delta \leq -2.5$. Segregated domains of water and oil separated by amphiphilic monolayers are observed in these configurations. We also note that the amphiphiles are almost always located at the water–oil interfaces and that there are hardly any amphiphiles dissolved in either

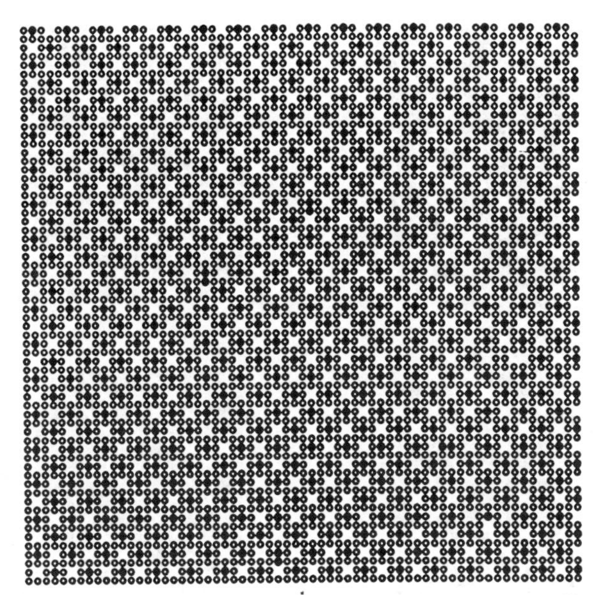

Figure 2.3. A typical configuration in the square phase corresponding to $\delta = 4$ and $T = 1$. The definitions of the dots are the same as in Fig. 2.2.

water or oil regions. As the temperature is increased more amphiphiles are dissolved in the water and oil domains and sections of the interfaces become unoccupied by amphiphiles. We also note that the average domain size decreases with increasing chemical potential, that is, as the amphiphile concentration increases, since amphiphiles create more interfaces. The same behavior was observed in experiments [14, 77].

We now study the nature of the transitions in the phase diagram displayed in Fig. 2.1 using the Ferrenberg–Swendsen extrapolation method. The specific heat is shown in Fig. 2.5 as a function of temperature for several system sizes at $\delta = 1$, that is, at a transition from the lamellar phase to the disordered phase. The maximum of the specific heat $C^{\max}$ is given as a function of L^2 in Fig. 2.6, which shows the linear dependence of $C^{\max}$ on L^2 implying that the transition is first order. The same behavior is seen in the amphiphile susceptibility χ. This behavior is in contrast to regions of the phase diagram where the phase transition is second order, as is the case for most of the transition line between the water–oil coexistence to the disordered phase. Here we observed that $C^{\max}$ increases for small values of L but becomes almost constant as L increases. Such behavior is expected for second-order phase transitions in two dimensions since the critical exponent for the specific heat is then zero. In Section III we show that the Ginzburg–Landau model is in the

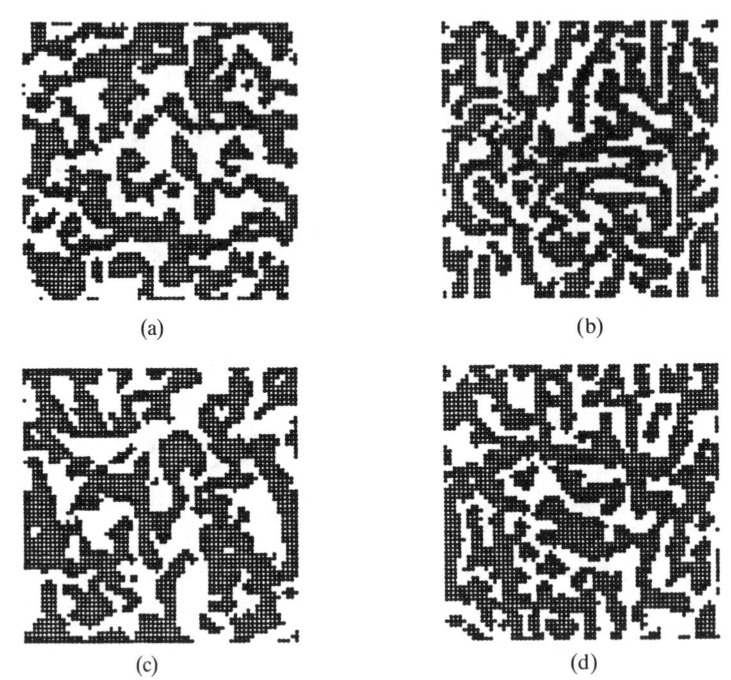

Figure 2.4. Configurations in the disordered phase between the water–oil coexistence and the lamellar phase at $T = 0.9$. The configuration (a) corresponds to $\delta = -2.9$, the configuration (b) corresponds to $\delta = -2.8$, the configuration (c) corresponds to $\delta = -2.7$ and the configuration (d) corresponds to $\delta = -2.5$. The definition of the dots is the same as in Fig. 2.2.

same universality class as the spin $\frac{1}{2}$ Ising model for small amphiphile chemical potentials.

We examined the behavior of χ at $\delta = 8$, that is, at the transition from the square phase to the disordered phase. Our calculations again show that the transition from the square phase to the disordered phase should be first order. We also calculated the free energy as a function of amphiphilic density for this transition. The value of the free energy at the transition point is shown in Fig. 2.7 for $\delta = 8$. This figure shows that as the system size increases, the depth of the double well structure increases, implying that the transition is first order [70]. This method therefore confirms the predictions made from the behavior of the specific heat and the susceptibility. The transition point occurs when the two minima have the same value.

We were not able to determine the order of the transition between the two-phase water–oil region and the disordered phase at low temperatures

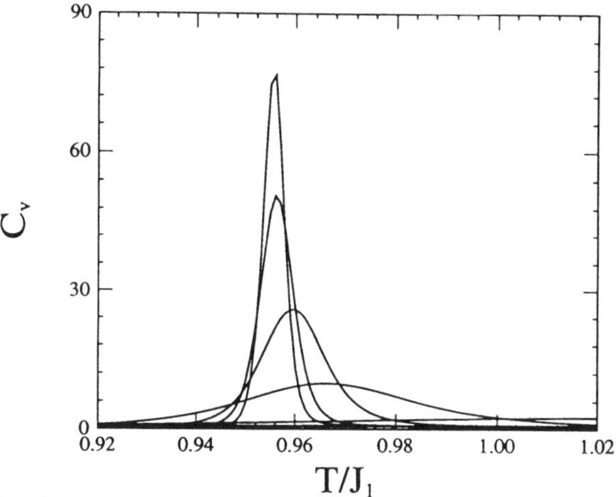

Figure 2.5. The specific heat in the square lattice as a function of temperature at $\delta = -1$ for several system sizes. Curves from lower to upper correspond to $L = 12$, 16, 20, 24, 32, and 40.

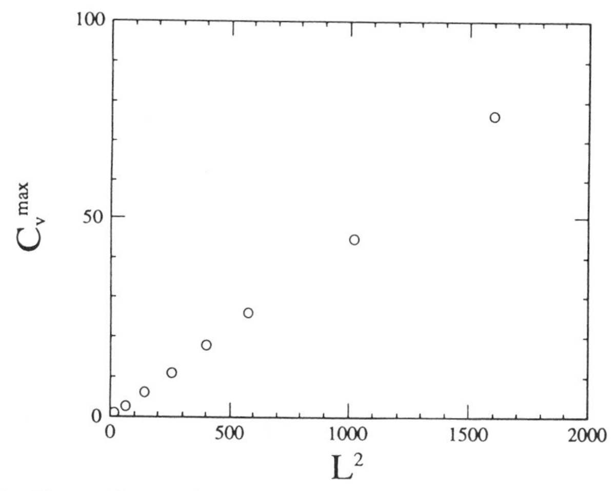

Figure 2.6. The maximum of the specific heat in the square lattice as a function the system size for $\delta = -1$.

in the simulations. This was due to the extremely small value of the interfacial tension between the water-rich (or oil-rich) phase and the disordered phase. It is observed experimentally that, when the microemulsion coexists simultaneously with water and oil, the interfacial tension between any two of the three phases is very small and corresponds to a few percent of the bare water–oil interfacial tension [10].

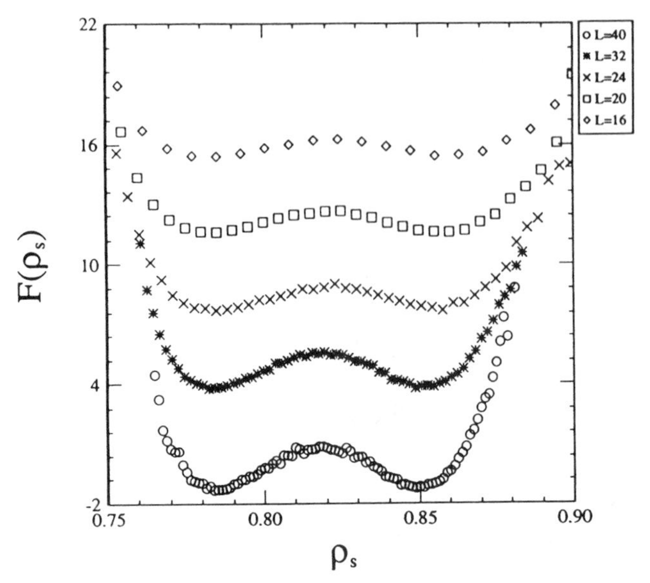

Figure 2.7. The free energy $\mathcal{F}$ as a function of amphiphile concentration for $\delta = 8$ in the square lattice model.

Finite size scaling behavior can therefore not be observed unless extremely large systems are studied. The only indication of a first-order transition is that the ground-state calculation shows that the transition at $T = 0$ is first order. The transition at $T = 0$ is also the transition to the lamellar phase. This implies that we could have a triple line that is separated from the second-order line by a tricritical point at low temperatures. In contrast Matsen and Sullivan [78] recently conjectured that no first-order transitions can take place in 2-D lattice models for mixtures containing surfactants.

Experiments show that for comparable densities of water and oil, the water–water scattering intensity exhibits a finite peak. The position of this peak increases and its height decreases as the amphiphilic concentration increases. It is also observed that the tail of the structure factor obeys Porod's law [79], that is, $S(q) \sim q^{-(d+1)}$, which is an indication that the width of the interface is very small compared to the size of the domains. For sufficiently strong amphiphiles, only a very small amphiphile concentration is expected to be dissolved in bulk water or bulk oil regions, that is, most of the amphiphiles are adsorbed at the interfaces. The total area of the interfaces should therefore be proportional to the amphiphilic density, implying that the average domain size decreases linearly with ρ_s.

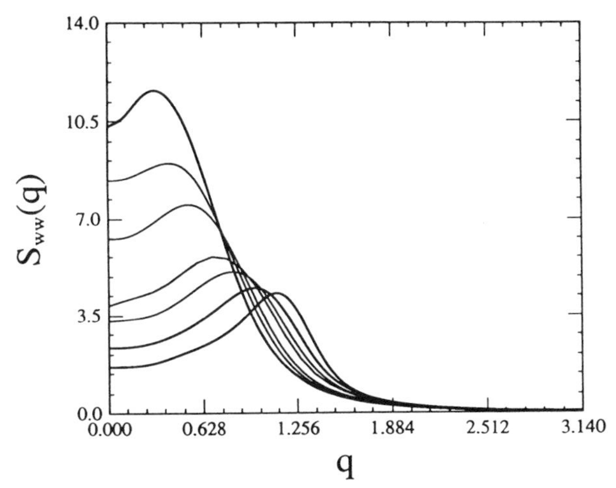

Figure 2.8. The evolution of the water–water structure factor in the microemulsion regime on the square lattice model at $T = 0.8$. Curves from upper to lower correspond to $\delta = -2.85$, -2.8, -2.75, -2.65, -2.6, -2.5, and -2.4, respectively.

We calculated the water–water and the amphiphile–amphiphile structure factor using Eq. (2.5) for $L = 64$ with a large number of averages. In Fig. 2.8, the water–water structure factor is shown for $T = 0.8$ in the microemulsion regime of the disordered phase and in the vicinity of the water–oil coexistence and the lamellar phase. As found experimentally, we observed a well-defined peak whose position shifts towards large wavenumbers as the amphiphilic concentration increases and whose height decreases with ρ_s. Figure 2.9 shows the variation of $q^{\max}$ with ρ_s which, as predicted, shows a linear dependence. For high temperatures, more diffuse domains of water and oil are expected and this prediction should then break down. We find that the tail of the water–water structure factor follows Porod's law $S(q) \sim q^{-3}$ at large wavenumbers (see Fig. 2.10).

Figure 2.11 shows the amphiphile–amphiphile structure factor $S_{ss}(q)$ in the microemulsion regime. In contrast to the water–water structure factor, $S_{ss}(q)$ is almost structureless. Moreover, we observe that the peak is at $q = 0$ for small chemical potentials. However, as the chemical potential increases, a peak develops and moves towards large wavenumbers, while the peak at $q = 0$ remains. Similar results have been observed in the amphiphile–amphiphile structure factor calculated by Gompper and Schick [31, 53].

We next present results for the phase diagram of the ternary mixture of water, oil, and surfactants for the 2-D triangular lattice when the average

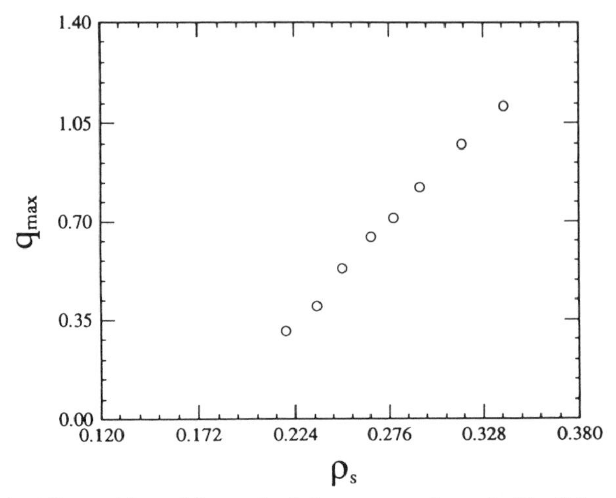

Figure 2.9. The position of the peak of the structure factor in Fig. 2.8 as a function of amphiphile concentration.

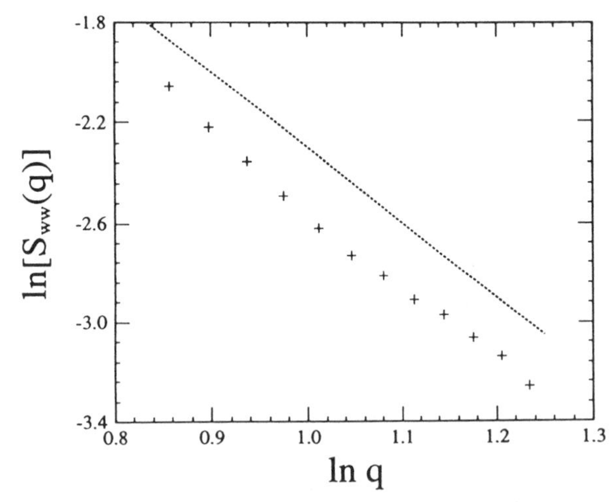

Figure 2.10. The tail of the water–water structure factor (+) in a log–log plot. The slope of the dotted line is −3.

concentrations of water and oil are equal. This phase diagram was calculated under the same conditions used for the square lattice case and the same number of phases were observed as is shown in Fig. 2.12. A broad water–oil coexistence region is found that extends to large negative δ. The concentration of amphiphiles is quite small in this region. The transition line separating the water–oil coexistence from the disordered

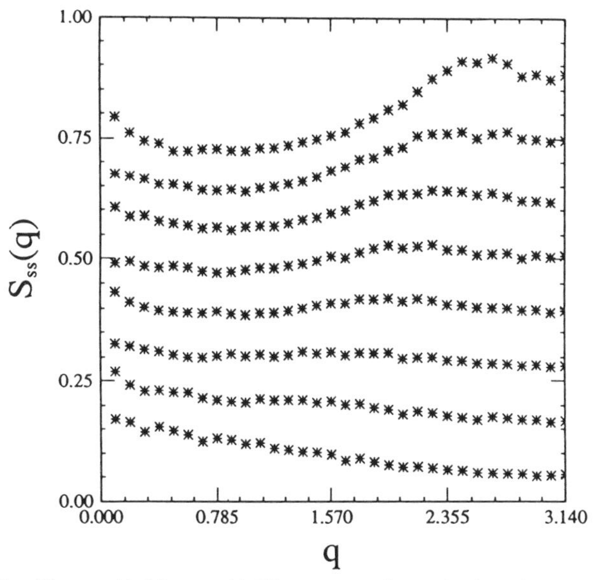

Figure 2.11. The amphiphile–amphiphile structure factor in the microemulsion regime of the disordered phase at $T = 0.8$. Curves from lower to upper correspond to $\delta = -2.85$, -2.8, -2.75, -2.7, -2.65, -2.6, -2.5, and -2.4, respectively.

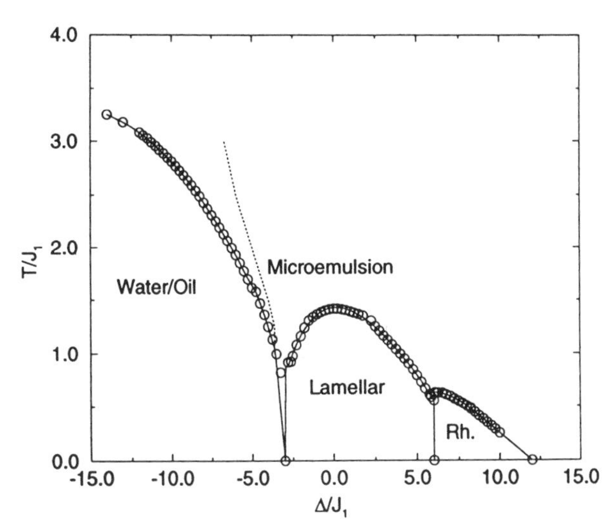

Figure 2.12. Monte Carlo phase diagram of the ternary mixture in the triangular lattice model with $\mu_w = \mu_o$ and $j_2 = 2$.

phase is mostly second order. We note that, as δ becomes very large but negative where there are essentially no amphiphiles in the system, the model is a spin 1 Ising model. In this limit the transition line approaches the critical temperature of the triangular lattice Ising model $T_c = 3.641$ [80].

At chemical potentials larger than $\delta \approx -4$, a triple-well structure is detected in the free energy versus the order parameter, $M = \Sigma_i \sigma_i / L^d$ as shown in Fig. 2.13. The three minima occur at $M = 0$ corresponding to the disordered phase and at $M = \pm 1$ corresponding to the water and oil phases. As the system size increases, the double-well structure becomes sharper leading to an uncertainty in the height of the free energy barrier. The presence of such a well is an indication that the phase transition from the region of water–oil coexistence to the disordered phase may well be first order. Indeed, the configuration shown in Fig. 2.14, obtained at $\delta = -3.5$ with fixed boundary conditions along one direction and periodic boundary conditions along the other, shows clearly a three-phase coexistence of the water, the oil, and the disordered phases. The presence of a triple line of coexistence and a line of second-order transitions imply the presence of a tricritical point. We were not, however, able to determine its exact location.

A lamellar phase is observed at low temperatures for chemical potentials larger than $\delta = -3$. The finite size scaling analysis of the phase transition from the lamellar phase to the disordered phase shows that this

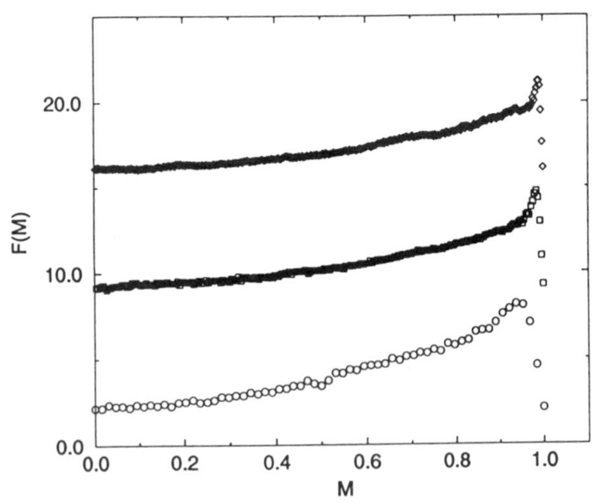

Figure 2.13. Free energy calculated from Eq. (2.10) as a function of $M = \Sigma\, \sigma_i / L^2$ with $\delta = -3.5$ for $L = 8$ (circles), $L = 16$ (squares), and $L = 20$ (diamonds).

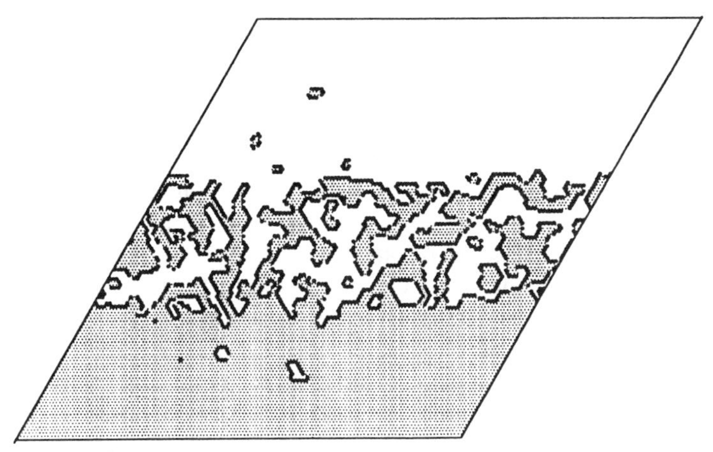

Figure 2.14. A configuration in a three-phase coexistence between the water, the oil, and the disordered phases at $\delta = -3.5$ and the closed circles represent amphiphiles, the open circles represent water and the white regions correspond to oil. $T = 1.01$ and the system size is 129×129.

phase boundary is always first order. This can be seen from Fig. 2.15 where the free energy as a function of amphiphile concentration is shown for two different values of δ. The increase in the free energy barrier with increasing linear system size L indicates that the transition between these two phases is first order [70]. It is interesting to note that there is no

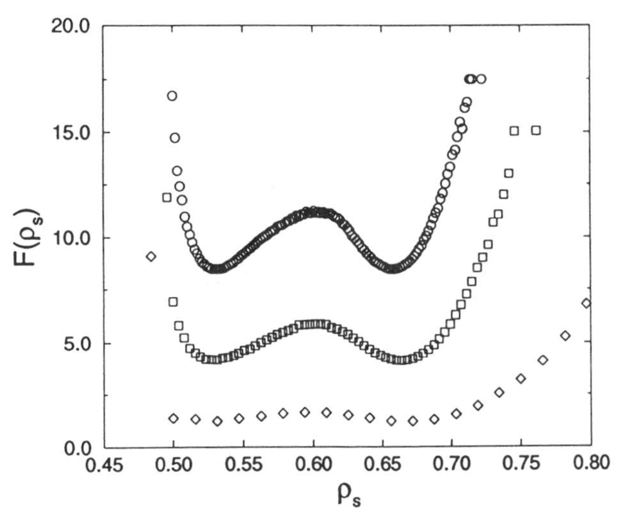

Figure 2.15. The free energy as a function of amphiphile concentration at $\delta = 4$ for the following system sizes: $L = 8$ (diamonds), $L = 16$ (squares), and $L = 24$ (circles).

three-phase coexistence between the lamellar, the water, and the oil phases except at $T = 0$. This allows the disordered phase to extend to zero temperature between the water–oil coexistence curve and the lamellar phase as was also observed in our simulations for the square lattice.

At higher amphiphile chemical potentials, the system undergoes a transition from the lamellar phase to the other liquid crystal phase with rhombic symmetry. A configuration obtained from the Monte Carlo simulation in this phase is displayed in Fig. 2.16. Again a finite size scaling analysis of the free energy barrier showed that this phase transition is first order.

A comparison of the phase diagrams for ternary mixtures using a square lattice (Fig. 2.1) and a triangular lattice (Fig. 2.12) show that both phase diagrams are very similar as expected from universality. The same structure of the disordered phase is found. The differences occur in the nonuniversal features such as the nature and extent of the liquid crystal phases.

2. *Phase Diagram for Binary Mixtures*

We now examine the limiting case for the triangular lattice where the concentration of oil is zero. The phase diagram for such a binary fluid mixture was determined in the same manner as the case of the ternary mixture discussed above. As before, the simulations were run for 4×10^6 Monte Carlo steps per site on several system sizes.

The phase diagram of the binary mixture is shown in Fig. 2.17. Only three phases are observed in this case. For $\delta < 0$ a pure water regime is observed at low temperatures. The dotted line in Fig. 2.17 shows where

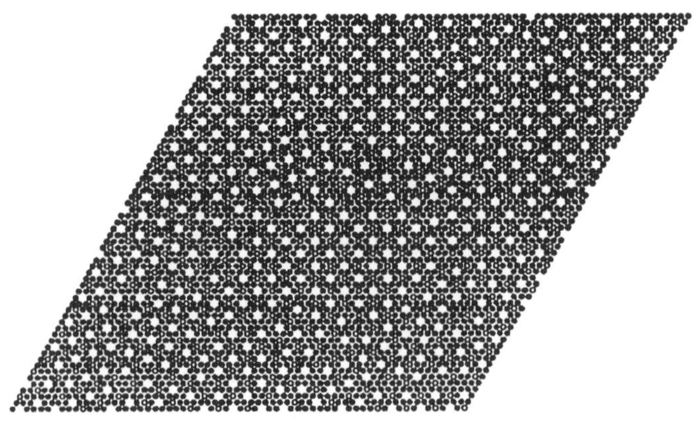

Figure 2.16. A configuration in the rhombic phase determined from Monte Carlo simulations at $\delta = 9$ and $T = 0.2$.

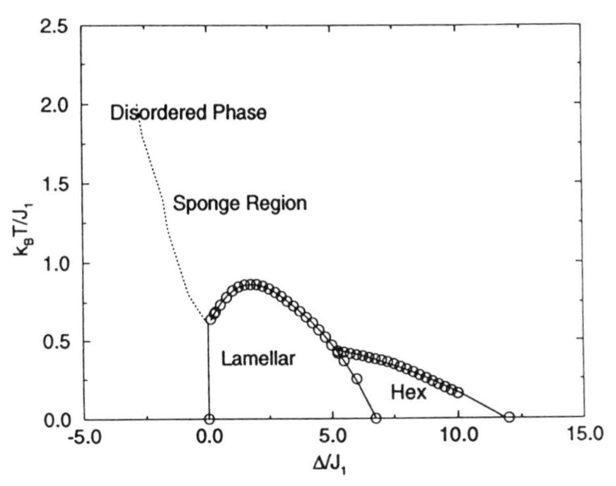

Figure 2.17. Monte Carlo phase diagram of the binary mixture of water and amphiphiles in the triangular lattice model with $j_2 = 2$.

this regime crosses to a disordered regime with nonzero amphiphile concentrations. This line does not, however, correspond to a phase transition. Although a peak in the heat capacity is observed at this line, the value of the maximum of the heat capacity does not increase with increasing system size. The mean-field calculation of this model also predicts a finite peak in the heat capacity in this region of the phase diagram. In this case the model reduces to the spin $\frac{1}{2}$ Ising model in a magnetic field that does not exhibit a phase transition. The disordered phase therefore extends to zero temperatures for chemical potentials smaller than $\delta = 0$.

The water-rich region is followed by a lamellar phase in which the amphiphiles form bilayers separated by a water layer. The transition separating the lamellar phase from the disordered phase was found to be first order by means of a finite size scaling analysis on the double-well structure of the free energy. As the amphiphile chemical potential is further increased, the lamellar phase crosses to another liquid crystal phase with a hexagonal symmetry, which can be identified with an inverted hexagonal phase. The nature of the phase transition from this phase to the lamellar and to the disordered phases is also first order.

The disordered phase in the phase diagram of Fig. 2.12 is divided into two subregions by a Lifshitz line. To the left of the Lifshitz line the structure factor of the disordered phase has a peak at $q = 0$. Again, to the right of the Lifshitz line, the structure factor of the disordered phase exhibits a well-defined broad peak with its position increasing as the

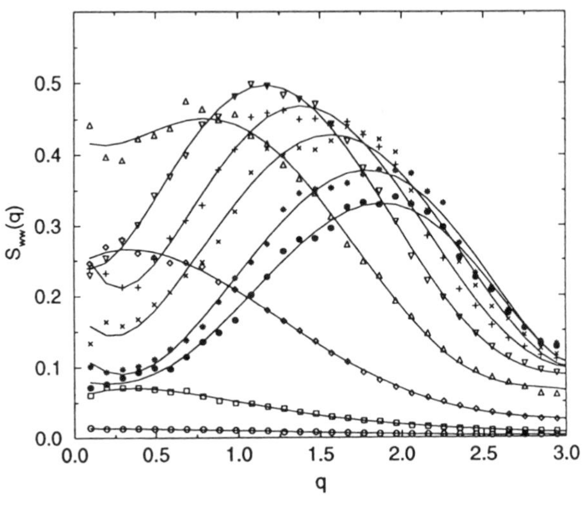

Figure 2.18. The water–water structure factor in the disordered sponge regime at $T = 1$ for the following amphiphile chemical potentials: $\delta = -2$ (open circles), $\delta = -1.6$ (squares), $\delta = -1.2$ (diamonds), $\delta = -0.8$ (triangles up), $\delta = -0.4$ (triangles down), $\delta = 0$ (+), $\delta = 0.4$ (×), $\delta = 1.2$ (stars), and $\delta = 2$ (closed circles). The solid lines are guides for the eye.

amphiphile concentration is increased, as shown in Fig. 2.18. The region to the right of the Lifshitz line of the phase diagram can be identified as a sponge phase. This behavior has been observed in experiments [81] and in previous mean-field calculations [52]. As for bicontinuous microemulsions, the presence of this peak implies that the water domains (or the amphiphilic bilayers) are now correlated. Indeed, the real space configurations corresponding to these structure factors are shown in Fig. 2.19 where we find a network of amphiphilic bilayers that exhibit short-range order.

D. Mean-Field Theory

1. *Phase Diagram*

For purposes of comparison we now present the phase diagram of the model of Eq. (2.3) for the square lattice and for equal concentrations of water and oil using a mean-field theory. At high temperatures the system equilibrates to a homogeneous disordered phase, whereas at low temperatures and small amphiphile chemical potentials, the system equilibrates to the water-rich or the oil-rich phases. The latter two phases are homogeneous and they are therefore characterized by a nonzero average value of the order parameter M, where $M = \Sigma\, \sigma_i / L^2$. The transition from

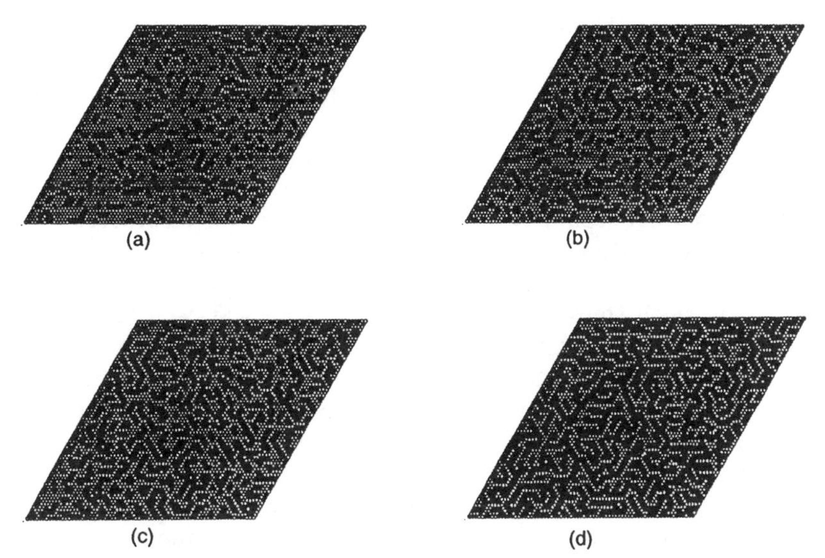

Figure 2.19. Monte Carlo configurations in the disordered phase at $T = 1$. (a–d) correspond, respectively, to $\delta = -0.4$, $\delta = 0$, $\delta = 0.4$, and $\delta = 1.2$. The closed circles represent amphiphiles, the open circles represent water and the white regions correspond to oil.

the disordered phase to the water–oil coexistence region can either be second or first order. This line is determined by expanding the mean-field free energy density around zero magnetization in a power series. Also, since we are dealing with homogeneous phases, we obtained the following free energy density,

$$\frac{\mathcal{F}(M)}{N} = \frac{\mathcal{F}(0)}{N} + \left(\frac{AT}{2} - 2\right)M^2 + \left[\left(\frac{32}{AT^3} - \frac{4A}{T}\right)\left(1 - \frac{1}{A}\right) + \frac{A^3T}{12}\right]M^4 + \mathcal{O}(M^6)\,, \tag{2.6}$$

where $A = 1 + 2\exp(\delta/t)$. A second-order transition line occurs when the prefactor of M^2 vanishes provided that the prefactor of M^4 stays positive and nonzero. When this prefactor becomes negative, the transition line becomes first order, as found by equating the free energies of water, oil, and the disordered phase. The first-order line is separated from the second-order line by a tricritical point. The equation of the second-order transition line is then given by,

$$\delta = T\ln\left(\frac{2}{T} - \frac{1}{2}\right) \tag{2.7}$$

When the homogeneous phases are examined using mean-field theory, the prefactors of M^n are independent of the amphiphilic interaction j_2. When $j_2 = 0$, we recover the mean-field phase diagram of the Blume–Capel model [82].

We calculated the transition lines that separate the lamellar phase from the water-oil coexistence, the disordered phase and the square phase, and the transition that separates the square phase from the disordered phase by using a novel Monte Carlo local mean-field method. We start from a completely disordered high-temperature state and then lower the temperature in small steps. The system is equilibrated at each temperature using a large number of iterations and N sites are chosen randomly (as in the standard Monte Carlo method) at each iteration. An attempt to change the local fields is made by using the mean-field equations and then the mean internal energy is calculated prior to and after this attempt. The acceptance or the rejection of these changes is controlled by the Metropolis Monte Carlo algorithm. This method allows the system to choose the configuration with the lowest free energy without imposing a real space structure of the phase as is usual in mean-field manipulations. The method described above resembles that proposed earlier by Soukoulis and Grest [83] to study the phase equilibria in spin glasses within the mean-field approximation.

The mean-field phase diagram is shown in Fig. 2.20 for $j_2 = 3$. This

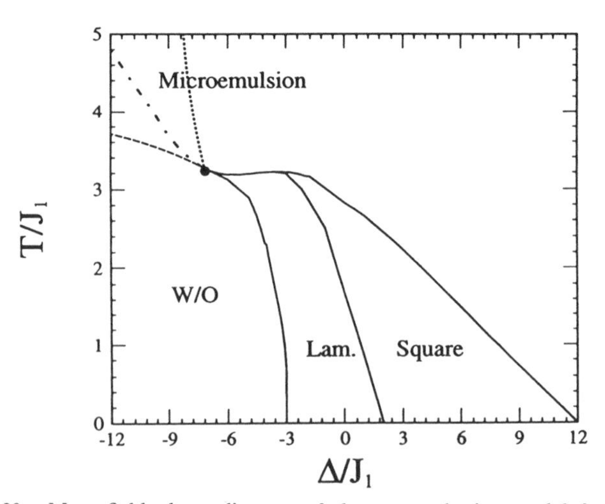

Figure 2.20. Mean-field phase diagram of the square lattice model for $j_2 = 3$. The dashed lines represent second-order phase transitions. The solid lines represent first-order phase transitions. The dotted line represents the Lifshitz line, the dash-dotted line represents the disorder line and the dot gives the location of the Lifshitz multicritical point.

particular value is chosen for purposes of comparison with the Monte Carlo results presented in Section II.D. The same phases were found: a water–oil coexistence at low temperatures and chemical potentials, followed by a lamellar phase, and then a square phase as the amphiphile chemical potential is increased. From the local mean-field calculations, we found that in the lamellar phase, the domains are in the (10) or the (01) direction. This finding is the case at low temperatures. However, the lamellae can also be in the (11) direction at higher temperatures and their period can be relatively large. In this case, amphiphiles at the water–oil interfaces act as if they are oriented along the diagonals. This result is a consequence of the mean-field treatment and is not allowed when the model is treated exactly or using the Monte Carlo method. Lamellar phases of different periods were also found in the mean-field picture.

Within mean-field theory, the disordered phase does not occur at very low temperature in contrast to the Monte Carlo results. We did not find three-phase coexistence between the water, oil, and disordered phases in mean field. The transition from the water–oil coexistence is always second order ending in a tricritical point, which is also the beginning of the phase transition from the water–oil coexistence to the lamellar phase, and the transition from the lamellar to the disordered phase. This multicritical point can be identified as a Lifshitz multicritical point.

2. *Structure of the Disordered Phase*

In this section, we give results for the structure factors and the correlation functions in the disordered phase as calculated from mean-field theory. The structure factor is calculated via Ornstein–Zernicke theory. To this purpose, we use the mean-field free energy and allow the local order parameter M_j and the local occupation number for the amphiphiles n_j to fluctuate around their equilibrium values in the disordered phases given, respectively, by $M_0 = 0$ and $n_0 = 1/[2\exp(\delta/t) + 1]$. The expansions of M_j and n_j in Fourier space around their equilibrium values are then given by

$$M_j = M_0 + \sum_{\mathbf{q}} M_{\mathbf{q}} \exp(i\mathbf{q} \cdot \mathbf{r}) \tag{2.8}$$

$$n_j = n_0 + \sum_{\mathbf{q}} n_{\mathbf{q}} \exp(i\mathbf{q} \cdot \mathbf{r}) \tag{2.9}$$

By substitution of these two equations into the mean-field expression for free energy, we obtain the following free energy density in Fourier space:

$$\frac{\mathcal{F}}{N} = \frac{\mathcal{F}_0}{N} + \sum_{\mathbf{q}} [A_{\mathbf{q}} M_{\mathbf{q}} M_{-\mathbf{q}} + B_{\mathbf{q}} n_{\mathbf{q}} n_{-\mathbf{q}} + C_{\mathbf{q}}(n_{\mathbf{q}} M_{-\mathbf{q}} + n_{-\mathbf{q}} M_{\mathbf{q}})] \tag{2.10}$$

where N is the total number of molecules. $A_{\mathbf{q}}$, $B_{\mathbf{q}}$, and $C_{\mathbf{q}}$ in Eq. (2.10) are given by

$$A_{\mathbf{q}} = -[\cos(q_x a) + \cos(q_y a)] - \frac{j_2^2}{T}(1 - n_0)[\sin^2(q_x a) + \sin^2(q_y a)] + \frac{n_0 T}{2(n_0^2 - M_0^2)} \tag{2.11}$$

$$B_{\mathbf{q}} = \frac{T}{2(1 - n_0)} + \frac{T n_0}{2(n_0^2 - M_0^2)} \tag{2.12}$$

and

$$C_{\mathbf{q}} = -\frac{M_0 T}{2(n_0^2 - M_0^2)} \tag{2.13}$$

Here a is the lattice spacing. The water–water structure factor is proportional to $T/A_{\mathbf{q}}$. We studied the evolution of the structure factor as the amphiphile chemical potential is increased for $T = 3.3$, which is immediately to the right of the Lifshitz tricritical point. As in the Monte Carlo results, we observe that the position of the peak, ($q_{\max}$) in the microemulsion regime increases as δ increases, whereas the height of the peak decreases. Furthermore, we calculated the dependence of $q_{\max}$ on amphiphilic density $\rho_s = 1 - n_0$. In contrast to the Monte Carlo results, the dependence of $q_{\max}$ on ρ_s is nonlinear. This occurs because amphiphiles are homogeneously distributed in space within the mean-field approximation and are not only found at the interfaces between water- and oil-rich regions. We see from Eq. (2.11), that the peak in the structure factor comes from the term coupled to j_2, implying the peak is at $q = 0$ as expected in the Blume–Capel model (i.e., for $j_2 = 0$).

The Lifshitz line, which separates the ordinary disordered fluid from the microemulsion can be obtained from the structure factor. This line is found by expanding $A_{\mathbf{q}}$ in a Taylor series at small wavenumbers up to fourth order. We then obtain the following expression for the structure factor for small wavenumbers

$$S(q) = \frac{T}{\left(\frac{T}{2n_0} - 2\right) + \left(1 - \frac{2j_2^2}{T}\rho_s\right)q^2a^2 + \left(-\frac{1}{6} + \frac{2j_2^2}{3T}\rho_s\right)q^4a^4 + \mathcal{O}(q^6a^6)} \tag{2.14}$$

If the prefactor of the q^2 term is positive, the peak of the structure factor is found at $q = 0$. However, when this prefactor becomes negative, the peak is found at $q_{\max} > 0$. The equation for the Lifshitz line is then given by

$$\delta = -T \ln\left(\frac{4j_2^2}{T} - 2\right) \tag{2.15}$$

This form implies that the maximum temperature at which the microemulsion can exist is less than $T_{\max} = 2j_2^2$. The Lifshitz line, drawn as a dotted line in Fig. 2.20, terminates at the multicritical point that belongs to a different universality class than usual tricritical points even within mean-field theory. This result has been discussed by Holyst and Schick [6].

The microemulsion can also be defined from the behavior of the density–density correlation function defined as

$$C_{ww}(\mathbf{r}) = \sum_{\mathbf{q}} S_{ww}(\mathbf{q}) \exp(i\mathbf{q} \cdot \mathbf{r}) \tag{2.16}$$

The correlation function of an ordinary disordered fluid does not exhibit oscillations (except those on the molecular scales). In contrast, a microemulsion exhibits oscillations on length scales much larger than intermolecular spacings. However, the correlation function can exhibit oscillations even when the structure factor has a peak at $q = 0$. The locus of the line that separates the oscillatory regime from the nonoscillatory one can also be taken as a limit of the microemulsion regime. This line is usually called the disorder line. Even though the accuracy of the Monte Carlo data is not sufficient to allow a determination of this line, it can easily be calculated within the mean-field approximation. We calculated this line in the (11) direction by numerically integrating the water–water structure factor. The resulting disorder line is shown as the dash–dotted line in Fig. 2.20. We note that both the disorder and the Lifshitz line meet at the Lifshitz tricritical point. We also note that the microemulsion region defined on the basis of the disorder line is broader than that defined on the basis of the Lifshitz line and extends to infinite temperature.

E. Conclusion

In this section, we determined the phase diagram of a lattice model for ternary mixtures of water, oil, and amphiphiles for equal concentrations of water and oil and for binary mixtures of water and oil in two dimensions. The phase diagram is determined by using both Monte Carlo simulations and in the case of ternary mixtures by using mean-field theory. The Monte Carlo method was combined with two numerical techniques: the method of Ferrenberg and Swendsen, which consists of extrapolating from one point in the phase diagram to other points, and the method of Lee and Kosterlitz, which consists of calculating the free energy at the transition. These two methods allow us to determine the nature of the phase transition.

For the ternary mixtures and using a square lattice, we found five phases, which are the water- and oil-rich phases, a lamellar phase, a square phase, and a disordered phase that behaves as a microemulsion in a broad region of the phase diagram. We found that the microemulsion region extends to extremely low temperatures, which leads to the absence of three-phase coexistence between water, oil, and the lamellar phase. For the triangular lattice, basically the same phases were found for the ternary case. However, instead of a square phase, the analogous liquid crystalline phase had the symmetry of a rhombic lattice.

The phase diagram of the model on the square lattice with the same parameters as determined from mean-field calculations shows the same phases. However, the nature of some of the transition lines is no longer the same. In particular, there is no three-phase coexistence between the water, the oil, and the disordered phase. Instead a three-phase coexistence between the water, the oil, and the lamellar phases is found. This result leads to a stability of the microemulsion at high temperatures only. The absence of fluctuations in the mean-field treatment makes the microemulsion unstable against the lamellar phase at low temperatures.

For the case of a binary mixture of water and amphiphiles on a triangular lattice, we found three phases. They correspond to a lamellar phase consisting of bilayers of surfactants separated by monolayers of water, a hexagonal phase with water organized into a hexagonal sublattice and coated by surfactant monolayers, and a disordered phase. The latter phase is observed for very low or very high surfactant concentration and for high temperatures. The structure of the hexagonal phase allows us to identify it as the 2-D analogy of the inverted hexagonal phase observed experimentally [81, 84]. The disordered phase is divided into a simple fluid for small concentrations of surfactants and a spongelike phase for high surfactant concentrations. The behavior of the water–water structure factor in this phase is in agreement with the experimental measurements, that is, a broad peak with its position increasing as the surfactant concentration increases. We note that this phase diagram is qualitatively similar to that observed in experiments for nonionic surfactants [81, 84].

We should mention that there are some limitations of the present model. Since hydrogen bonding between water molecules and surfactants is not included in this work, the two-phase coexistence between the disordered and water-rich phases seen experimentally in binary mixtures is not found from the present model. Also a hexagonal phase that usually occurs before the lamellar phase was not observed.

Since the model produces phase behavior in qualitative agreement with the experimental observations, we are confident that it captures the essential physics of fluid mixtures containing surfactants. A 3-D version of this model is at present under investigation by Matsen, et al. [85]. The effects of hydrogen bonding between surfactant polar heads and water molecules, as suggested by Matsen et al. [62], is under study in 3-D for the binary mixture using Monte Carlo and other simulation methods. A recent article by Schmid and Schick [86] extends the methods described in Section II.B to the calculation of the interfacial tension.

III. GINZBURG–LANDAU MODELS

A. Introduction

In Section II, we discussed the application of microscopic lattice models to the study of ternary mixtures of water, oil, and surfactants. We showed that these models enabled us to understand the universal behavior of such systems and to analyze experimental results for microemulsions. However, they suffer from several restrictions due to the use of a lattice. For example, water, oil, and surfactant molecules have the same size in lattice models, whereas in reality water molecules are smaller than oil or surfactants molecules. Furthermore, the lamellar phases predicted by lattice models are strongly influenced by the underlying lattice. For example the lamellae are usually oriented along lattice directions and their period is often only a few lattice spacings. Continuum models for ternary mixtures do not suffer from these difficulties. In this section we first discuss the available continuum models for ternary mixtures containing surfactants and then present the Ginzburg–Landau model used here in Section III.B. Section III.C contains a mean-fields analysis of this model and the structure factor for these systems is analyzed in Section III.D. The effect of thermal fluctuations on the phase diagram is studied in Section III.E. Interfacial behavior and wetting are dealt with in Section III.F and a brief conclusion is presented in Section III.G.

Continuum models may be divided into two groups:

Interfacial Models. The first group of continuum models only consider the interfacial surfactant film and their Hamiltonians are based on the Helfrich Hamiltonian [87]. In these models, the surfactants form a 2-D incompressible fluid lying on a minimal surface. Water and oil are assumed to be completely segregated, and the surfactants are not allowed to dissolve into either water or oil domains [88, 89]. Furthermore, the interfaces are completely covered by the surfactants. Such interfacial models should therefore be applicable to the case of strong surfactants.

Ginzburg–Landau Models. The second group of the continuum models are of the Ginzburg–Landau type and their Hamiltonians are written in terms of a set of spatially varying fields, which include one or more order parameters. In contrast to the lattice models where the variables are microscopic, the local fields in the Ginzburg–Landau models are coarse-grained variables averaged over small regions. These regions are characterized by length scales that are larger than the molecular sizes.

Gompper and Schick proposed a Ginzburg–Landau model that is based on a single local field corresponding to the difference in the water and oil concentrations. This model was used to investigate the interfacial properties along the triple line of coexistence between the water phase, the oil phase, and the microemulsion. In this model, the effect of surfactants is implicit in the coefficients of the free energy functional [28]. The authors showed that this model exhibits a wetting transition that is closely related to the short-range order in the disordered phase. More recently, Gompper and Kraus calculated the phase diagram of this model for equal concentrations of water and oil using both mean-field analysis and Monte Carlo simulations [90]. The resulting phase diagram shows the same features as those observed in both lattice models and experiments.

A more detailed description of ternary mixtures can be made by using a Ginzburg–Landau model, which are based on two local fields instead of a single field. Chen et al. [91] proposed a model of this type, which is based on the usual scalar field corresponding to the difference between the water and oil concentrations, and a vector field whose orientation and magnitude represent the local orientation and the local concentration of surfactants respectively. The phase diagram of this model agrees qualitatively with experiment. However, the concentration of surfactants was controlled by varying an interaction parameter rather than a chemical potential.

Another way of representing the surfactants in a Ginzburg–Landau model is by a scalar field. This representation appears at first sight to be inconsistent with a vector description of surfactants. However, in a coarse-grained model, the local concentration and the orientation of surfactants is integrated over small regions. This implies that one is able to describe the orientation of the surfactants in terms of the gradient of the local field corresponding to the difference in the concentrations of water and oil. This results in the two scalar fields describing the Ginzburg–Landau model of Section III.B.

Finally, there are also more involved Ginzburg–Landau models such as the "mixed" model proposed by Kawasaki and Kawakatsu [92], which treats the surfactants as discrete entities and the model of Gompper and Klein [93] for binary mixtures.

B. The Model

The Ginzburg–Landau model used here is formulated in terms of two scalar local fields: a ψ field, which is proportional to the difference in the water and oil concentrations; and a ρ field, which is proportional to the local concentration of surfactants. We write the free energy of our model as a sum of two parts. The first part describes a ternary mixture in which the third component is not interfacially active, that is,

$$\mathscr{F}_1\{\psi, \rho\} = \int d\mathbf{r} \left[-\frac{r}{2}\psi^2 + \frac{u}{4}\psi^4 + \frac{a}{2}\rho^2 + \frac{g}{2}\rho^2\psi^2 + \frac{c}{2}(\nabla\psi)^2 + \frac{h}{2}(\nabla^2\psi)^2 - H\psi - \mu\rho \right] \tag{3.1}$$

The prefactor of ψ^2 can be written as $r \sim (1 - T)$, where $T = 1$ corresponds to the mean-field critical temperature in the absence of the third component [94]. If ρ does not represent surfactants, the term $(\nabla^2\psi)^2$ becomes irrelevant. The fourth term in Eq. (3.1) ensures that the concentration of the third component (surfactants) in the bulk regions (i.e., when $|\psi| \neq 0$) remains small. As discussed below, the quadratic power in the ρ field is essential for three-phase coexistence between the water, the oil, and the disordered phases. In Eq. (3.1), the effective field H represents the difference in the chemical potentials of water and oil, and μ represents the chemical potential of surfactants. All the parameters of the free energy functional of Eq. (3.1), except r and H, are always positive. The phase diagram corresponding to this free energy functional has been calculated by Ohta et al. [95], and it shows the same features as the spin 1 model [49].

The second part of the free energy describes the effect of surfactants at the water–oil interfaces and can be written as follows:

$$\mathscr{F}_2\{\psi, \rho\} = -\int d\mathbf{r}\, (\rho\nabla\psi) \cdot \left[\frac{c'}{2}\nabla\psi + \frac{h'}{2}\nabla(\nabla^2\psi) \right] \tag{3.2}$$

Here we assume that surfactants, adsorbed at an interface, are oriented *on the average* along the normal of this interface, which is given by $\nabla\psi(\mathbf{r})$. This assumption is a reasonable approximation since we are dealing with a coarse-grained model in which the orientations of single surfactant molecules are averaged out. The expression in the square bracket of the right-hand side of Eq. (3.2) gives the expansion of the gradient of ψ to lowest orders. The first term leads to a decrease in the interfacial tension between water and oil domains when the interface is occupied by surfactants, and the second term relates to the bending elasticity of surfactant films. This term is therefore responsible for the occurrence of liquid crystalline phases such as the lamellar and hexagonal phases, as well as the microemulsion regime. Finally, we assume that the surfactant films have no spontaneous curvature. Thus, in the case of equal average concentrations of water and oil, the chemical potentials of water and oil can be taken to be equal, that is, $H = 0$. The total free energy functional is then given by

$$\mathcal{F}\{\psi, \rho\} = \mathcal{F}_1\{\psi, \rho\} + \mathcal{F}_2\{\psi, \rho\} \tag{3.3}$$

As in the lattice model, we limit ourselves to the case of equal average concentrations of water and oil.

C. Mean-Field Analysis

1. *Phase Diagram*

In this section, we present a mean-field analysis of the model described by the free energy functional (3.3). The phase diagram is obtained by minimizing the free energy for a given set of system parameters. The functional derivatives of $\mathcal{F}$ with respect to ψ and ρ fields can be derived from Eqs. (3.1–3.3):

$$\begin{aligned}\frac{\delta\mathcal{F}}{\delta\psi} = & -r\psi + u\psi^3 + g\rho^2\psi - (c - c'\rho)\nabla^2\psi + c'\nabla\rho\cdot\nabla\psi \\ & + (h + h'\rho)\nabla^4\psi + \frac{h'}{2}\nabla^2\rho\nabla^2\psi + \frac{3h'}{2}\nabla\rho\cdot\nabla(\nabla^2\psi) \\ & + \frac{h'}{2}\nabla\cdot[(\nabla\rho\cdot\nabla)\nabla\psi + (\nabla\psi\cdot\nabla)\nabla\rho]\end{aligned} \tag{3.4}$$

and

$$\frac{\delta\mathcal{F}}{\delta\rho} = a\rho + g\rho\psi^2 - \frac{c'}{2}(\nabla\psi)^2 - \frac{h'}{2}\nabla\psi\cdot\nabla(\nabla^2\psi) - \mu \tag{3.5}$$

Both derivatives vanish at equilibrium.

We first examine the homogeneously ordered phases. These correspond to the water- and oil-rich phases, as well as the disordered phase. For homogenous phases all the terms in Eqs. (3.4) and (3.5) that involve gradients are set to zero. We then obtain the following homogeneous mean-field equations:

$$\frac{\delta \mathcal{F}}{\delta \psi} = -r\psi + u\psi^3 + g\rho^2\psi = 0 \tag{3.6}$$

and

$$\frac{\delta \mathcal{F}}{\delta \rho} = a\rho + g\rho\psi^2 - \mu = 0 \tag{3.7}$$

In order to obtain the transition lines between the water–oil coexistence region and the disordered phase, ρ is expanded in terms of ψ up to sixth order using Eq. (3.7). Substituting the result into Eq. (3.6) we obtain the following expression for the free energy density of the homogeneous phases:

$$f(\psi) = \left(-\frac{r}{2} + \frac{g\mu^2}{2a^2}\right)\psi^2 + \left(\frac{u}{4} - \frac{g^2\mu}{2\alpha^3}\right)\psi^4 + \frac{g^3\mu^2}{2a^4}\psi^6 + \mathcal{O}(\psi^8) \tag{3.8}$$

Equation (3.8) has the required form for systems exhibiting tricritical behavior. Note that the term $(g/2)\rho^2\psi^2$ of Eq. (3.1) is essential for obtaining this form since the lower order term $(g'/2)\rho\psi^2$, would only give a term in ψ^4 in the free energy density that does not describe tricritical behavior. Although the presence of the higher order term usually requires the inclusion of the lower order term $(g'/2)\rho\psi^2$, it is straightforward to show that a change $\rho' = \rho + (g'/2g)$ leads to exactly the same form as found in Eqs. (3.1) and (3.2). We can therefore omit this term without altering the model.

A coexistence region of the water and oil phases occurs when the prefactor of ψ^2 is negative. A second-order transition separates the water–oil coexistence from the disordered phase when this prefactor vanishes while the prefactor of ψ^4 remains positive, that is, when

$$r = \frac{g\mu^2}{a^2}$$

A tricritical point occurs when the prefactors of ψ^2 and ψ^4 simultaneously vanish, which leads to $(\mu_{trc}, r_{trc}) = (\sqrt{\frac{ua^3}{2g^2}}, \frac{ua}{2g})$. Furthermore, the remainder of the transition line is a triple line along which the water, the oil, and the disordered phases coexist. This line is obtained by equating the free energies of the disordered and the water or the oil phases. The equation of the triple line is therefore,

$$r = \frac{g\mu^2}{a^2}\left[1 - \left(\frac{1}{2} - \frac{ua^3}{4\mu^2 g^2}\right)^2\right]$$

We have so far only considered homogeneous phases, but we are cognizant of the fact that ternary mixtures of water, oil, and surfactants exhibit long-range modulated phases such as lamellar and other liquid crystalline phases. When the concentrations of water and oil are comparable, the most probable phase would be the lamellar phase.[2] In this case, the system exhibits oscillations with a nonzero wavenumber q_0 which is obtained by linearizing the free energy in Fourier space, and which corresponds to (see below)

$$q_0 = \sqrt{\frac{c'\dfrac{\mu}{a} - c}{2\left(h'\dfrac{\mu}{a} + h\right)}} \tag{3.9}$$

It is of interest to note that q_0 increases monotonically as the surfactant chemical potential increases, implying that the wavelength decreases as the surfactant concentration increases. The simplest form for the ψ field is $\psi(\mathbf{r}) = \psi_0 \cos(q_0 x)$ assuming that the modulations are along the x direction. Therefore, the ρ field is also sinusoidal but with one-half the period of that of ψ,[3] that is,

$$\rho(\mathbf{r}) = \rho_0 - \rho_1 \cos(2q_0 x) \tag{3.10}$$

with

$$\rho_0 = \frac{\mu}{a} + \alpha\psi_0^2 + \beta\psi_0^4$$

[2] For equal concentrations of water and oil, a cubic phase might be observed as was seen in Section II, where it is located for higher surfactant concentrations. However, we have not detected this phase in our Ginzburg–Landau calculations. We therefore do not consider its stability here.

[3] The ρ field has in principle higher mode terms. For simplicity, we only retained the lowest order mode, that is, the one with $q = 2q_0$.

and

$$\rho_1 = \gamma\psi_0^2 + \delta\psi_0^4$$

and the constants α, β, γ, and δ are given by

$$\alpha = \frac{1}{a}\left(-\frac{g\mu}{2a} + \frac{q_0^2 c'}{4} - \frac{h' q_0^4}{4}\right)$$

$$\beta = \frac{1}{a}\left(\frac{3\mu d^2}{8a^2} - \frac{gc' q_0^2}{16a} + \frac{gh' q_0^4}{16a}\right)$$

$$\gamma = \frac{1}{a}\left(\frac{g\mu}{2a} + \frac{c' q_0^2}{4} - \frac{h' q_0^4}{4}\right)$$

and

$$\delta = -\frac{g^2\mu}{2a^3}$$

after integration over one period, we obtain the following free energy density in the lamellar phase

$$f_{\text{lam}} = \left[-\frac{r}{2} + \frac{g}{4}\left(\rho_0^2 + \frac{\rho_1^2}{2} - \rho_0\rho_1\right) + q_0^2\left(c - c'\rho_1 - \frac{c'}{2}\rho_1\right) + \frac{q_0^4}{4}\left(h + h'\rho_0 + \frac{h'}{2}\rho_1\right)\right]\psi_0^2 + \frac{3u}{32}\psi_0^4 + \frac{g}{2}\left(\rho_0^2 + \frac{\rho_1^2}{2}\right) - \mu\rho_0 \tag{3.11}$$

In order to find the transition lines between the lamellar phase and the water–oil coexistence region or the disordered phase, we computed the free energy densities of the various phases and then compared them. The equilibrium phase is the phase corresponding to the minimum value of the free energy.

In Figs. 3.1 and 3.2, we show two phase diagrams predicted from the model for two sets of parameters. Figure 3.1 corresponds to the case when $u = 1$ and Fig. 3.2 corresponds to $u = 0.15$. The other parameters are the same for both figures and correspond to $a = 2$, $g = 0.5$, $c = 2$, $c' = 2$, $h = 0$, and $h' = 0.5$. When u is large, we do not find a phase coexistence between the water, the oil, and the disordered phases. In this case, the Lifshitz and disorder lines meet at a multicritical Lifshitz point where the water, oil, lamellar, and disordered phases occur simultaneous-

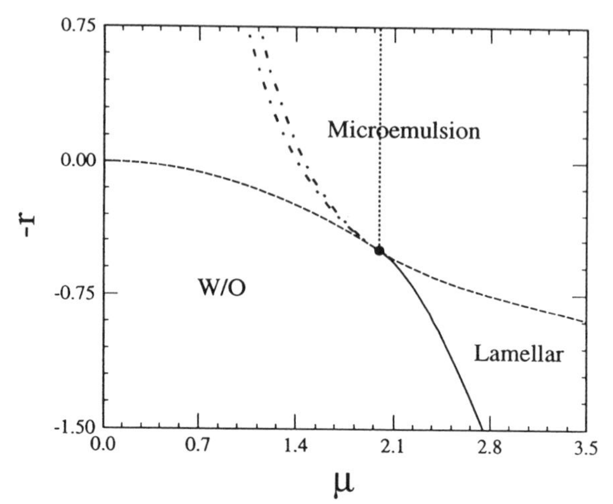

Figure 3.1. Phase diagram of the Ginzburg–Landau model corresponding to $u = 1$. The dashed lines are second-order lines, the solid line is a first-order line, the dotted line is the Lifshitz line, the left and the right dot–dashed lines are, respectively, the 3- and the 2-D disorder lines, and the dot gives the location of the Lifshitz multicritical point.

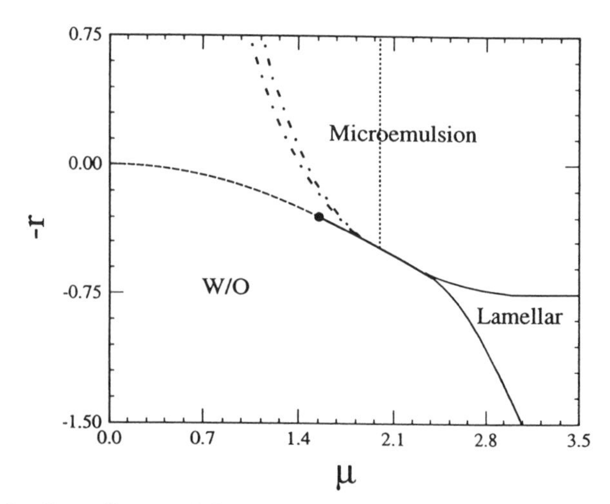

Figure 3.2. Phase diagram of the Ginzburg–Landau model corresponding to $u = 0.15$. See Fig. 3.1 for the explanation of the lines. The dot gives the location of the tricritical point.

ly. This point has many interesting features, as discussed for the case of diblock copolymer surfactants mixed with a binary blend of two immiscible homopolymers [6]. The locus and the nature of this transition is confirmed by the divergence in the peak of the structure factor. It is interesting to note that the phase diagram of Fig. 3.1 closely resembles that found by Holyst and Schick [6] for ternary mixtures of two homopolymers and diblock copolymers. This result suggests that a large value of u is also appropriate to describe surfactants such as diblock copolymers. In contrast, Fig. 3.2 shows that, when u is small, there is a three-phase coexistence between water, oil, and the disordered phases along a triple line. This is the usual case found for ternary mixtures of water, oil, and amphiphilic surfactants. In this case, we found that the transition from the lamellar to the disordered phase is first order. The transition from the water–oil coexistence to the lamellar phase is also first order. In Figs. 3.1 and 3.2, the dot–dashed lines represent the disorder line and the dotted line represent the Lifshitz line. These lines separate the simple disordered fluid from the microemulsion. The calculational details of these lines and their nature will be discussed in further detail in Section III.D.

D. Structure Factor

As seen in Section II, the microemulsion can be distinguished from simple (unstructured) fluids by the presence of short-range correlations. In particular, for bicontinuous microemulsions, the position of the peak increases and the height of the peak decreases with increasing surfactant concentration.

The structure factor is obtained by expanding the ψ field in Fourier space about its equilibrium value $\psi_0 = 0$ in the disordered phase. The Fourier transform of the free energy in this phase can be written as follows in terms of ψ when the ρ field has been integrated out:

$$\mathcal{F} = \int d\mathbf{q}\, \psi(\mathbf{q})\psi(-\mathbf{q})(A + Bq^2 + Cq^4) \tag{3.12}$$

The water–water structure factor is then given by

$$S_{ww}(q) = \langle \psi(\mathbf{q})\psi(-\mathbf{q}) \rangle = \frac{T}{A + Bq^2 + Cq^4} \tag{3.13}$$

which has the same form as proposed by Teubner and Strey [13] to describe bicontinuous microemulsions. The coefficients A, B, and C are given by

$$A = -\frac{r}{2} + \frac{g\mu^2}{2a^2}$$

$$B = \frac{1}{2}\left(c - c'\frac{\mu}{a}\right)$$

$$C = \frac{1}{2}\left(h + h'\frac{\mu}{a}\right)$$

When the prefactor of q^2, B, is positive, the structure factor exhibits a peak at $q = 0$ as usually observed in simple fluids. However, when B becomes negative, a peak in the structure factor appears at $q_0 > 0$, with q_0 is given by Eq. (3.9). The regime with $q_0 > 0$ can be identified as a microemulsion and the line separating the two regimes is the Lifshitz line shown as a dotted line in Figs. 3.1 and 3.2. We note that in this model the Lifshitz line does not depend on temperature.

In Fig. 3.3, we display the water–water structure factor in the microemulsion regime for $u = 0.15$ and $r = 0.5116$ just after the triple line. Beyond the Lifshitz line, the position of the peak moves to larger values while the height of the peak decreases, as the surfactant con-

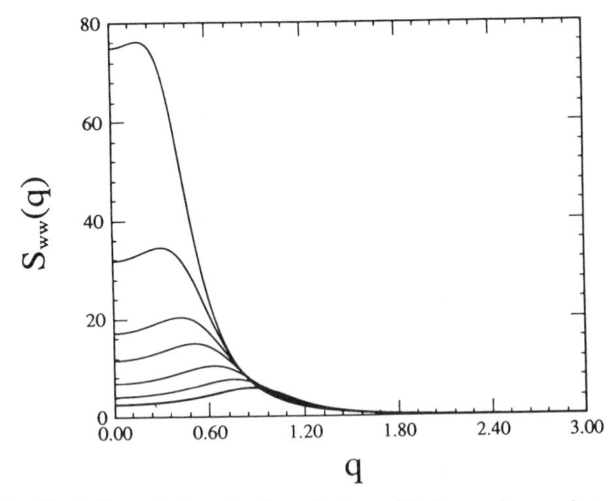

Figure 3.3. Evolution of the structure factor with increasing surfactant chemical potential right after the triple line at $r = 0.5116$. The curves from upper to lower correspond to $\mu = 2.03, 2.1, 2.2, 2.3, 2.5, 2.8, 3.2$, respectively. The other parameters are the same as those for Fig. 3.1.

centration is increased. This result implies a decrease in the size of the water and oil microdomains. This implication is reasonable since an increase in the surfactant concentration tends to create more interfaces, and therefore smaller domains, as discussed earlier. As the temperature increases the structure factor becomes broader, indicating an increase in disorder.

As seen in Section II, the definition of the bicontinuous microemulsion using the water–water structure factor can be used for comparison with experiment, since structure factors are proportional to the corresponding small angle neutron scattering intensity. We can also use the real space correlation function to define microemulsions. The water–water correlation function [54] is given by the inverse Fourier transform of $S_{ww}(q)$. The isotropic nature of the disordered phase allows us to obtain the correlation function in a simple manner. In three dimensions, we find that if $(B/A)^2 - 4(C/A) > 0$, the correlation function at large distances behaves as

$$G_{ww}(\mathbf{r}) \sim \frac{1}{r} \exp\left(-\frac{r}{\xi}\right)$$

where the correlation length is given by

$$\xi = \frac{\sqrt{2}}{[(B/C) - \sqrt{(B/C)^2 - 4(A/C)}]^{1/2}}$$

However, when $(B/A)^2 - 4(C/A) < 0$, the correlation function becomes

$$G_{ww}(\mathbf{r}) \sim \frac{1}{\tau} \exp\left(-\frac{r}{\xi}\right) \sin\left(2\pi \frac{r}{\lambda}\right)$$

where the two length scales ξ and λ are given by

$$\xi = \frac{\sqrt{2}}{(\sqrt{A/C} + (B/2C)]^{1/2}}$$

and

$$\lambda = \frac{2\pi\sqrt{2}}{[\sqrt{A/C} - (B/2C)]^{1/2}}$$

respectively. This shows that the correlation function can exhibit long-wavelength oscillations, even if the structure factor has a peak at $q = 0$, that is, when $B > 0$. However, these oscillations have a period that is larger than the correlation length ξ, and therefore are very difficult to be detected in experiments or in numerical simulations. A change from the nonoscillatory regime to the oscillatory regime occurs at the disorder line, the equation of which is given by

$$r = \frac{g\mu^2}{a^2} + \frac{\left(c - c'\frac{\mu}{a}\right)^2}{4\left(h + h'\frac{\mu}{a}\right)}$$

The disorder line depends on the spatial dimension d, whereas the Lifshitz line is independent of d. The disorder lines in two and three dimensions are shown as the left and right dot–dashed lines, respectively, in Figs. 3.1 and 3.2. We observe that the disorder line in two dimensions is slightly to the right of the disorder line in three dimensions. However, the two lines become almost identical when the phase transition from the water–oil to the disordered phase is approached.

When the phase diagram exhibits three-phase coexistence between the water, oil, and the disordered phase, the disorder line meets the triple line at a smaller chemical potential than the Lifshitz line. By contrast, when a Lifshitz multicritical point is present, the Lifshitz and the disorder lines meet at this point. It is important to note that the Lifshitz and the disorder lines are not phase transition lines in the thermodynamic sense since no singularities are encountered in the free energy along these lines.

E. Effect of Fluctuations

The mean-field analysis presented in Section III.C neglects thermal fluctuations. This is usually a good approximation sufficiently far away from phase transitions. For systems involving surfactants, a large number of fluctuating interfaces may exist due to a significant reduction of the interfacial tension, as mentioned in Section II for the lattice model. It is thus important to find out if thermal fluctuations alter the mean-field phase diagram to any significant degree. Our approach to this problem is to allow the system to evolve in time using Langevin equations derived from the free energy functional given by Eqs. (3.1)–(3.3). This will lead us to the equilibrium phases for a given system parameters, which can then be compared with those derived from the mean-field analysis. An additional advantage of this approach is that the dynamics can also be studied. Although this is not the focus of the present section (see Section IV), it is of interest to study the dynamical evolution of the system. In this section, we also analytically consider the stability of the interfaces in the lamellar phase, using an approach similar to previous work [96]. We show that, close to the transition from water–oil coexistence region to the lamellar phase, the lamellar phase becomes unstable against the formation of a microemulsion. This finding is confirmed by the Langevin calculations.

1. *Solution of the Langevin Equations*

The Langevin equations for this model are derived from the free energy functional given by Eqs. (3.1) and (3.2),

$$\frac{\partial\psi(\mathbf{r},t)}{\partial t} = -M_\psi \frac{\delta\mathcal{F}}{\delta\psi(\mathbf{r},t)} + \eta_\psi(\mathbf{r},t) \tag{3.14}$$

and

$$\frac{\partial\rho(\mathbf{r},t)}{\partial t} = -M_\rho \frac{\delta\mathcal{F}}{\delta\rho(\mathbf{r},t)} + \eta_\rho(\mathbf{r},t) \tag{3.15}$$

where η_ψ and η_ρ are thermal noise terms that satisfy the fluctuation–dissipation relation

$$\langle\eta(\mathbf{r},t)\eta(\mathbf{r},t')\rangle = 2TM\delta(\mathbf{r}-\mathbf{r})\delta(t-t') \tag{3.16}$$

The two functional derivatives of the free energy in Eqs. (3.14) and (3.15) are given by Eqs. (3.4) and (3.5), respectively. The Boltzmann constant has been set to unity. The kinetic coefficients M_ψ and M_ρ usually depend on the conservation laws for the two concentration fields. Since in this section we only concentrate on the equilibrium phase behavior that is independent of the dynamical evolution of the system; we used the fastest dynamics leading to equilibrium. This corresponds to model A in critical dynamics [97] where M_ψ and M_ρ are both positive constants. We note that model A does not describe the true dynamics in these ternary mixtures since in real mixtures, the concentrations of the three components should remain constant during the dynamics. A realistic study of the dynamics thus requires the use of conserved Langevin equations as will be seen in Section IV.

We first consider the dynamics at the very early times. We begin by taking the initial condition for the two fields to be $\psi(\mathbf{r},t=0)=0$ and $\rho(\mathbf{r},t=0)=\rho_0$, which can be different from zero. This finding could, for example, correspond to a state at an infinite temperature at which the system is completely disordered. We then follow the early dynamics after a quench at a low temperature (i.e., within an ordered phase). We also assume that the two fields are initially subject to infinitesimal fluctuations perhaps induced by thermal noise. We can then write

$$\psi(\mathbf{r}, t) = \frac{1}{(2\pi)^d} \int \psi_{\mathbf{q}} \exp[i\mathbf{q} \cdot \mathbf{r} + \Omega(\mathbf{q})t]$$

and

$$\rho(\mathbf{r}, t) = \frac{1}{(2\pi)^d} \int \rho_{\mathbf{q}} \exp[i\mathbf{q} \cdot \mathbf{r} + \Omega(\mathbf{q})t]$$

If we retain only the linear terms in ψ and ρ, we obtain the following linear equation for the ρ field when $\mathbf{q} \neq 0$ using Eq. (3.15)

$$\frac{\partial \rho_{\mathbf{q}}}{\partial t} = - M_\rho a \rho_{\mathbf{q}}$$

The ρ field will then decay as follows:

$$\rho_{\mathbf{q}}(t) = \rho_{\mathbf{q}}(0) \exp(- M_\rho a t)$$

When $q = 0$, we have

$$\rho_{\mathbf{q}=0}(t) = \frac{\mu}{a} + \rho_0 \exp(- M_\rho a t)$$

This implies that all fluctuations in the surfactant field decay exponentially leading to a homogeneous distribution of surfactants.

The ψ field will now have the following time dependence,

$$\psi_{\mathbf{q}}(t) = \psi_{\mathbf{q}}(0) \exp[\Omega(\mathbf{q})t]$$

where the dispersion relation is given by

$$\Omega(\mathbf{q}) = -M_\psi \left[\left(-r + g \frac{\mu^2}{a^2} \right) + \left(c - c' \frac{\mu}{a} \right) q^2 + \left(h + h' \frac{\mu}{a} \right) q^4 \right] \quad (3.17)$$

The parameter $\Omega(\mathbf{q})$ has a peak at $q = 0$ when $\mu < ac/c'$, but peaks at a finite wavenumber given by Eq. (3.9) when $\mu > ac/c'$, as shown in Fig. 3.4. The results above show that the ρ field is completely decoupled from the ψ field at very early times and that the ψ field grows exponentially for large wavelengths but decays exponentially for short ones. Another interesting result is that there is a wide region of the water–oil coexistence in the phase diagram (see Figs. 3.1 and 3.2), where the linear dispersion relation shows a finite peak implying that at early times the structure factor should exhibit a peak, whereas at late times and at

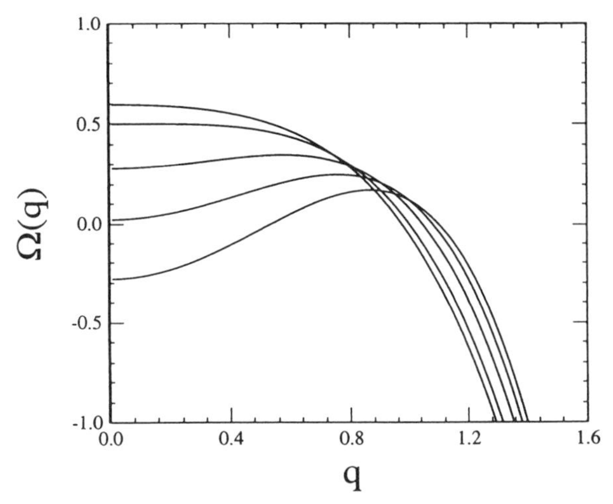

Figure 3.4. The linear dispersion relation versus q. The curves from upper to lower correspond, respectively, to $\mu = 1.8$, 2, 2.4, 2.8, and 3.2.

equilibrium, the peak should be at $q = 0$, since a homogeneous ordered phase is expected. This occurs in all regions of the water–oil coexistence with $\mu > ac/c'$. For the lamellar phase, the wavelength selected linearly is expected to remain the same at equilibrium.

We solved the Langevin equations given by Eqs. (3.14) and (3.15) numerically on a 2-D grid with the parameters corresponding to both Figs. 3.1 and 3.2 using the standard Euler technique and the finite difference scheme.

The structure factor can be calculated from the following expression:

$$S(\mathbf{q}, t) = \left\langle \frac{1}{L^2} \left| \sum_{\mathbf{r}_i} [\psi(\mathbf{r}_i, t) - \psi_0] \exp(i\mathbf{q} \cdot \mathbf{r}_i) \right|^2 \right\rangle \qquad (3.18)$$

where the summation is over the grid points and L is the grid size. The results are presented for the circularly averaged structure factor $S(q, t) = \Sigma' S(\mathbf{q}, t)/\Sigma' 1$, where the Σ' denotes a sum over circular shells defined by $n - \frac{1}{2} \leq qL/(2\pi) \leq n + \frac{1}{2}$.

The time evolution of the structure factor is shown in Figs. 3.5 and 3.6 for the ψ field, when $\mu = 2.3$. This value of μ falls within the water–oil coexistence but is close to the transition to the lamellar phase. We observe a finite peak at very early times with a value close to that predicted by the linear theory. This peak shifts towards smaller wavenumbers as time increases, and then completely disappears at late times. The structure factor narrows at late times, indicating that the system orders

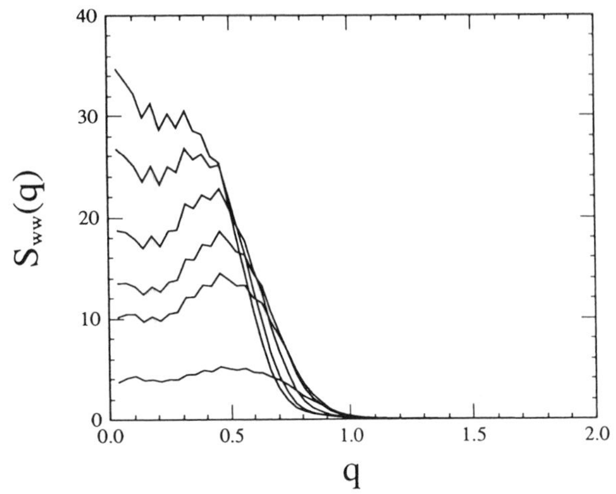

Figure 3.5. The early time evolution of the structure factor within the water–oil coexistence for $\mu = 2.3$ and $r = 1$. The curves from lower to upper correspond to $t = 10$, 20, 30, 60, 110, and 160, respectively.

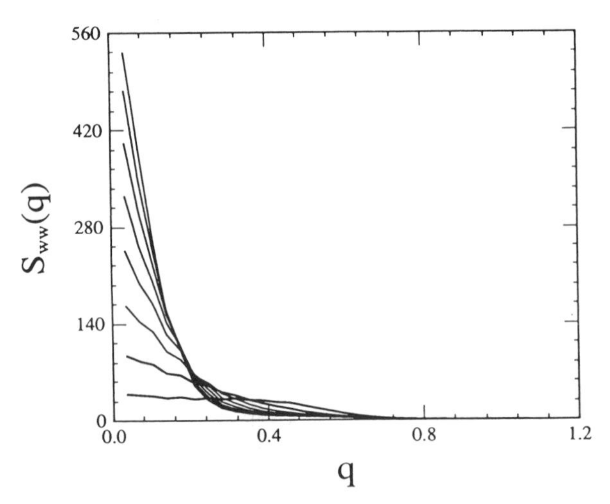

Figure 3.6. The intermediate and late times evolution of the structure factor within the water–oil coexistence for $\mu = 2.3$ and $r = 1$. Curves from lower to upper correspond to $t = 180$, 440, 700, 960, 1220, 1480, 1740 and 2000, respectively.

towards a homogeneous equilibrium state corresponding to either water or oil phases. The average domain size is then calculated from the second moment of the structure factor,

$$R(t) = \frac{2\pi}{q_2(t)} = 2\pi\sqrt{\frac{\sum_{q=0}^{q_c} S(q,t)}{\sum_{q=0}^{q_c} q^2 S(q,t)}} \tag{3.19}$$

In Fig. 3.7, we show this average domain size as a function of time for $\mu = 2.3$ and $r = 1$. At early times, when the structure factor exhibits a finite peak, the domains size grows very slowly. However, when the peak disappears, the average domains size grows fast and algebraically, $R(t) \sim t^n$, where the exponent n is very close to $\frac{1}{2}$. Allen and Cahn [98, 99] showed that the average domain size grows algebraically with an exponent equal to $\frac{1}{2}$ during equilibration to a homogeneous phase in systems with a nonconserved order parameter. Since our asymptotic value is very close to this prediction, we conclude that the system in this region of the phase diagram equilibrates to a homogeneous state corresponding to a water or an oil phase. The Langevin approach thus agrees with the mean-field predictions in this part of the phase diagram.

We also performed Langevin calculations in the lamellar region of the mean-field phase diagram. Figure 3.8 gives the spatial dependence of the ψ field and the ρ field for $u = 1$, $r = 1$, and $\mu = 2.5$, that is, very close to the water–oil coexistence. This figure clearly shows that the equilibrium

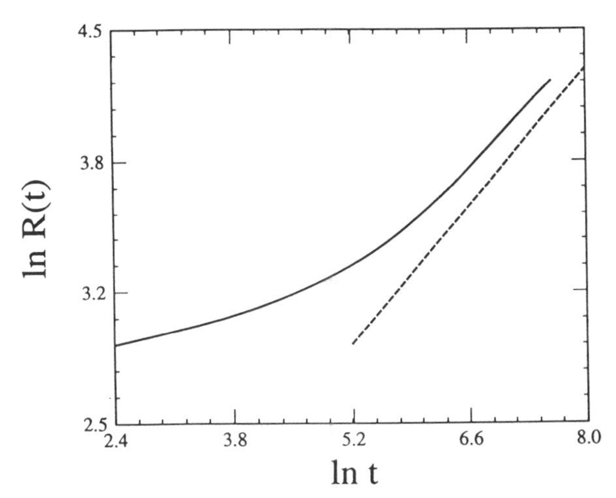

Figure 3.7. The time dependence of the average domain size for $\mu = 2.3$ and $r = 1$. The other parameters are the same as in Fig. 3.2. The dashed line has a gradient of $\frac{1}{2}$.

Figure 3.8. The left configuration shows the spatial dependence of the ψ field at $u = 1$, $r = 1$, and $\mu = 2.5$. The right configuration shows the corresponding distribution of surfactants. The surfactants have a higher density at the interfaces and a lower density in the bulk.

state is disordered implying that the lamellar phase is unstable against the microemulsion in this region of the phase diagram.

We then performed Langevin simulations deep inside the lamellar phase. Figure 3.9 shows the spatial distribution of the ψ field and the surfactant field for $u = 1$, $r = 1$, and $\mu = 3.1$. In contrast to the case of $\mu = 2.5$, we observe a well-correlated lamellar structure, implying that the Langevin calculations are in agreement with the mean-field predic-

Figure 3.9. The left configuration shows the spatial dependence of the ψ field at $u = 1$, $r = 1$, and $\mu = 3.1$. The right configuration shows the corresponding distribution of surfactants.

tions in this region. However, we note that the lamellae are not perfect as predicted by the mean-field approximation. The Langevin simulations therefore suggest the existence of a region of the lamellar phase, which becomes unstable against the microemulsion when fluctuations are included. In Section III.E.2 we predict this region of instability from a stability analysis of the lamellar interfaces. The structure factor for the case of $\mu = 2.5$ exhibits a broad peak at a nonzero value of q indicating the microemulsion character. By contrast, the structure factor for the case of $\mu = 3.1$ exhibits a narrow Bragg-like peak at the value of q corresponding to the inverse separation of the stripes thus showing the long-range order expected for a lamellar phase. In both cases, we found that the position of the peak is the same as predicted by linear stability analysis. We also found that the position of the peak of the surfactant–surfactant structure factor is twice that of the water–water structure factor. This finding is reasonable since all surfactants are correlated within a distance that is twice that of water or oil domains.

The equilibration was found to be very fast in the disordered region of the mean-field phase diagram. Configurations for this region close to the triple line are shown in Figs. 3.10–3.12 for parameters corresponding to the phase diagram in Fig. 3.2. The corresponding simulations were performed in the three regions of the disordered phase, that is, to the left of the disorder line (Fig. 3.10), between the disorder line and the Lifshitz line (Fig. 3.11), and to the right of the Lifshitz line (Fig. 3.12). The

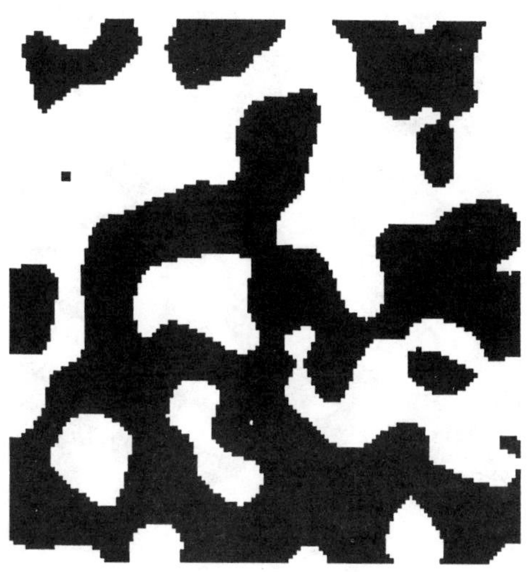

Figure 3.10. A configuration obtained from Langevin simulations in the disordered phase at $(r, \mu) = (1.6, 0.3)$.

Figure 3.11. A configuration obtained from Langevin simulations in the disordered phase at $(r, \mu) = (1.9, 0.42)$.

Figure 3.12. A configuration obtained from Langevin simulations in the disordered phase at $(r, \mu) = (2.2, 0.42)$.

structure is completely disordered before the disorder line is reached, with the structure factor peaks at $q = 0$ as expected for simple fluids. Almost the same behavior is observed between the disorder and the Lifshitz line. Beyond the Lifshitz line, the fluid exhibits short-range order, as expected in microemulsions. In these regions the Langevin simulations again agree with the mean-field calculations.

2. *Stability of the Lamellar Phase*

We now study the stability of the lamellar interfaces with respect to infinitesimal fluctuations. To this end, we describe the ψ field in terms of an oscillating profile with a single spatial frequency $\psi(\mathbf{r}) = \psi_0 \cos(q_0 x)$, as mentioned in Section III.A. The surfactant field is also assumed to have the same type of profile as seen in Eq. (3.10). We assume for simplicity that ψ only varies along the x direction.

Assuming the function $u(\mathbf{r}, t)$ to be the displacement of an interface around its equilibrium position as predicted by the mean-field theory (see Section III.B). The ψ field can then be written as $\psi(\mathbf{r}, t) = \psi_0 \cos[q_0(x + u(\mathbf{r}, t))]$ and the ρ field as $\rho(\mathbf{r}, t) = \rho_0 - \rho_1 \cos[2q_0(x + u(\mathbf{r}, t))]$. After integrating out the ρ field, as in Section III.A, we obtain the following differential equation for the displacement $u(\mathbf{r}, t)$:

$$\frac{\partial u}{\partial t} = A_x \frac{\partial^2 u}{\partial x^2} + A_z \frac{\partial^2 u}{\partial z^2} - D\nabla^4 u \tag{3.20}$$

where A_x, A_z, and D are given by

$$A_x = \left(c - c'\frac{\mu}{a}\right) + 6q_0^2\left(h + h'\frac{\mu}{a}\right) - [c'(\alpha + \gamma) - 6q_0^2 h'(\alpha + 3\gamma)]\psi_0^2 \\ - [c'(\beta + \delta) - 6q_0^2 h'(\beta + 3\delta)]\psi_0^4 \tag{3.21}$$

$$A_z = \left(c - c'\frac{\mu}{a}\right) + 2q_0^2\left(h + h'\frac{\mu}{a}\right) - [c'(\alpha + \gamma) - 2q_0^2 h'(\alpha + 3\gamma)]\psi_0^2 \\ - [c'(\beta + \delta) - 2q_0^2 h'(\beta + 3\delta)]\psi_0^4 \tag{3.22}$$

and

$$D = \left(h + h'\frac{\mu}{a}\right) + h'(\alpha + \gamma)\psi_0^2 + h'(\beta + \delta)\psi_0^4$$

After somewhat lengthy calculations, we obtained the following free energy in the lamellar phase in terms of the displacements of the lamellar interfaces $u(\mathbf{r}, t)$:

$$\mathscr{F}\{u(\mathbf{r})\} = \int dx\, dz \left[\frac{A_x}{2}\left(\frac{\partial u}{\partial x}\right)^2 + \frac{A_z}{2}\left(\frac{\partial u}{\partial z}\right)^2 + \frac{D}{2}(\nabla^2 u)^2\right] \quad (3.23)$$

Assuming that the fluctuations are very small, one can study the growth of different modes in the linear approximation. This can be done by writing

$$u(\mathbf{r}, t) = \int dq_x\, dq_z\, u_{\mathbf{q}} e^{i\mathbf{q}\cdot\mathbf{r} + \Omega_u(\mathbf{q})t}$$

Using Eq. (3.20) we obtain the following dispersion relation for the u field[4]

$$\Omega_u(q_x, q_z) = -A_x q_x^2 - A_z q_z^2 - Dq^4 \quad (3.24)$$

The dispersion relation along the x axis is then given by

$$\Omega u(q_x) = -A_x q_x^2 - Dq_x^4 \quad (3.25)$$

and the dispersion relation along the z axis is given by

$$\Omega_u(q_z) = -A_z q_z^2 - Dq_z^4 \quad (3.26)$$

We found that $\Omega_u(\mathbf{q})$ is always negative along the x direction, which is the direction perpendicular to the lamellar interfaces, for all chemical potentials within the lamellar phase. This is shown in Fig. 3.13 and implies that the lamellae are linearly stable along this direction. Figure 3.14 shows that the behavior of the dispersion relation in the z direction depends on the value of μ. In particular, $\Omega_u(\mathbf{q})$ is always negative for $\mu > \mu_0$, where μ_0 is temperature dependent. However, when $\mu < \mu^0$, $\Omega_u(\mathbf{q})$ is positive for small wavenumbers and negative for large wavenumbers. This finding implies that small fluctuations at the interfaces grow along the z direction for $\mu < \mu_0$, thereby destroying the long-range order. This leads to a disordered microemulsion phase. In contrast, the fluctuations decay for $\mu < \mu_0$, and therefore the lamellar phase is stable in this region (we recognize that the lamellar phase is expected to have only quasilong-range order as usually observed in layered systems in two dimensions [96, 100]). The behavior of $\Omega_u(\mathbf{q})$ then allows us to calculate the new transition line between the microemulsion and the lamellar phase

[4] Note that this dispersion relation is different from that of the ψ field given by Eq. (3.17).

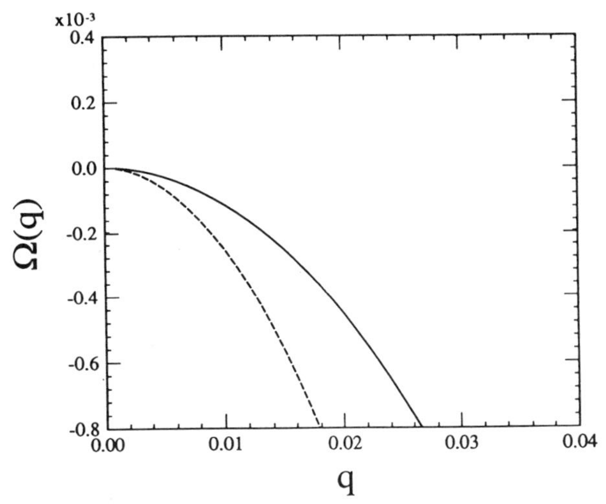

Figure 3.13. The dispersion relation for $u(\mathbf{r}, t)$ along the x axis. The solid line corresponds to $\mu = 2.5$ and the dashed line corresponds to $\mu = 3.1$. The parameters are the same as for Fig. 3.1.

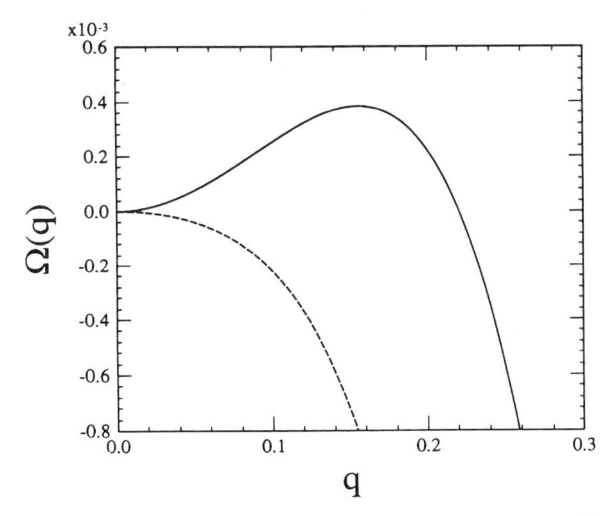

Figure 3.14. The dispersion relation for $u(\mathbf{r}, t)$ along the z axis. The solid line corresponds to $\mu = 2.5$ and the dashed line corresponds to $\mu = 3.1$. The parameters are the same as for Fig. 3.1.

at low temperatures. This line corresponds to the set of points where $\Omega_u(\mathbf{q})$ first exhibits a peak at nonzero values of q along the z direction. This agrees with the Langevin calculations. The result that the lamellar phase becomes unstable against the microemulsion close to the water–oil coexistence when fluctuations are included has been discussed within the lattice models in Section II.[5]

F. Interfacial Behavior along the Triple Line

Recent experimental work provides evidence for the existence of a wetting transition for intermediate chain length surfactants at, for example, the triple line between water, oil, and the disordered phases. It has been observed that this wetting transition changes character when the disordered fluid begins to exhibit short-range order, that is, when the simple fluid becomes a bicontinuous microemulsion [25, 27–29, 102]. Here we investigate the nature of wetting in ternary systems containing surfactants using the continuous model presented in Section III.B. This is achieved by solving the Langevin equations (3.14) and (3.15) on a square grid with fixed boundary conditions along the x axis. These equations were solved numerically for parameters corresponding to the triple line of Fig. 3.2. Two types of boundaries are considered: a water–oil interface, and a water–disordered interface.

Profiles of the water–oil interface corresponding to different values of μ along the triple line are shown in Figs. 3.15–3.17. These profiles give the change in value of the order parameter ψ across the interfaces from the water phase ($\psi > 0$) to the disordered phase ($\psi = 0$) and to the oil phase ($\psi < 0$). We observe that the interface between the water and the disordered phases has a monotonic profile if the disordered phase is a simple fluid, that is, before the disorder line is reached. However, small oscillations appear in the profile for values of μ beyond the disorder line where the disordered phase becomes structured. The oscillations become more pronounced as the temperature decreases and the microemulsion becomes more structures. Since $\psi > 0$ represents water, and $\psi < 0$ represents oil, a dip in ψ at the interface (see Figs. 3.16 and 3.17) indicates the presence of a thin oil layer in contact with the water phase. Likewise, a bump in the profile at the oil side indicates the presence of a

[5] Recently, Levin et al. [101] studied the effects of fluctuations on the mean-field phase diagram of the Widom's model [44] by a renormalization group (RG) method. In contrast to the mean-field calculations, which predict that the water, the oil, and the lamellar phase coexist along a triple line, the RG calculations show that the lamellar phase does not exist with water or oil allowing the disordered phase to extend to very low temperatures between the water–oil coexistence and the lamellar phase.

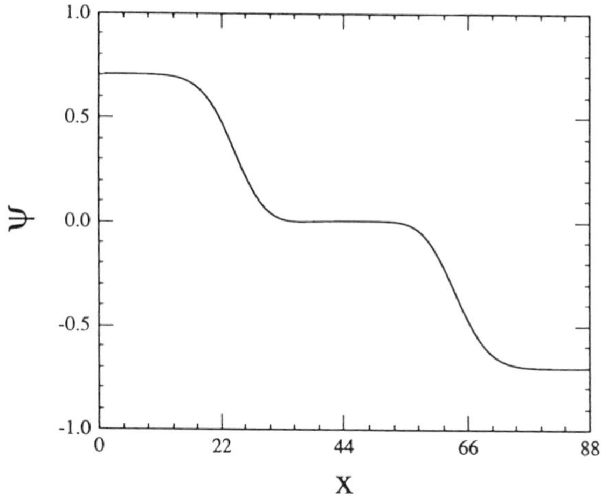

Figure 3.15. Profile of the ψ field along the x axis when the water, the disordered and the oil phases coexist along the triple line when $\mu = 1.6$. The other parameters are the same as for Fig. 3.2.

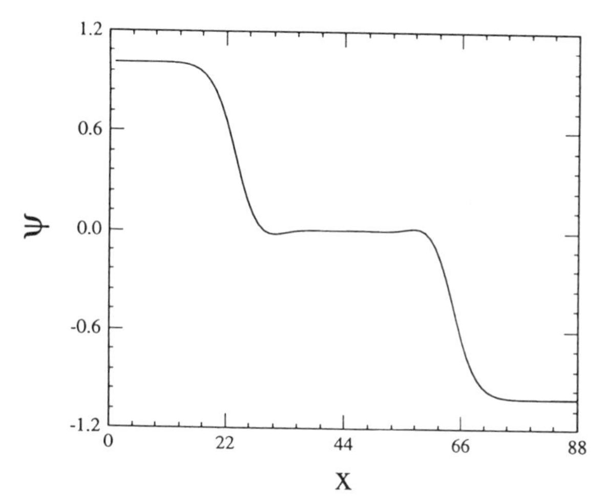

Figure 3.16. As for Fig. 3.15 but with $\mu = 1.9$.

thin layer of water in contact to the oil phase. Similar behavior was observed in previous studies [28, 29].

Wetting phenomena are usually discussed in terms of interfacial tension and contact angles. Let σ_{WO}, σ_{WD}, and σ_{OD} be the interfacial tensions of water–oil, water–disordered, and oil–disordered interfaces, respectively. As is usual for wetting phenomena [103], the disordered

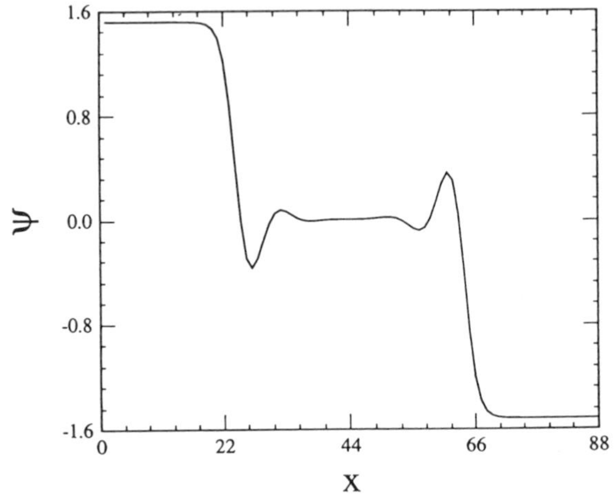

Figure 3.17. As for Fig. 3.16 but with $\mu = 2.3$.

phase wets the water–oil interface if $\sigma_{WO} = \sigma_{WD} + \sigma_{OD}$; if $\sigma_{WO} < \sigma_{WD} + \sigma_{OD}$, there is only a partial wetting. In our case water and oil are symmetric, and therefore $\sigma_{WD} = \sigma_{OD}$. Hence, complete wetting occurs when $\sigma_{WO} = 2\sigma_{WD}$.

If we assume that the system is uniform along the z direction, the interfacial tensions of water–oil and water–disordered interfaces can then be calculated by solving the Langevin equations numerically on a 2-D grid in the (x, y) plane. To this purpose we used periodic boundary condition along the x direction. The boundary conditions at the two boundaries in the y direction were fixed in such a way that they correspond to water and oil, respectively. The relevant interfacial tensions were then calculated in terms of the difference between the free energy for the system in the presence and in the absence of an interface, respectively, since σ_{WO} and σ_{WD} can be written

$$\sigma_{WO} = \frac{1}{L} (\mathcal{F}_{WO} - \mathcal{F}_W) \tag{3.27}$$

and

$$\sigma_{WD} = \frac{1}{L} (\mathcal{F}_{WD} - \mathcal{F}_W) \tag{3.28}$$

where L is the system size in the x direction. We observed that for

$\mu < 1.86$

$$\sigma_{WO} = 2\sigma_{WD} \tag{3.29}$$

implying a wetting regime. For $\mu > 1.86$, however, we found that

$$\sigma_{WO} < 2\sigma_{WD} \tag{3.30}$$

implying that the water–oil interface is only partially wetted by the disordered fluid. Since the disorder line meets the triple line at $\mu = 1.86$ (see Figs. 3.1), the wetting transition changes character from complete wetting to partial wetting when the disordered fluid becomes a microemulsion, as discussed above. We should note that for high temperatures when we are in the wetting regime, the interfacial tension is extremely small but positive. Such a small value of the interfacial tension might be due to the closeness of the system to the tricritical point.

The nature of this wetting transition can be described in terms of the contact angle [103] θ, as schematically shown in Fig. 3.18. The parameter θ is calculated from the following equation:

$$\theta = 2\cos^{-1}\left(\frac{\sigma_{WO}}{2\sigma_{WD}}\right) \tag{3.31}$$

where θ is plotted as a function of chemical potential in Fig. 3.19. For $\mu < 1.86$ this angle is zero indicating a complete wetting regime. The parameter θ increases rapidly by contrast after $\mu > 1.86$. Note that the contact angle should be a continuous quantity, since the interfacial

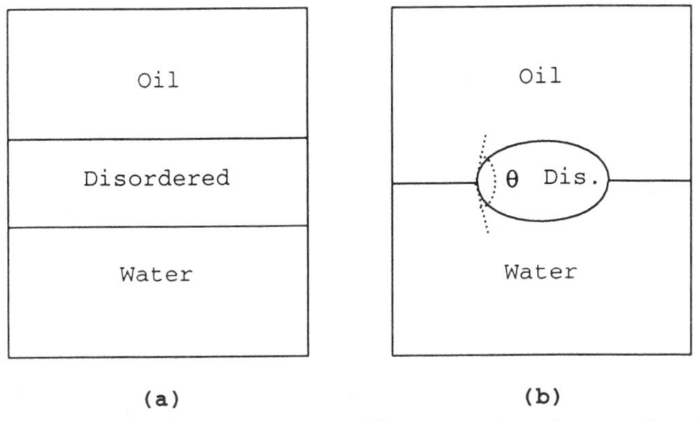

Figure 3.18. Three phases in coexistence: (*a*) corresponds to the case where the middle (disordered) phase wets the water–oil interface and (*b*) corresponds to the case where the disordered phase does not wet the water–oil interface.

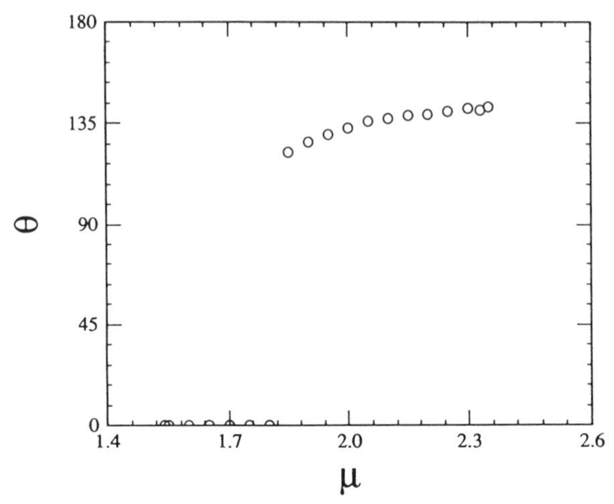

Figure 3.19. The contact angle versus the chemical potential along the triple line.

tension is always continuous. Therefore, the apparent jump in θ is in effect a rapid increase in the interfacial tension.

G. Conclusion

In this section we proposed and studied in detail a continuous model of ternary mixtures of water, oil, and surfactants for equal concentrations of water and oil. The model is based on two scalar local fields that represent the difference in the concentrations of water and oil and the concentration of surfactants. The mean-field phase diagram of the model shows the existence of four phases: the water, oil, lamellar, and disordered phases. The disordered phase possesses a microemulsion regime in a large region of the phase diagram. The water–water structure factor shows qualitatively the same behavior as that observed in experiments, that is, an increase in the position of the peak and a decrease in peak height when the surfactant concentration is increased. The effects of fluctuations on this model are analyzed via Langevin equations, which are derived from the free energy functional of the model and then iterated on a 2-D grid. The Langevin simulations show real space structures in both the disordered and the lamellar phases. The equilibrium states generated by the Langevin simulations are largely consistent with those from the mean-field analysis. However, in some parts of the lamellar phase region of the mean-field phase diagram, the Langevin iterations show that the lamellar phase becomes unstable against the microemulsion due to the inclusion of fluctuations. We also performed a linear stability of the interfaces in the lamellar phase. It was found that within the mean-field lamellar phase, the lamellae are unstable against infinitesimal fluctuations for small surfactant concentrations, that is, close to the water–oil coexistence. This finding implies that the microemulsion is the equilibrium structure in this region. For higher surfactant concentrations, the lamellar phase becomes stable. These findings are consistent with the calculations for the lattice model presented in Section II.

The Ginzburg–Landau model is extremely useful for the examination of a variety of phenomena that are difficult to examine using lattice models. For example, water–oil–surfactant systems often exhibit a hexagonal phase when the concentrations of water and oil are quite different. We therefore performed Langevin simulations for unequal concentrations of water and oil and we were able to find the hexagonal phase for relatively high surfactant chemical potentials as shown in Fig. 3.20. For lower surfactant concentrations, the simulations give a configuration that corresponds to microemulsions with a globular structure. Although we did not examine the phase boundaries between the various phases, this model indeed exhibits many of the structures observed experimentally.

Another application of the Ginzburg–Landau approach is the calculation of the interfacial behavior of mixtures of oil, water, and surfactant when the water, the oil, and the disordered phases coexist. We found that the wetting transition changes character at the point where the simple fluid becomes a microemulsion. This behavior was previously observed in both experimental and theoretical studies.

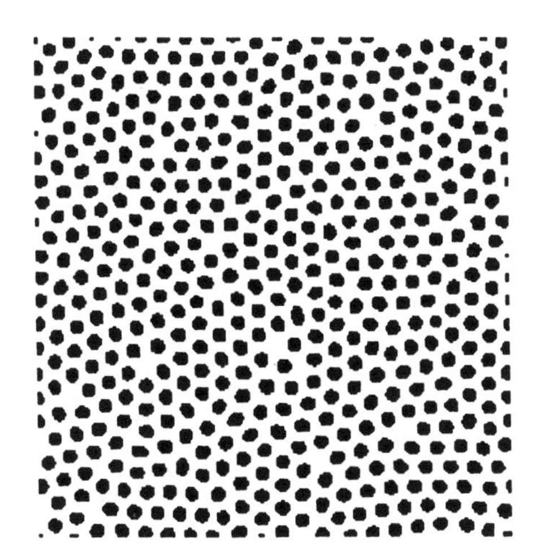

Figure 3.20. A configuration in the hexagonal phase obtained from Langevin simulations. The parameters are the same as those of Fig. 3.1 with $r = 1$, $\mu = 3.1$, and $\mu_w - \mu_0 = 0.1$.

IV. DYNAMICS OF A MODEL OF A DISORDERED MICROEMULSION

A. Introduction

In this section, the dynamics of phase separation in 2-D binary systems containing surfactants is studied by means of a time-dependent Ginzburg–Landau model. This is a new area that offers many opportunities for future theoretical and experimental work. Since relatively little work has been done to date, we review our own work on this subject, which has been reported in several papers [33–35].

We focus on the case where a quench is made from a completely disordered high-temperature phase to a microemulsion with short-range order, but no long-range order. We term this process *microphase separation* to distinguish it from phase separation. We have made use of time-dependent Ginzburg–Landau models of the type discussed in Section III where two scalar fields represent the local order parameter and the local surfactant concentration, respectively. The model is constructed so that the interfacial tension that drives phase separation in binary systems vanishes if order develops beyond some length scale set by the surfactant density ρ_0.

During the growth of order in *macroscopic* phase separation [104], the average domain size of one phase $R(t)$ grows to a macroscopic size following a power law $R(t) \sim t^n$, where $n = \frac{1}{3}$ if the process is controlled by a conserved mode and $n = \frac{1}{2}$ if no conservation laws limit the process. Growth occurs so as to minimize the interfacial energy following a quench, where the driving force is proportional to the product of interfacial tension and local curvature. In contrast, for microphase separation in a microemulsion, we consider the limit of domain growth as the interfacial tension becomes vanishingly small due to the presence of surfactants. The domains are then characterized by small length scales [105]. We find that domain growth, regardless of the presence or absence of conservation laws, is characterized by anomalously slow dynamics in the intermediate time regime and the structure factor shows approximate dynamic scaling. We furthermore estimate a cross-over scaling form numerically for the effect of surfactants on phase separation and derive its form from arguments based on interface dynamics. These arguments also indicate the origin of the slow growth at intermediate times.

B. A Model for Microphase Separation

We have used the most basic Ginzburg–Landau free energy functional essentially given by Eq. (3.2), which incorporates the above considerations. As discussed in Section III, the free energy is a functional of two local fields dependent on field point $\mathbf{x}$: $\psi(\mathbf{x})$ and $\rho(\mathbf{x})$, where ψ is the local order parameter corresponding to the difference in the local densities of water and oil, and ρ represents the local concentration of surfactants. Our dynamic model is based on the coupled-variable model of Hohenberg, and co-workers [97, 106], which was introduced to study critical dynamics. Their free energy functional[6] is given by Eq. (3.1) with $h = H = 0$,

$$\mathcal{F}[\psi, \rho] = \int d\mathbf{x} \left[\frac{c}{2} (\nabla\psi)^2 - \frac{r}{2} \psi^2 + \frac{u}{4} \psi^4 + \frac{g}{2} \rho^2 \psi^2 + \frac{a}{2} \rho^2 - \mu\rho \right] \quad (4.1)$$

where c, r, u, g, a, and μ are positive constants for temperatures below the critical temperature T_c. This model is in the same equilibrium universality class at the Ising model, which can easily be found on integrating away the variable ρ. The double-well structure below T_c ensures phase separation to the two bulk phases given by $\psi = \pm 1$. The novelty of the mode-coupling contribution of the ρ variable arises from the time t dependence. The equations of motion are given by Eqs. (3.14) and (3.15), with the thermal noise set to zero,

$$\frac{\partial \psi}{\partial t} = -M_\psi \frac{\delta \mathcal{F}}{\delta \psi} \quad (4.2)$$

and

$$\frac{\partial \rho}{\partial t} = -M_\rho \frac{\delta \mathcal{F}}{\delta \rho} \quad (4.3)$$

We do not expect the inclusion of noise terms to change the dynamics significantly, as shown in studies of spinodal decomposition. Following Hohenberg et al., we consider models A, B, C, and D, which are defined by the presence or absence of conservation laws: the kinetic coefficients M_ψ and M_ρ are constants (implying no conservation laws) except in model B, where $M_\psi \propto -\nabla^2$ (implying a conservation law); in model C where $M_\rho \propto -\nabla^2$; and model D where both $M_\psi \propto -\nabla^2$ and $M_\rho \propto -\nabla^2$.

To this we add the most basic term that preferentially forces a concentration ρ of surfactants to the interface between different values of

[6] Our form for $\mathcal{F}$ adds an extra term proportional to $(\rho - \langle \rho \rangle)^2$ to the free energy of Halperin et al. [106] This is convenient for the numerical study, though not essential.

ψ, that is, the first term of Eq. (3.2):

$$\mathcal{F}_{\text{surfactant}} = -s \int d\mathbf{x}\, \rho(\nabla\psi)^2 \tag{4.4}$$

where s is a constant. This is a reduced version of the free energy that was studied in Section III. The free energy of Eq. (4.4) causes the interfacial tension between the two bulk phases of ψ to vanish, if there is a nonzero average concentration of surfactant ρ_0, thus inhibiting macroscopic phase separation. Hence, the dynamics of this system and its short-range order are different in several important respects from those of a simple disordered system.

Typical results obtained from numerically integrating the Langevin equations for model D, where both fields are conserved, are shown in Fig. 4.1.[7] Results for models A, B, and C are similar and are reported in our original papers [33, 34, 35]. Figure 4.1 shows the spatial configura-

Figure 4.1. Final configurations (after 100,000 iterations) for the values of $\rho_0 = 0.1$, 0.15, 0.17, and 0.2, from upper to lower. The lower row of snapshots shows the spatial distribution of the ψ field, in which black regions correspond to positive ψ and white regions to negative ψ. The upper row shows the spatial distribution of the ρ field. White regions correspond to small values of ρ (<0.04) and black regions correspond to positive values (close to 1).

[7] Euler's method was used on a 2-D grid of linear size $L = 128$ with periodic boundary condition, a spatial mesh of $\Delta x = 0.7$, and a time mesh size of $\Delta t = 0.02$; smaller mesh sizes gave essentially the same results. Parameters used in the simulations were $c = \frac{1}{2}$, $r = 1$, $u = 1$, $g = 5$, $a = \frac{1}{2}$, and $s = \frac{1}{4}$. The proportionality constants in the mobilities M_ψ and M_ρ were set to $\frac{1}{2}$. The equations were integrated up to a time of 100,000 iterations, and 50 independent runs were averaged.

tions of the system after 100,000 iterations, for $\rho_0 = 0.1$, 0.15, 0.17, and 0.2. The lower row of snapshots illustrates the ordering field $\psi(\mathbf{x})$ and the upper row displays the surfactant field $\rho(\mathbf{x})$. It is clear that the surfactants accumulate at the interfaces, as expected. We also observe that the domain sizes decrease considerably as ρ_0 is increased.

Our quantitative results involve the structure factor, which is the Fourier transform of the real space pair correlation function

$$S(k, t) = \left\langle \frac{1}{L^2} \left| \sum_{\mathbf{x}_i} \psi(\mathbf{x}_i, t) e^{ik \cdot x_i} \right|^2 \right\rangle \tag{4.5}$$

in a system of size L^2, where the angular brackets denote an average over initial conditions and angular orientations, as discussed in Section III. The wavenumber k is defined by $k^2 = 2/\Delta x^2 [2 - \cos(k_x \Delta x) - \cos(k_y \Delta x)]$, where $k_x = (2\pi/L)m$, $k_y = (2\pi/L)n$, in units Δx, and $m, n = 0, 1, 2, \ldots, L$. The average domain size is estimated from the second moment of the structure factor using Eq. (3.19).

The process of microphase separation was found to be exceedingly slow, in contrast to the more familiar case of the phase separation in a first-order transition, where one finds power-law growth in the domain size. Figure 4.2 shows $R(t)$ as a function of t for different values of ρ_0. We observe that, for values $\rho_0 \geq 0.1$, the growth appears logarithmic in time, and often becomes even slower for very late times. It is worth noting that the final domain size decreases as ρ_0 increases, in agreement with the

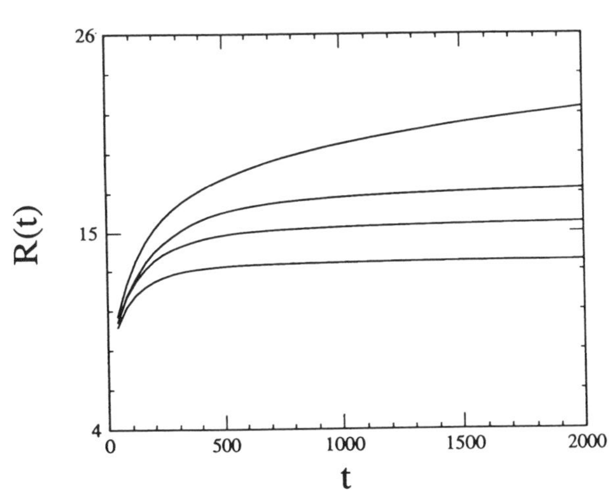

Figure 4.2. Time evolution of the domain size for the different values of ρ_0. Curves from upper to lower correspond to $\rho_0 = 0.1$, 0.15, 0.17, and 0.2.

configurations shown in Fig. 4.1. No power-law growth was observed for any of the surfactant concentrations considered in this study. The reason of such a slow growth is due to the accumulation of surfactants at the interfaces, which reduces the interfacial tension. This results in a dramatic decrease of the driving force for the growth, which is proportional to the product of interfacial tension and the local curvature.

The structure factor exhibits a peak at a nonzero wavenumber k for either ψ or ρ in model D. This peak moves to smaller values of k as time increases indicating coarsening, as shown in Fig. 4.3. At very late times the system no longer coarsens and the peak stays at a fixed value of $k = k_e$, which corresponds to the inverse of the equilibrium domain size. Our calculation shows that k_e increases its value as ρ_0 increases, consistent with the behavior of the domain size mentioned above. This behavior is due to the conservation law; from the study of models A, B, and C, we have found that if a field is nonconserved, the structure factor only peaks at $k = 0$.

Systems undergoing phase separation often exhibit self-similar behavior, which is reflected in the dynamic scaling of the structure factor [104]. This occurs when the average domain size becomes the only relevant length scale. Although the domain size saturates at late times in our case, implying that the width of the interface can play a role as a second length scale, this width is still much smaller than the domain size. It is therefore natural to make the usual scaling ansatz [104] $S(k, t) =$

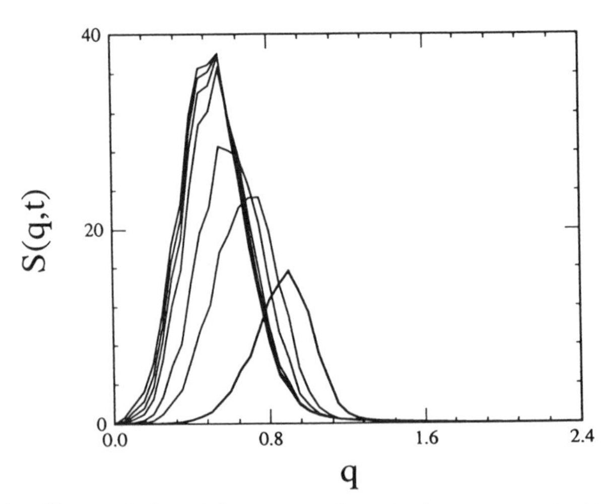

Figure 4.3. Time evolution of the structure factor in the case of model D for $\rho_0 = 0.17$. The curves from lower to upper correspond to $t = 40$–2000.

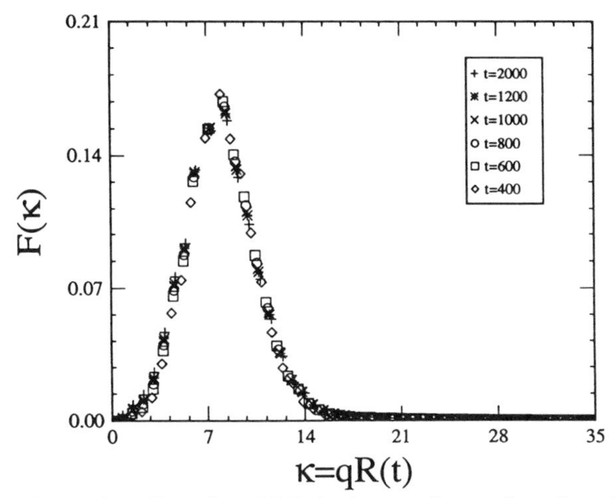

Figure 4.4. Dynamic scaling of model D for intermediate to late times for $\rho_0 = 0.17$. Data for $t = 400$–2000 and wave vector values from $k = 0$ to $k = \pi$ are collapsed onto a single curve.

$R(t)^d F(\kappa)$ where $\kappa = kR(t)$, and $d = 2$ is the dimension of the system. The function $F(\kappa)$ is shown in Fig. 4.4 for $\rho_0 = 0.15$ and good scaling is observed. Similar results were obtained for other values of ρ_0. We have also made a direct calculation of the real space two-point correlation function for the ψ field and observed the same dynamic scaling.

This scaling can only be approximate here, however, since R does not diverge as it does in phase separation. Indeed, we expect that there is a cross-over form since we should obtain the usual $t^{1/3}$ growth law for $\rho_0 = 0$, with nonzero ρ_0 leading to the slow dynamics mentioned above. We also expect the equilibrium domain size to be proportional to $1/\rho_0$ for nonzero ρ_0, since the surfactants mostly accumulate at the interfaces. This implies that the total length of the interface is proportional to ρ_0. These considerations suggest the following cross-over form

$$R(t)t^{-1/3} = h(\rho_0^3 t) \tag{4.6}$$

The function $h(\tau)$ is equal to a constant for $\tau = 0$ and is proportional to $\tau^{-1/3}$ for large τ. Our data for ρ_0 ranging from 0.15 to 0.20 are consistent with this form, as shown in Fig. 4.5. However, a detailed confirmation of this cross-over form requires data for smaller values of ρ_0, and longer time regimes, than we have studied so far.

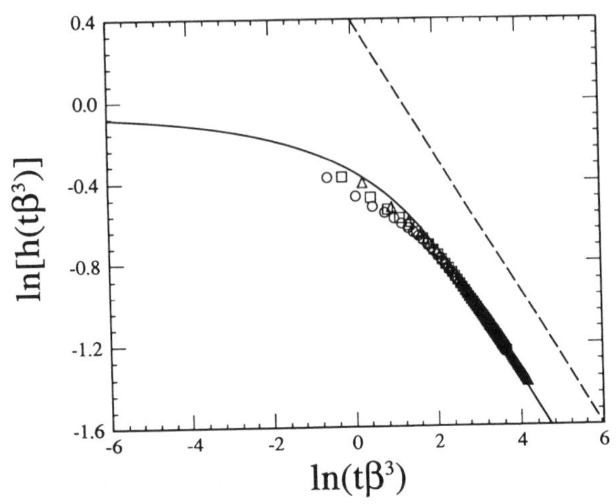

Figure 4.5. The ln–ln plot of cross-over scaling function h versus its argument for different ρ_0 for model D. Circles, squares, and triangles are for $\rho_0 = 0.15$, 0.17, and 0.2, respectively. The dashed line has a gradient of $-\frac{1}{3}$. The theory from reference [35] gives the solid line. The parameter $\beta \propto \rho_0$.

C. Interface Theory for Microphase Separation

One can give an indication of the origin of these results by generalizing standard interface analysis for domain growth. Consider the nonconserved case of model A first. In the absence of surfactants, growth occurs so as to minimize surface area, to which the thermodynamic force of curvature $1/R$ is conjugate. Hence, we expect that the velocity of the interface is proportional to that thermodynamic force

$$\frac{dR}{dt} \propto \frac{1}{R} \tag{4.7}$$

where the asymptotic solution is

$$R \sim t^{1/2} \tag{4.8}$$

for $\rho_0 = 0$.

However, in the surfactant system, the driving force eventually vanishes as the surfactants are adsorbed on the interfaces. Therefore, the proportionality constant above that may be thought of as an interfacial tension, is decreased by an amount determined by the number of surfactants, which is proportional to the interface area. That is, the equation above becomes

$$\frac{dR}{dt} \propto (1 - (\beta R)^2)\frac{1}{R} \tag{4.9}$$

in three dimensions, where β is the maximum curvature for a given concentration of surfactants (hence $\beta \propto \rho_0$). The asymptotic solution has the form,

$$R^2 \sim \beta^{-2}[1 - \exp(-\beta^2 t)] \tag{4.10}$$

which gives anomalously slow growth, similar to what we observed in our numerical work. Indeed, the result can be written in the scaling form

$$Rt^{-1/2} = f(\beta^2 t) \tag{4.11}$$

where the cross-over scaling function is

$$f(x) \propto x^{-1/2}\sqrt{[1 - \exp(-x)]} \tag{4.12}$$

for the nonconserved case of models A or C.

This calculation has previously been performed for conserved systems such as models B or D in the limit in which one of the two phases has negligible amount, that is, the domain structure consists of many widely separated droplets [35]. Again, on assuming the interfacial tension is decreased proportional to the amount of interface, one has

$$R(t)t^{-1/3} = h(\beta^3 t) \tag{4.13}$$

where we note again that $\beta \propto \rho_0$, and

$$h(x) \sim x^{-1/3}(1 - e^{-x}) \tag{4.14}$$

The shape of this cross-over scaling function and its asymptotic behavior is quite similar to that observed in the numerical simulations, as shown in Fig. 4.5. One can also calculate the cluster distribution function and find its dependence on the amount of surfactants.

To conclude this section, we would like to emphasize that experiments on this dynamical process would be of great value. Most of the results above are for 2-D systems, but we expect qualitatively similar behavior in 3-D systems. It is worth noting that although extensive experimental studies have been performed on complex fluids, very little is known about the dynamics of formation of micellar and bicontinuous microemulsions, or other colloidal systems starting from a completely disordered state. We believe such experimental work would be very timely, and that there are a number of issues to address—the anomalously slow growth, the approximate scaling, and the cross-over scaling form reported above—which are fundamental to the dynamical fluctuations of these systems.

V. SUMMARY

The theoretical approaches reviewed in this chapter were shown to be successful in describing the phase equilibria of ternary mixtures of water, oil, and surfactants. Other more complicated phenomena, such as the relaxational dynamics of this mixture, can also be studied in order to give an overall picture of the physical process. These models are phenomenological and are well suited for the study of universal features such as the existence of certain phases. Due to advances in computational techniques, the lattice based models can now be studied in detail to give reliable information. For instance the order of the phase transitions can be determined unambiguously using the new method of Lee and Kosterlitz [70]. The quality of the numerical work is also enhanced with the help of a new extrapolation technique of Ferrenberg and Swendsen [69].

For the 2-D lattice model studied in Section II, we found that the transition from water–oil coexistence to the disordered phase is mainly second order and that it may become weakly first order at low temperatures. This implies a three-phase coexistence region between the water, oil, and disordered phases, which is also observed experimentally. The water and oil phases do not coexist with the lamellar phase since the disordered phase extends to very low temperatures. The transition from the lamellar to the disordered phase and that from the square to the disordered phase are always found to be first order.

The disordered phase does not have the same character in all parts of the phase diagram. For low surfactant concentrations, the disordered phase was found to be a simple fluid, since a peak in the structure factor is observed at $q_{\max} = 0$. However, for higher surfactant concentrations, we found that the water–water structure factor exhibits a peak at $q_{\max} > 0$, implying that the water domains are correlated within distances inversely proportional to $q_{\max}$. This region of the disordered phase was identified as a microemulsion. As the surfactant concentration increases, the position of the peak increases, but the height of the peak decreases. This occurs because the surfactants adsorb at the water–oil interface so that an increase in surfactant concentration leads to an increase in the total area of the interfaces. The structure factor obeys Porod's law at high wavenumbers, that is, $S_{ww}(q) \sim q^{-(d+1)}$. This behavior is a consequence of well-defined thin interfaces. It was shown in previous theoretical studies on microemulsions that fluctuations have a major effect on the phase diagram [52]. In particular, we found that the lamellar phase becomes unstable against the microemulsion close to the water–oil coexistence.

The coarse grained Ginzburg–Landau model discussed in detail in Section III is based on two local fields: a ψ field, which corresponds to the difference in the local concentrations of water and oil and a ρ field, which is proportional to the surfactant concentration. In contrast to the lattice model, the surfactant field here is a scalar field as the microscopic orientational degrees of freedom for the surfactants have been integrated out and replaced by the gradient in the ψ field. Our mean-field analysis shows that, for strong surfactants, the transition from the water–oil coexistence region to the disordered phase is second order for low surfactant concentrations and eventually becomes first order at higher concentrations. The two transitions are again separated by a tricritical point. The water–oil coexistence region is separated from the lamellar phase by a first-order transition line. The transition from the lamellar phase to the disordered phase is also found to be first order. By tuning the parameters of the model, it is possible to decrease the extent of the triple line for coexistence between the water, oil, and disordered phases so as to exhibit a Lifshitz multicritical point. This occurs when the surfactants have weak interactions with the two other components. In this case, only the transition from the water–oil coexistence to the lamellar phase remains first order.

As in the lattice model, the disordered fluid region can be divided into two subregions: an ordinary simple fluid and a microemulsion. The two regimes are separated by either a Lifshitz line or a disorder line. When the phase diagram exhibits a triple line of coexistence between the water, the oil, and the disordered phases, the disorder line and the Lifshitz lines meet the triple line at different points. However, when there is no triple line, the Lifshitz line and the disorder line meets at the tricritical point, which then becomes a Lifshitz tricritical point. We also found that the Lifshitz line is independent of the spatial dimension, while the disorder line depends weakly on the spatial dimension.

The Ginzburg–Landau model also allows a study of the effects of thermal fluctuations. To this purpose, two Langevin equations that describe the local dynamics of the two fields were derived. These equations include fluctuations, and therefore lead to results that are closer to an exact solution than the mean-field theory. We found that the Langevin solutions, which were obtained using a *2-D* grid, mostly agreed with the mean-field results for the phase diagram except for a region of the lamellar phase close to the water–oil coexistence where the lamellar phase becomes unstable against the microemulsion. This finding is similar to results found using the Monte Carlo simulations for the lattice model. We also note that the lamellar domains in this model can have any orientation due to the absence of an underlying lattice. In contrast to the lattice model, this can lead to quasilong-range order.

The most interesting feature of the ternary mixtures is the extremely low value found for the interfacial tension between the water and the oil phases when these phases coexist with the microemulsion. To this purpose, we studied the interfacial properties along the triple line using the Ginzburg–Landau model. We found that the disordered phase wets the water–oil interface provided it is not a microemulsion, that is, along the triple line but to the left of the point where the disorder line meets the triple line. A partial wetting behavior is, however, found when the disordered phase becomes a microemulsion, that is, to the right of the disorder line and along the triple line. These results are in agreement with experimental data [25]. From this point of view, the disorder line seems to be more appropriate than the Lifshitz line as a definition of the microemulsion. Unfortunately, the disorder line is not easily accessible to experiment. In the partial wetting regime along the triple line, decaying oscillations were found in the ψ field at the water–microemulsion and the oil–microemulsion interfaces. The oscillations become stronger as the average size of the microemulsion domains decreases. These results are also supported by experimental evidence [107].

An extremely interesting topic concerning these ternary mixtures is the dynamics of formation of microemulsions. This dynamics has not been studied carefully in the literature. In comparison with the case of binary mixtures, the presence of surfactants drastically slows down the dynamics. This result leads to a microphase separation instead of a complete phase separation. The dynamics is found to be slower for higher surfactant concentrations due to the accumulation of surfactants at the water–oil interfaces during growth. As time increases the total interfacial area decreases leading to an increase in the interfacial surfactant concentration, thereby reducing the driving force of the growth of domains. The growth eventually stops when the driving force vanishes, that is, when the effective surface tension becomes zero. The structure factor is found to exhibit an approximate dynamic scaling in the intermediate time regime before growth stops. We also numerically estimated the dynamical cross-over scaling that governs the crossover of the growth dynamics from lower to higher surfactant concentrations. Finally, on the basis of interface dynamics we can account for the slow growth and for the form of the cross-over scaling function. Our numerical results have been verified by a modified Lifshitz–Slyozov theory recently proposed by Yao and Laradji [35]. As mentioned in Section I, Kawakatsu and Kawasaki

[36–38] proposed a hybrid model in which water and oil are treated in a coarse-grained manner, whereas surfactants are treated individually, and their motion obeys Newton's laws of motion. Using this model, they investigated the dynamics of microphase separation in two dimensions and found results similar to ours.

A question that we have not addressed is the effect of spontaneous curvature of the surfactant film on the phase diagram of ternary systems containing surfactants. Experimental evidence shows that the phase diagram is symmetric only for a narrow temperature range in the case of nonionic surfactants and for a narrow range of salinity in the case of ionic surfactants. To explore these and other effects on the phase diagram, a spontaneous curvature term should be included, which should depend on external parameters such as temperature and salinity for either the lattice or the continuum model. Another interesting area of study concerns the effect of cosurfactants on the phase diagram of ternary systems containing surfactants. Cosurfactants are usually alcohols that adsorb at the water–oil interfaces in the same way as surfactants. These cosurfactants tend to decrease the bending modulus of the surfactant film, thereby increasing the region of the phase diagram over which the microemulsion is stable. Finally, we stress that all our numerical results are based on simulations in two dimensions. Extension to three dimensions is of great importance and is in the process of investigation.

ACKNOWLEDGMENTS

This work was supported by the Natural Sciences and Engineering Research Council of Canada and le Fonds pour la Formation de Chercheurs et l'Aide á la Recherche de la Province de Québec.

REFERENCES

1. T. Witten, *Phys. Today* **43**, 21 (1990).
2. J. Meunier, D. Langevin, and N. Boccara. *Physics of amphiphillic layers, Part VI.* Springer-Verlag, 1987.
3. G. Gompper and M. Schick, "Self-Assembling Amphiphilic Systems" to appear in *Phase Transitions and Critical Phenomena*, C. Domb and J. Lebowitz, Eds.,
4. Z. Wang and S. Safran, *J. Phys.* (*Fr.*) **51**, 185 (1990).
5. Z. Wang and S. Safran, *J. Chem. Phys.* **94**, 679 (1991).
6. R. Holyst and M. Schick, *J. Chem. Phys.* **96**, 7728 (1992).
7. E. T. K. Winey and L. Fetters, *Macromolecules*, **24**, 6182 (1991); K. Winey, E. Thomas, and L. Fetters, *J. Chem. Phys.* **95**, 9367 (1991); *Macromolecules*, **25**, 422 (1992).

8. T. Hoar and J. Schulman. *Nature* (*London*), **152**, 102 (1943).
9. J. Schulman, W. Stockenius, and L. Prince, *J. Phys. Chem.*, **63**, 1677 (1959).
10. E. Kaler, K. Benett, H. Davis, and L. Scriven, *J. Chem. Phys.*, **79**, 941 (1984).
11. M. Kotlarchyk. S. H. Chen, J. S. Huang, and M. W. Kim, *Phys. Rev. Lett.* **53**, 941 (1984); *Phys. Rev. A*, **29**, 2054 (1984).
12. R. O. L. Auvray, J. Cotton, and C. Taupin. *J. Phys.* (*Fr.*) **45** 913 (1984); *J. Phys. Chem.* **88**, 4586 (1984); *Physica B*, **136**, 281 (1986).
13. M. Teubner and R. Strey, *J. Chem. Phys.* **87**, 3195 (1987).
14. C. Alba-Samionesco, J. Teixeira, and C. Angell, *J. Chem. Phys.* **91**, 395 (1989).
15. B. Barago. D. Richter, J. S. Huang, S. A. Safran, and S. T. Milner, *Phys. Rev. Lett.* **65**, 3348 (1990).
16. P.-G. Nilsson and B. Lindman. *J. Phys. Chem.* **88**, 4764 (1984).
17. W. Brown, R. Rymdén, J. van Stam, M. Almgren, and G. Svensk, *J. Phys. Chem.* **93**, 2512 (1989).
18. H. D. E. W. Kaler, and L. Scriven. *J. Chem. Phys.* **79**, 5685 (1983).
19. M. Kotlarchyk, *Physica* **136B**, 274 (1986).
20. J. Wilcoxon and E. Kaler, *J. Chem. Phys.*, **86**, 4684 (1987).
21. P. Kekicheff and G. Tiddy, *J. Phys. Chem.*, **93**, 2520 (1989).
22. D. S. J. P. Wilcoxon and E. Kaler, *J. Chem. Phys.*, **90**, 1909 (1989).
23. Y. Yan and J. Clarke, *J. Chem. Phys.*, **93**, 4501 (1990).
24. S. Chen, S. L. Chang, R. Strey, J. Samseth, and K. Mortensen, *J. Phys. Chem.*, **95**, 7427 (1991).
25. O. Abillon, L. Lee, D. Langevin, and K. Wong, *Physica A*, **172**, 209 (1991).
26. M. López-Quintela, *J. Non-Equilib. Thermodyn.*, **14**, 279 (1989); 287 (1989).
27. L. Chen, L. Feng, M. Robert, and K. Shukla, *Phys. Rev. A*, **42**, 4716 (1990).
28. G. Gompper and M. Schick, *Phys. Rev. Lett.*, **65**, 1116 (1990).
29. M. Matsen and D. Sullivan, *J. Phys. II* (*Paris*), **2**, 93 (1992).
30. V. Degiorgio and M. Corti, *Chem. Phys. Lett.*, **151**, 349 (1988).
31. G. Gompper and M. Schick, *Chem. Phys. Lett.*, **163**, 475 (1989).
32. M. Borkovec, *J. Chem. Phys.*, **91**, 6268 (1989).
33. M. Laradji, H. Guo, M. Grant, and M. Zuckermann, *J. Phys. A: Math. Gen.* (Letter to the Editor), **24**, 629 (1991).
34. M. Laradji, H. Guo, M. Grant, and M. Zuckermann, *J. Phys. Cond. Matter*, **4**, 6715 (1992).
35. J. Yao and M. Laradji, *Phys. Rev. E.*, **47**, 2695 (1993).
36. T. Kawakatsu and K. Kawasaki, *Physica A*, **167**, 690 (1990).
37. T. Kawakatsu and K. Kawasaki, *J. Coll. Int. Sci.*, **145**, 413 (1991).
38. T. Kawakatsu and K. Kawasaki, *Physica A*, **145**, 420 (1991).
39. A. Ciah, J. Hoye, and G. Stell, *J. Chem. Phys.*, **90**, 1215 (1989).
40. A. Ciah and J. Hoye, *J. Chem. Phys.*, **90**, 1222 (1989).
41. J. Wheeler and B. Widom. *J. Am. Chem. Soc.*, **90** 3064 (1968).
42. B. Widom, *J. Chem. Phys.* **81**, 1030 (1984).

43. B. Widom, *J. Chem. Phys.*, **84**, 6943 (1986).
44. K. Dawson, M. Lipkin, and B. Widom, *J. Chem. Phys.*, **88**, 5149 (1988).
45. M. S. A. Hansen and D. Stauffer, *Phys. Rev. A*, **44**, 3686 (1991).
46. D. Stauffer, and N. Jan, *J. Chem. Phys.*, **87**, 6210 (1987).
47. N. Jan and D. Stauffer, *J. Phys. (Fr.)*, **49**, 623 (1988).
48. D. Stauffer and H. F. Eicke, *Physica A*, **182**, 29 (1992).
49. M. Blume, V. Emery, and R. Griffiths, *Phys. Rev. B*, **4**, 1071 (1971).
50. M. Schick and W. Shih, *Phys. Rev. Lett.*, **59**, 1205 (1987).
51. G. Carneiro and M. Schick, *J. Chem. Phys.* **89**, 4638 (1988).
52. G. Gompper and M. Schick, *Phys. Rev. Lett.*, **62**, 1647 (1989).
53. G. Gompper and M. Schick, *Phys. Rev. A*, **42**, 2137 (1990).
54. G. Gompper and M. Schick, *Phys. Rev. B*, **41**, 9148 (1990).
55. M. S. J. Lerczak and G. Gompper, *Phys. Rev. A*, **46**, 985 (1992).
56. R. Larson, *J. Chem. Phys.*, **89**, 1642 (1988).
57. J. Halley and A. Kolan, *J. Chem. Phys.*, **88**, 3313 (1988).
58. M. Matsen and D. Sullivan, Phys. Rev. A, **41**, 2021 (1990).
59. M. Laradji, H. Guo, M. Grant, and M. Zuckermann, Phys. Rev. A, **44**, 8184 (1991).
60. M. Laradji, H. Guo, M. Grant, and M. Zuckermann, Mater. Res. Soc. Proc., **248**, 23 (1992).
61. J. R. Gunn and K. A. Dawson, *J. Chem. Phys.*, **96**, 3152 (1992).
62. M. Matsen, M. Schick, and D. E. Sullivan, *J. Chem. Phys.*, **98**, 2341 (1993).
63. S. Alexander, *J. Phys. (Paris) Lett.*, **39**, L1 (1978).
64. C. J. K. Chen, C. Ebner, and R. Pandit, *J. Phys. C*, **20**, L361 (1987): *Phys. Rev. A*, **38**, 8184 (1988).
65. T. P. Stockfisch and J. C. Wheeler, *J. Chem. Phys.*, **92**, 3292 (1988).
66. B. Smit, Ph.D. thesis, University of Utrecht, 1992.
67. N. Metropolis, A. Rosenbluth, T. Teller, and R. Teller, *J. Chem. Phys.*, **12**, 1087 (1953).
68. N. Bartelt, T. Einstein, and L. Roelofs, *Phys. Rev. B*, **34**, 1616 (1986).
69. A. Ferrenberg and R. Swendsen, *Phys. Rev. Lett.*, **63**, 1195 (1988).
70. J. Lee and J. Kosterlitz, *Phys. Rev. Lett.*, **65**, 137 (1990).
71. P. Bowen, L. Burke, P. Corsten, K. Crowell, K. Farrell, J. MacDonald, R. Macdonald, A. MacIsaac, P. Poole, and N. Jan, *Phys. Rev. B*, **40**, 7439 (1990).
72. R. Toral and A. Chakrabarti, *Phys. Rev. B*, **40**, 2445 (1990).
73. Z. Zhang, M. Laradji, H. Guo, O. Mouritsen, and M. Zuckermann, *Phys. Rev. A*, **45**, 7560 (1992).
74. E. Corvera, M. Laradji, and M. Zuckermann, *Phys. Rev. E.*, **47**, 696 (1993).
75. J. Cardy, *Current Physics, Sources and Comments*, Vol. 2. North-Holland, New York, 1988.
76. K. Binder, *Phys. Rev. Lett.*, **47**, 119 (1981).
77. M. Kotlarchyk, S.-H. Chen, J. Huang, and M. Kim, *Phys. Rev. Lett.*, **53**, 941 (1984).
78. M. Matsen and D. E. Sullivan, *Phys. Rev. E*, submitted.

79. H. Tomita, *Prog. Theor. Phys.*, **72**, 656 (1984).
80. R. Baxter, in *Exactly Solved Models in Statistical Mechanics*, Academic, 1982, pp. 276–321.
81. V. Degiorgio and M. Corti, *Chem. Phys. Lett.*, **151**, 349 (1988).
82. M. Blume, *Phys. Rev.*, **141**, 517 (1966).
83. K. L. C. M. Soukoulis and G. Grest, *Phys. Rev. B*, **28**, 1495 (1983); **28** 1510 (1983).
84. J. C. Lang and R. D. Morgan, *J. Chem. Phys.*, **73**, 5849 (1980).
85. M. Matsen, D. E. Sullivan, and M. J. Zuckermann, private communication.
86. F. Schmid and M. Schick, *Phys. Rev. E*, **48**, 1882 (1993).
87. W. Helfrich, *Z. Naturforschung A*, **33**, 305 (1981).
88. S. Safran and L. Turkevick, *Phys. Rev. Lett.*, **50**, 1930 (1983).
89. M. Cates, D. Roux, A. Andelmann, S. T. Milner, and S. A. Safran, *Europhys. Lett.*, **5**, 733 (1988).
90. G. Gompper and X. Kraus, *Phys. Rev. E*, **47**, 4301 (1993).
91. K. Chen, C. Jayaprakash, R. Pandit, and W. Wenzel, *Phys. Rev. Lett.*, **65**, 2736 (1990).
92. K. Kawasaki and T. Kawakatsu, *Physica A*, **164**, 549 (1990).
93. G. Gompper and S. Klein, *J. Phys. II Fr.*, **2**, 1725 (1992).
94. D. Amit, *Field Theory, the Renormalization Group, and Critical Phenomena*, World Scientific, Singapore, 1984.
95. T. Ohta, K. Kawasaki, A. Sato, and Y. Enomoto, *Phys. Lett. A*, **126**, 93 (1987).
96. J. Toner and D. Nelson, *Phys. Rev. B*, **23**, 316 (1981).
97. P. Honenberg and B. Halperin, *Rev. Mod. Phys.*, **49**, 435 (1977).
98. S. Allen and J. Cahn, *Acta Metall.*, **27**, 1085 (1979).
99. T. Ohta, D. Jasnow, and K. Kawasaki, *Phys. Rev. Lett.*, **49**, 1223 (1981).
100. K. Elder, J. Vinals, and M. Grant, *Phys. Rev. Lett.*, **65**, 316 (1992).
101. Y. Levin, C. Mundy, and K. Dawson, *Phys. Rev. A*, **45**, 7309 (1992).
102. K. Skukla, B. Payandah, and M. Robert, *J. Stat. Phys.*, **63**, 1053 (1991).
103. P. de Gennes, *Rev. Mod. Phys.* **57**, 827 (1985).
104. J. Gunton, M. S. Miguel, and P. Sahni, *Phase Transitions and Critical Phenomena*, Vol. 8. Academic, London, 1983.
105. O. G. Mouritsen and P. J. Shah, *Phys. Rev. B*, **40**, 11445 (1989).
106. B. Halperin, P. Hohenberg, and S. Ma, *Phys. Rev. B*, **10**, 139 (1974).
107. J. Meunier, *J. Phys. Lett. Fr.*, **46**, 1005 (1985).

COLORED NOISE IN DYNAMICAL SYSTEMS

PETER HÄNGGI AND PETER JUNG

Department of Physics, University of Augsburg, Memminger Str. 6, D-86135 Augsburg, Germany

CONTENTS

Advances in Chemical Physics, Volume LXXXIX, Edited by I. Prigogine and Stuart A. Rice.
ISBN 0-471-05157-8

I. INTRODUCTION

The general subject of colored noise driven dynamical flows is rooted in the study of the motion of small particles suspended in a fluid and moving under the influence of random forces that result from collisions with molecules of the fluid induced by thermal fluctuations. In short, the phenomenon of *Brownian motion* [1]. In the earliest studies of Brownian motion, the damping of the motion of the suspended particles was very large compared to that of the fluid molecules, so that inertial effects could be neglected. Moreover, the thermal fluctuations occur on a time scale that is very much shorter than that of the Brownian particle. It is then a good approximation to assume that the random forces are uncorrelated delta functions as perceived by the particle on its own, much slower time scale. This assumption considerably simplifies the problem, because it allows one to treat the stochastic dynamical motions as a Markovian process for which many methods and approximation schemes are available. The fluctuations that can be treated under this assumption have often been termed "white noise". Thus, white noise fluctuations $\xi(t)$, are those for which the autocorrelation function is given by

$$\langle \xi(t)\xi(s) \rangle = 2D\delta(t-s) \tag{1.1}$$

where we designate the noise intensity as D. This noise has no time scale and exists independently of any other physical system. A large body of literature on white noise applications exists, and appropriate starting points are the now classic reviews of Chandrasekhar, Uhlenbeck and Ornstein, and others in the collection of Wax [2]; the texts by Stratonovich [3], van Kampen [4], Risken [5], and Horsthemke and Lefever [6]; or the reports by Hänggi and Thomas [7], and Fox [8].

In the physical world this idealization, however, is never exactly realized. What must be done is to consider the noise and the physical system within which, or upon which, it is operating together. Specifically, the time scales of the two systems must be taken into account. Therefore, we seek, in the first instance, a noise with a well-defined characteristic time. One of the simplest examples for an introductory discussion of time-correlated noise is the Ornstein–Uhlenbeck process, which exhibits an exponential correlation function,

$$\langle \xi(t)\xi(s)\rangle = (D/\tau_n)\exp(-|t-s|/\tau_n) \tag{1.2}$$

with noise correlation time τ_n. This fluctuation process is called "colored noise" in analogy with the effects of filtering on white light. The terms "white" and "colored noise", are, of course, jargon words. Nevertheless, because they are widely recognized and understood, even though somewhat imprecise, we shall use them throughout. Now it is important to understand what clock is being used to measure the noise correlation time. It is the physical system itself, which is either generating the noise internally or is subject to the noise as an external forcing which, according to its own characteristic response time, perceives the time scale of the noise. The physical system is frequently and simply modeled with a stochastic differential equation termed a Langevin equation. A damped oscillator subject to a one-dimensional (1-D) deterministic potential $U(x)$, and additive noise serves as a simple example:

$$m\ddot{x} + m\gamma\dot{x} = -\frac{dU(x)}{dx} + \xi(t) \tag{1.3}$$

where m is the mass and γ is the damping factor. By dividing through by $m\gamma$, neglecting the inertial term in the limit of large γ, and properly scaling the system coordinate x to be dimensionless, that is, $x \to \alpha x$ and $D \to D(\alpha)$, we have

$$\dot{x} = \left[\frac{1}{\tau_s}\right]\left\{-\left[\frac{dU(x)}{dx}\right] + \xi(t)\right\} \tag{1.4}$$

where τ_s is the system characteristic time. Now for simplicity, we can scale time in the Langevin equation so that τ_s is removed by letting $t' = t/\tau_s$. But we must measure time in Eq. (1.2) on the same scale, where with $\tilde{D} = D(\alpha)/\tau_s$

$$\langle \xi(t')\xi(s')\rangle = \left[\frac{\tau_s}{\tau_n}\right]\tilde{D}\exp\left[-|t'-s'|\left(\frac{\tau_s}{\tau_n}\right)\right] \tag{1.5}$$

It has become customary to write the Langevin equation in this scale, and the noise correlation function in terms of a dimensionless time $\tau \equiv \tau_n/\tau_s$, so that

$$\dot{x} = -\frac{dU(x)}{dx} + \xi(t) \tag{1.6}$$

and

$$\langle \xi(t)\xi(s) \rangle = (D/\tau) \exp(-|t-s|/\tau) \tag{1.7}$$

where $D \equiv \tilde{D}$ is now a dimensionless noise intensity. Though this is a convenient scale, which we shall adopt throughout this paper, it has sometimes obscured the role of colored noise in real physical systems by hiding the "clock" with which the system measures time. Next, we shall discuss various approximations, some valid only for small values of τ. In order to successfully apply these theories to any real physical system, it is essential to understand what "small" means. As the above discussion indicates, this means that $\tau_n \ll \tau_s$ (or $\tau \ll 1$), but how much smaller depends not only on the particular approximations used but also on the problem or system. A related question is one of measurability and distinguishability. The analogue simulators and associated measurements, which mimic real physical systems, are not accurate enough to convincingly distinguish any colored noise approximate theory from the white noise predictions for the same system for $\tau \cong 0.1$. Digital simulations can, of course, achieve much greater accuracy and further improvements are currently forthcoming. Even so, it is probably not practical (with finite computing time) to expect distinguishability for $\tau \cong 0.001$.

Moving toward larger τ, approximations that are based on perturbations of the white noise theory become progressively less accurate once again for values of τ that depend both on the system and on the particular approximation used. These so-called "small τ" approximate theories have roots that date to the original work of Stratonovich [3] cited above. A different approach is expounded by Risken [5] who has pioneered the use of matrix continued fraction expansions, which offer solutions to the colored noise problem. These expansions are in principle exact, but are in practice rendered approximate by the necessity to truncate and numerically invert a final matrix of infinite dimension. Moreover, in the absence of supercomputers, the matrix continued fraction method is practically limited to systems with a low number of state variables.

One of the earliest definitive results, which indicated that colored noise plays an important role in nature was Kubo's explanation of motional narrowing of the observed magnetic resonance line shapes induced by thermal fluctuations [9]. Kubo's model is exactly solvable and applicable to all ranges of τ, since it treats a linear system: an oscillator with a noisy frequency. The observed statistical properties of the fluctuations of dye laser light [10] offered the next solid evidence that the noise found in some physical systems is colored. Initially, the evidence was provided by numerical simulations of the nonlinear laser Langevin equations using

Ornstein–Uhlenbeck noise, compared to measurements of the correlation function of the actual laser light fluctuations [11]. Soon after, an early success of the so-called "small τ" theory resulted from its application to the same experimental dye laser data [12]. Recently, it has been demonstrated that the pump parameter in dye lasers can be adjusted close enough to the laser transition that the laser fluctuations are driven by (pump) noise of moderate values of τ [13]. The intensity fluctuations in all pumped lasers originate from two sources: the spontaneous emission, or quantum, noise that derives from the statistics of photon emission from the inverted population within the laser cavity, and the pump noise that derives from fluctuations in the intensity of the pump. The pump noise is governed by a much slower time scale than the emission noise, and so has been treated as colored noise, while the emission noise has until recently been assumed to be white. Colored spontaneous emission noise has been shown to have a strong influence on the properties of the proposed correlated spontaneous emission laser [14]. Noise color also has a strong effect on the systematics of noise induced bifurcations among ordered and turbulent states in nematic liquid crystals [15, 16].

II. USE AND ABUSE OF COLORED NOISE

This section first reviews the development of the field of systems driven by noise starting from the early work on Brownian motion around the turn of the century, then continues with the pioneering studies and applications to physical systems during the decade of the 1950s, and concludes with the more recent theoretical developments through the early 1980s.

A. The Role of White Noise

As already mentioned in Section I, the most well-known application of a noisy differential equation for a state variable $x(t)$ is the theory of Brownian motion, described first in 1828 [17], with the first precise experiments carried out in 1888 by Gouy [18]. The description in terms of a noisy differential equation wherein one splits the motion into two parts, a slowly varying systematic part and a rapidly varying random part, is due to Langevin [19], who first wrote the familiar expression for the damped motion of a randomly forced particle, with $\dot{x} = v$,

$$m\dot{v} = -m\gamma v + \xi(t) \tag{2.1}$$

with

$$\langle \xi(t)\xi(t') \rangle = \frac{2\gamma kT}{m} \delta(t-t') \tag{2.2}$$

A study of the solution of this equation, however, had to await Ornstein's early work on Brownian motion (see the historical discussion on colored noise given below). It is important to note that a general study of Eq. (2.1) is nontrivial. In order to make progress, one necessarily must specify the properties of the random force. It is often justifiable to assume that the random forces, which sometimes derive from the environment, are correlated on a very small time scale τ_n compared to the characteristic relaxation time for the system τ_s, around a locally stable state. An idealized treatment then assumes the random force to have zero correlation time, that is $\xi(t)$ is approximated by a (generalized) δ-correlated process:

$$\langle \xi(t)\xi(s) \rangle = 2D\delta(t-s) \tag{2.3}$$

where all the frequencies of its power spectrum $S_\xi(\omega) = \int_{-\infty}^{\infty} \langle \xi(t)\xi(s) \rangle e^{-i\omega t}\, dt \equiv 2D$, are present with equal weight. Obviously, there exist several classes of such white noise processes, all of which are completely understood [20]. The classes are defined in terms of the derivative $\xi(t) = dz(t)/dt$ of processes with stationary, independent increments. For example, the derivative of the Wiener process [21, 22] defines *Gaussian white noise*, whereas the derivative of the Poisson process yields *white shot noise*. These two elementary noise processes form the building blocks for the theory of Markov processes [20, 23–26]. Stochastic differential equations composed of nonlinear drift flows $f_\alpha(x)$ and multiplicative noise forces $g_{\alpha i}(x)\xi_i(t)$, that is,

$$\dot{x}_\alpha = f_\alpha(x) + \sum_{i=1}^{n} g_{\alpha i}(x)\xi_i(t) \qquad \alpha = 1, \ldots, n \tag{2.4}$$

where $\xi_i(t)$ denotes a white (generally) non-Gaussian random force, thus describe a multidimensional Markov process $\mathbf{x}(t)$. The corresponding master equation, which describes the rate of change of the probability, as well as the statistical properties of the nonlinear noise forces, has been discussed in the literature [27–29].

From an historical point of view, the statistical consequences of Eq. (2.1) have first been studied by Ornstein [30, 31], implicitly assuming *Gaussian* white noise (see also [32, 33] and the bibliographical notes given in [34]). For the mean-square displacement in thermal equilibrium, he obtained from Eq. (2.1) the central result [30, 31],

$$\langle (x - x_0)^2 \rangle_{x_0} = \left(\frac{2kT}{m\gamma^2}\right)[\gamma t - 1 + e^{-\gamma t}] \tag{2.5}$$

where an average over $x(t_0) = x_0$ is implied by the subscript on the left-hand side. In Eq. (2.5) T denotes the temperature and k is the Boltzmann constant. This result, which accounts also for the inertia effects $m\dot{v}$, generalizes the celebrated result by Einstein [35].

$$\langle x^2 \rangle = \left(\frac{2kT}{m\gamma}\right)t \tag{2.6}$$

The inertia induced shift obtained by Ornstein in Eq. (2.5), given by the terms $[-1 + \exp(-\gamma t)]/\gamma$ inside the bracket is, of course, the result of the two-dimensional 2-D stochastic motion in phase space, which is equivalent to a colored noise driven dynamics in configuration space. The passage from Eq. (2.1) to a partial differential equation for the probability[1]

$$\frac{\partial}{\partial t} p_t(v) = \gamma \frac{\partial}{\partial v} [v p_t(v)] + \frac{kT\gamma}{m} \frac{\partial^2}{\partial v^2} p_t(v) \tag{2.7}$$

has been achieved by Fokker [36, 37], Smoluchowski [38], and Planck [39]. Actually, Eq. (2.7) was obtained earlier by Lord Rayleigh [40, 41] who employed a limiting procedure from a discrete state Brownian motion model for a heavy particle (the Rayleigh model). This connection between the Langevin equation driven by Gaussian white noise and the parabolic partial differential equation, Eq. (2.7), commonly known as the "Fokker–Planck equation", was subsequently generalized to account for the Brownian motion of the configuration coordinate of a particle moving in an external potential field by Smoluchowski [42, 43] and Fürth [44]. The generalization to the full phase space, that is, $\partial p_t(x, v)/\partial t$, has been obtained first by Klein [45] (see also [46]).

Useful applications of the theory of Brownian motion to the calculation of other statistical quantities, such as the probability density of first passage times or absorption and escape probabilities, had been considered as early as 1915 by Schrödinger [47] and others [42–44]. For nonlinear flows, interesting applications, such as calculations of the stationary probability density of a 2-D noisy limit cycle, and the exact

[1] We will refer throughout this article to the "probability density" as simply the "probability".

quadrature formulas for the mean first-passage time of 1-D Fokker–Planck processes, have been obtained as early as 1933 by Pontryagin et al. [48] (translated by Barber in [49]).

Interesting as these applications are, however, the fact is that noise with zero correlation time leads to stochastic realizations, generated by the noncontinuous noise-sample paths, which are in reality nonphysical. For example, for the state variable $x_\alpha(t)$ driven by a white Gaussian noise source, as given by Eq. (2.4), the sample paths are not of bounded variation, nor are they continuous (as it is the case with white shot noise) or differentiable [21, 22, 50]. Thus, any results obtained from white noise theory that make predictions about the dynamics on time scales approximately equal to τ_n clearly do not lie within its regime of validity. Nevertheless, the results of measurements on real physical systems for which noise forces with very large effective bandwidths are encountered (at least a factor of 10 larger than the deterministic system bandwidth) are for almost all practical purposes indistinguishable from the predictions of the white noise theory. Of course, all actual noise encountered in nature has some nonzero (though perhaps small) correlation time. In Section III, corrections to the white noise theory that are necessary to describe systems driven by noise with nonnegligible correlation time, commonly known as "colored noise", are considered.

B. The Role of Colored Noise

Statistical fluctuations always reflect a lack of knowledge about the exact state of the system. Usually, the system behavior is modeled in terms of two classes of variables: state variables that change on a slow time scale, which are most often monitored directly in experiments, and those that are generally more rapidly varying and more closely related to the random forces. Moreover, the random forces themselves can be classified into two groups as "internal noise" or "external noise", though this distinction is often ambiguous depending, as it does, on how the boundary between the "system" and the "external world" is drawn. Generally, external noise can be thought of as imposed on some subsystem by a larger fluctuating environment in which the subsystem is immersed. In laboratory experiments, external noise with well characterized and immediately controllable statistical properties, such as the stationary probability density, intensity, and correlation time, is imposed by the experimenter on the system whose response he then measures. In laboratory experiments, as well as in many naturally occurring instances of nonequilibrium noise driven systems, the external noise can take on correlation times that are much smaller than, comparable to, or much larger than the characteristic relaxation times of the system. Furthermore, random forces of moderate-to-large correlation times, $\tau_n \cong \tau_s$, can also emerge with internal noise as was already clearly shown in 1962 by Kubo in the case of spin relaxation in magnetic systems [9].

In practice, for a complex system, any strongly colored noise implies a significant deviation from Markovian behavior. In the theoretical treatment of such systems it is often the case that strongly colored internal noise emerges as the result of *coarse graining over a hidden set of slow variables*. In this context, we touch upon a major problem that sooner or later confronts nearly every perplexed modelist of noisy stochastic flows: Given a nonlinear system, which and how many slow variables are needed to adequately describe the system dynamics? One generally hopes to monitor only a few, and preferably just one physical variable. There is a price to be paid for this simplification, however, precisely because such a resulting low-dimensional flow implies a loss of the Markovian properties of the original higher dimensional system. Systems that exhibit noise of moderate or large correlation time are often intrinsically high dimensionally, and can be reduced in dimension only at the expense of the Markovian character. Because multidimensional Markovian objects, of the form given by Eq. (2.4) with $n > 1$, present a rather complicated dynamics that is already difficult to study in analytical form, the study of colored noise driven flows even in one-coordinate dimension, such as

$$\dot{x} = f(x) + g(x)\xi(t) \tag{2.8}$$

where $\xi(t)$ is a stationary noise with correlation function,

$$\langle \xi(t)\xi(s) \rangle = D\gamma(t - s) \tag{2.9}$$

is thus expected to be challenging as well. However, any modeling in terms of colored noise is expected to be more physically realistic, since when a nonzero correlation time is explicitly accounted for, the realizations become differentiable as they must be for all real macroscopic systems. The white noise limits of such theories can then be compared to purely white noise theories as well as to the results of experiments performed with wide bandwidth noise. Often, the fluctuations $\xi(t)$ represent the cumulative effects of many weakly coupled environmental degrees of freedom. Outside critical neighborhoods, and in the absence of long-range correlations that induce large scale collective effects, one can invoke the *central limit theorem* and thus treat the fluctuations as Gaussian. In particular, if $\xi(t)$ is *in addition Markovian*, then Doob's theorem[2] [50, 52] states that $\xi(t)$ is necessarily an *Ornstein–Uhlenbeck*

[2] For a generalization of Doob's theorem to nonstationary processes (see [51]).

process, with exponential correlation function,

$$\langle \xi(t)\xi(s)\rangle = \left(\frac{D}{\tau}\right)e^{-|t-s|/\tau} \tag{2.10}$$

with the Lorentzian power spectrum $S_\xi(\omega) = 2D/(\tau^2\omega^2 + 1)$ [Fourier transform (FT) of the correlation function]. In the following discussion, we will often restrict the discussion to Gaussian processes with exponential correlation functions, as given by Eq. (2.10), unless stated otherwise. This exponentially correlated Gaussian noise source has been widely used in numerous recent studies. One of the earliest application dates back to 1966 by Berne et al. [53], where it has been used to model transport in simple liquids.

Pioneering studies of stochastic, nonlinear flows of the type given by Eq. (2.8), and applied to problems in electrical engineering and radiophysics, date to the late 1950s and were developed primarily by the school surrounding Stratonovich and co-workers [3, 54]. They considered corrections to the white noise theory valid for small τ, meaning, of course, that their approximate theory would apply in the range $\tau_n \ll \tau_s$, and succeeded to obtain an approximate Fokker–Planck like evolution for the probability [55],

$$\frac{\partial p_t(x,\tau)}{\partial t} = -\frac{\partial}{\partial x}[f(x)p_t(x,\tau)] + D\frac{\partial}{\partial x}g(x)\frac{\partial}{\partial x} \times \left\{g(x)\left[1 + \tau g(x)\left(\frac{f(x)}{g(x)}\right)'\right]p_t(x,\tau)\right\} \tag{2.11}$$

where the prime ($'$) indicates differentiation with respect to x.

This celebrated result is now commonly known as the "small τ approximation". Over the last two decades it has been applied to many different systems, rederived, commented on, and extended by many authors using a variety of methods. In particular, we mention here the method of cumulant expansions [56–63], expansions in functional derivatives [64–66], singular perturbation methods [67–71], the method of moments [72], adiabatic elimination procedures [73, 74], and projector operator techniques [75–77]. We wish to emphasize that this list of references is intended to be representative only and certainly is not complete. These different methods will not be further reviewed, instead we will return, in Section IV.A, to the small τ approximation with a discussion of the regime of its validity.

Colored noise of arbitrarily long correlation time has been considered in Kubo's cornerstone paper on the theory of line shapes and relaxation in magnetic resonance systems [9, 78]. He employed a modified Bloch equation, that is, the so-called "Kubo oscillator"[3]

$$\dot{x} = i[\omega_0 + \xi(t)]x(t) \tag{2.12}$$

Because this is inherently a linear system, and because $\xi(t)$ is Gaussian, Ornstein–Uhlenbeck noise in Eq. (2.10), Eq. (2.12) can be solved exactly for the first moment [9, 78]

$$\langle x(t) \rangle = \langle x(0) \rangle \exp\{i\omega_0 t - D[t - \tau(1 - \exp(-t/\tau))]\} \tag{2.13}$$

The transformation, $u = \ln x$ yields a linear stochastic flow with additive noise,

$$\dot{u} = i[\omega_0 + \xi(t)] \tag{2.14}$$

which in turn yields an exact master equation for $p_t(x) = p_t(u)|x|^{-1}$; see Eqs. (3.28–3.34). Defining the relaxation function $\phi(t)$,

$$\phi(-t) = \phi(t) \equiv \left\langle \exp \int_0^t \xi(s)\, ds \right\rangle = \frac{\langle x(t) \rangle}{\langle x(0) \rangle} e^{-i\omega_0 t} \tag{2.15}$$

we obtain, in the white noise limit, from Eq. (2.13) (termed "fast modulation limit" in [9]),

$$\lim_{\tau \to 0} \phi(t) = \phi_0(t) = \exp(-Dt) \qquad \tau \to 0 \tag{2.16}$$

whereas in the case of large τ we have the Gaussian (termed "slow modulation limit" in [9]), that is, with $(D\tau)^{1/2} \gg 1$

$$\lim_{\tau \to \infty} \phi(t) = \phi_\infty(t) = \exp[-(Dt^2/2\tau) + O(\tau^{-2})] \qquad \tau \to \infty \tag{2.17}$$

For the absorption spectrum,

$$I(\omega - \omega_0) = \frac{1}{2\pi} \int_{-\infty}^{\infty} dt\, \phi(t) e^{-i(\omega - \omega_0)t} \tag{2.18}$$

one obtains, using Eq. (2.16), a Lorentzian in the white noise limit,

$$I_0(\omega - \omega_0) = \frac{1}{\pi} \frac{D}{(\omega - \omega_0)^2 + D^2} \tag{2.19}$$

[3] Note that the Kubo oscillator, Eq. (2.12), is not overdamped, although it is described by a differential equation of first order. Eliminating the real or the imaginary part of the complex variable x reveals the undamped harmonic oscillator if no noise is present.

In contrast, for the limit of large τ, or the "slow modulation limit", Eq. (2.17) yields a Gaussian line shape,

$$I_0(\omega - \omega_0) = \left(\frac{2\pi D}{\tau}\right)^{-1/2} \exp\left[-\frac{(\omega - \omega_0)^2}{2D/\tau}\right] \tag{2.20}$$

Thus, with $\sigma^2 = D/\tau$ held constant, and with $D \ll \sigma$ (i.e. $D \ll \sqrt{D/\tau}$, as $\tau \ll 1$), the line shape in Eq. (2.19) becomes *narrowed* as compared with that given by Eq. (2.20). In nuclear magnetic resonance (NMR), this experimentally well-known effect is called "motional narrowing " [9, 78], whereas in paramagnetic resonance it is called "exchange narrowing" [79, 80]. It is worth emphasizing that *Kubo's* explanation of these line shapes represents an exact theory, valid for arbitrarily long correlation time τ, successfully applied to nonsubtle, experimentally observable features of many-body spin systems, which are naturally subject to *internal* colored noise. It is, historically, one of the best known examples elucidating the effect of fluctuations with nonzero τ in a macroscopic physical system.

Prior to the advent of this theory, Anderson [81] and Kubo [82] considered a Markovian modulation $\xi(t)$ in Eq. (2.12), which became known as the "Kubo–Anderson process". The noise model consisted of a discontinuous Markovian process $z(t)$, which was made up of independently occurring steps with random amplitudes $m_i(t)$ during the interval $t_i \leq t < t_{i+1}$. The amplitudes were distributed with density $\rho(m)$, and the jump times $\{t_i\}$ were determined by a Poissonian distribution. This process, too, has an exponential correlation.

$$\langle z(t)z(s)\rangle = \langle m^2\rangle e^{-\lambda|t-s|} \tag{2.21}$$

with $\langle m_i\rangle = 0$, where λ denotes the Poisson parameter in $P[n(t) = k] = [(\lambda t)^k/k!]\exp(-\lambda t)$, with $n(t)$ describing the number of jumps. When $\rho(m) = (\frac{1}{2})[\delta(m + a) + \delta(m - a)]$ one obtains as a special case the (symmetric) two-state Markov process $z(t)$ discussed, for example, in [7]

$$z(t) = a(-1)^{n(t)} \tag{2.22}$$

with

$$\langle z(t)z(s)\rangle = a^2 e^{-\lambda|t-s|} \tag{2.23}$$

This process is also known as "telegraphic noise" and "dichotomous noise", and it plays an important role in applications in radiophysics [83, 84] and in noise-induced transition phenomena [6, 85]. In particular, for nonlinear colored noise flows of the form

$$\dot{x}(t) = f(x) + g(x)z(t) \tag{2.24}$$

one obtains an exact, retarded but closed master equation, which results in stationary probabilities [83–85], and non-Markovian mean first-passage times [86–90], which can be calculated exactly. An interesting application of nonexponential correlated colored noise has been brought forth by Brissaud and Frisch [91, 92] in order to explain noise-induced Stark broadening. They make use of the "Kangaroo process", that is, a Kubo–Anderson like noise with a correlation function $\langle z(t)z(s)\rangle \propto |t-s|^{-1}$. This noise, however, is not always realistic, at least for short times (or high frequencies), since it clearly does not have an integrable FT. In this chapter, we confine the discussion largely to Gaussian noise. For the many results and applications of stochastic flows driven by, for example, two-state or "dichotomous" noise as given by Eq. (2.24), we refer the reader to Section IV.

III. COLORED NOISE THEORY

A. Characterization of Colored Noise

In the following sections we shall elaborate on various theoretical methods being tailored to investigate stochastic differential equations driven by colored noise sources, Eq. (2.8). These dynamical flows are rather difficult to study because the statistical properties of such flows depend at least on two intrinsic parameters which, apart from the statistical nature of the random force (i.e., Gaussian versus non-Gaussian noise) characterize the correlation function of the noise. The first parameter is its overall *noise intensity* D, which we identify with the zero-frequency part of the power spectrum of the (stationary) noise source $\xi(t)$,

$$2D \equiv \int_{-\infty}^{\infty} |\langle \xi(t)\xi(0)\rangle|\, dt \equiv S_{\xi}(\omega = 0) \tag{3.1}$$

The second parameter refers to the intrinsic *correlation time* τ of $\xi(t)$,

$$\tau \equiv \frac{\int_0^\infty |\langle \xi(t)\xi(0)\rangle|\, dt}{\langle \xi^2\rangle} \tag{3.2}$$

Thus, the complete theoretical analysis of the noisy dynamical flow involves a study in terms of a two-parameter space (D, τ); certainly a

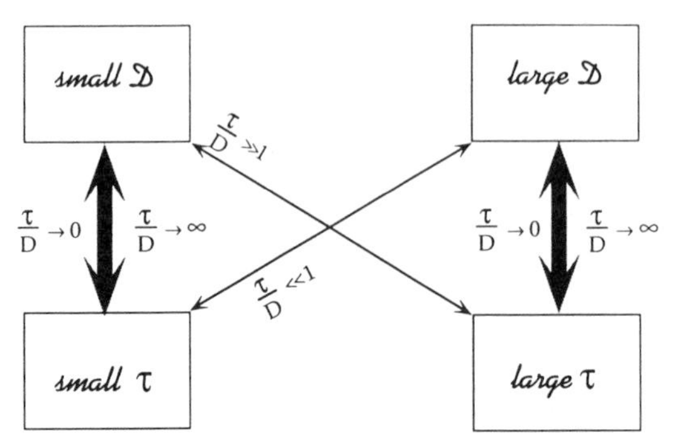

Figure 3.1. Study of colored noise: The various asymptotic regimes in parameter space (τ, D) are indicated by the set of boxes. The arrows connect mutually asymptotic regimes.

rather formidable task. Accurate approximation schemes for colored noise are thus expected only in the asymptotic limits of one or both parameters D and/or τ. These possible asymptotic regimes are depicted in Fig. 3.1. Note, that for the asymptotic regimes indicated in Fig. 3.1 by vertical double arrows, one must also distinguish between the limiting behaviors $\tau/D \to 0$, or $\tau/D \to \infty$, that is, one has to account for the relative change of one parameter compared to the relative change of its accompanying second parameter. Because most of the practical applications are driven by weak noise intensities our primary goal has been to develop workable approximation schemes that hold for weak noise D (see Section V). For a state vector $\mathbf{x} = (x_1, \ldots, x_n)$ we shall assume in the following a noisy, multidimensional dynamical law, which is of the form

$$\dot{x}_\alpha = f_\alpha(\mathbf{x}, \lambda) + \sum_{i=1}^{n} g_{\alpha i}(\mathbf{x}, \lambda)\xi_i(t) \qquad \alpha = 1, \ldots, n \tag{3.3}$$

wherein λ denotes a set of external control parameters, and $\{\xi_i(t)\}$ are *colored* noise forces. In particular, we restrict ourselves to memory-free dynamical laws for $\dot{\mathbf{x}}(t)$, which generally model nonequilibrium phenomena. For the vast literature on colored thermal equilibrium fluctuations obeying a fluctuation–dissipation theorem in terms of a memory friction we refer the reader to the references given in Section VIII. The dynamics in Eq. (3.3) constitute a *non-Markovian process* $\mathbf{x}(t)$; this occurs because $\mathbf{x}(t)$ is with colored noise $\{\xi_i(t)\}$ *not* made up of independent, infinitesimal increments that are statistically uncorrelated [93]. Moreover, it should be noted that the drift vectors $\{f_\alpha(\mathbf{x}, \lambda)\}$ are generally *not* identical with the deterministic flow F_α

$$\dot{x}_\alpha = F_\alpha(\mathbf{x}, \lambda) \tag{3.4}$$

which is the result of the motion of the conditional average $\langle x_\alpha(t)|\mathbf{x}(t) = \mathbf{x}\rangle$, with the noise intensity approaching zero; but $f_\alpha(x, \lambda)$ contain, in general, nonlinear, noise induced effects of the fluctuations $\xi_i(t)$. This constitutes one of the major problems of any phenomenological modeling if partially internal noise sources are involved: The deterministic flow does not even determine the drift term uniquely [94]. The basis for the form in Eq. (3.3) is related to the notion already discussed in Section II. Usually, one assumes that the variables of the system separate into two classes: One class of slowly varying macroscopic variables $\mathbf{x}(t)$, the motion of which is determined by the drift vector $\mathbf{f}(\mathbf{x}, \lambda)$ varying on a time scale τ_s, and the small perturbation around this relevant motion, that is, its irrelevant motion varying on a much shorter time scale $\tau \ll \tau_s$. In this case, the random forces $\xi_i(t)$ are almost white noise forces. This limit is referred to in the literature as (see Section II) short-correlation time limit, "off-white noise", or "pink noise", respectively. In general, however, the experimenter does not monitor the complete set of all slowly varying variables. Then the noise term $\xi_i(t)$ can be correlated on the same, or on an even larger time scale, that is, $\tau \geq \tau_s$. We will refer to this situation as "moderate-to-strong noise color". Before we engage in the study of colored noise driven flows, we first elaborate on general properties of propagating non-Markovian processes

B. Time Evolution of Non-Markovian Processes

The stochastic flow generated by Eq. (3.3) is non-Markovian, whose conditional probability $R(\mathbf{x}t\,|\,\mathbf{y}s), t > s$, depends on the previous history. Thus, the operator $\mathbf{R}(t\,|\,s)$ with the kernel $R(\mathbf{x}t\,|\,\mathbf{y}s)$ is not a linear operator, but depends in a nonlinear way on the initial probability $p_0(x)$ at time t_0 of preparation [95]. Consequently, this operator fails to satisfy the celebrated Bachelier–Smoluchowski–Kolmogorov–Chapman equation [95]. Therefore, an important question is whether it is possible to construct an operator $\mathbf{G}(t\,|\,s)$ for the single-event probability $p(t)[= p(\mathbf{x}, t)]$,

$$p(t) = \mathbf{G}(t\,|\,s)p(s) \tag{3.5}$$

which satisfies the property of a propagator,

$$\mathbf{G}(t\,|\,t_1) = \mathbf{G}(t\,|\,s)\mathbf{G}(s\,|\,t_1) \qquad t \geq s \geq t_1 \tag{3.6}$$

with

$$\mathbf{G}(t^+\,|\,t) = \mathbf{1} \tag{3.7}$$

This concept allows the derivation of a master equation

$$\dot{p}(t) = \mathbf{\Gamma}(t)p(t) \tag{3.8}$$

with the operator $\mathbf{\Gamma}(t)$ given by

$$\mathbf{\Gamma}(t) = \frac{d}{du}\mathbf{G}(u \mid t)|_{u=t^+} \tag{3.9}$$

The existence of such propagators $\{\mathbf{G}(s \mid t)\}$, or corresponding (pseudo-Markovian) generators $\mathbf{\Gamma}(t)$, which *yield the identical propagation behavior of the non-Markovian single event probability has been studied some time ago* [96–99]. It has been shown that Eq. (3.5) does not determine the propagator set $\{\mathbf{G}(t \mid s)\}$ uniquely, but there exists many such sets, which in general depend on the initial probability p_0. We must stress that the kernels $\mathbf{G}(\mathbf{x}t \mid ys)$ of these propagator sets are in general *different* from the conditional probability $R(\mathbf{x}t \mid \mathbf{y}s)$; thus they cannot be used to calculate correlation functions such as $C_{ij}(t, s) = \langle x_i(t)x_j(s)\rangle$, or conditional averages, $\langle \mathbf{x}(t) \mid \mathbf{x}(s) = \mathbf{x}\rangle$. There is one important exception, however: If the system has been prepared at time t_0 without any memory of the past, and for time sets for which $R(t \mid t_0)$ is nonsingular for $t > t_0$, the operator

$$\mathbf{G}(t \mid s) = \mathbf{R}(t \mid t_0)\mathbf{R}(s \mid t_0)^{-1} \tag{3.10}$$

or

$$\mathbf{\Gamma}(t) = \dot{\mathbf{R}}(t \mid t_0)\mathbf{R}(s \mid t_0)^{-1} \tag{3.11}$$

is independent of p_0; that is, $\mathbf{G}(t \mid t_0) = \mathbf{R}(t \mid t_0)$. Therefore, $\mathbf{G}(t \mid t_0)$ can be invoked to calculate *initial*, non-Markovian correlations, such as $\langle \mathbf{x}(t)\mathbf{x}(t_0)\rangle$. For the dynamical flow in Eq. (3.3) we shall *further* assume for the following that the system has been prepared at time t_0 without any memory of the past and *without correlations between system and environment*, that is, $p_0(x, \mathit{environment}) = p_0(x)p_0(\mathit{environment})$. This preparation scheme (π) will be termed "correlation-free preparation", with the conditional preparation function $W_\pi[\text{environment} \mid \mathbf{x}(t_0)] = p_0(\text{environment})$ [97].

This concept can be generalized [100] for any preparation scheme characterized by a preparation function $W_\pi[\text{environment} \mid \mathbf{x}(t_0)]$, characterizing the distribution of microstates of the environment for given macrostate $\mathbf{x}_0 = \mathbf{x}(t_0)$ [100]. Of particular importance is the stationary preparation W_s for which $W_s[\text{environment} \mid \mathbf{x}_0]p_0(\mathbf{x}_0)$ represents the *stationary* probability of the total system (system plus bath). Then

$\mathbf{R}_s(t\,|\,t_0)$ becomes time homogeneous, that is, $\mathbf{R}_s(t\,|\,t_0) = \mathbf{R}_s(t+s\,|\,t_0+s)$, and thus can be used to calculate stationary correlation properties [95–100].

C. Correlation Formulas between Noise Functionals

In this section we restrict for the sake of clarity and simplification only, the further discussion to 1-D stochastic flows in Eq. (2.8), that is,

$$\dot{x} = f(x) + g(x)\xi(t) \tag{3.12}$$

Moreover, for a multiplicative noise function $g(x)$, which does not vanish, we can use the transform: $x \rightarrow y = \int^x [dz/g(z)]$, $h(x) = f(x)/g(x)$, to obtain the simplified, additive noise equation

$$\dot{y} = h(y) + \xi(t) \tag{3.13}$$

Its single event probability $p_t(x)$ is given in terms of an average over the noise realizations of $\xi(t)$, that is,

$$p_t(x) = \langle \delta(x(t) - x) \rangle \tag{3.14}$$

The rate of change of $p_t(x)$ then obeys

$$\dot{p}_t(x) = -\frac{\partial}{\partial x}\langle \delta(x(t)-x)\dot{x}(t)\rangle = -\frac{\partial}{\partial x}[f(x)p_t(x)] - \frac{\partial}{\partial x} g(x) \times \langle \xi(t)\delta(x(t)-x)\rangle \tag{3.15}$$

We note that a colored noise master equation for the probability $p_t(x)$ introduces a correlation between the noise $\xi(t)$ and the functional $\mathscr{F}\{\xi\} = \delta(x(t)-x)$ of the colored noise source $\xi(s)$, $t \geq s \geq t_0$ (t_0 is the time of preparation). The expression in Eq. (3.15) can only be disentangled further if we explicitly invoke the statistical properties of the noise $\xi(t)$. We now give (without proof) some important relations that are needed for the derivation of a colored noise master equation. For an explicit derivation of these relations the readers are referred to the original paper [101] and the reviews [102, 103]. Moreover, we shall explicitly assume that the random force $\xi(t)$ is of vanishing mean,

$$\langle \xi(t) \rangle \equiv C_1(t) = 0 \tag{3.16}$$

Let $F\{\xi\}$, $G\{\xi\}$ denote two functionals of $\xi(t)$. Then we have with

$F\{\xi\} = \xi(t)$ and $G\{\xi\} = \delta(x(t) - x)$ the important relation [101, 102]

$$\langle \xi(t)\delta(x(t) - x)\rangle = \sum_{n=1}^{\infty} \left(\frac{1}{n!}\right) \int_{t_0}^{t} \cdots \int_{t_0}^{t} dt_1 \cdots dt_n C_{n+1}(t, t_1, \ldots, t_n) \times \left\langle \frac{\delta^n[\delta(x(t) - x)]}{\delta\xi(t_1) \cdots \delta\xi(t_n)} \right\rangle \tag{3.17}$$

Here, $C_m(t_1, \ldots, t_m)$ denotes the mth order cumulant of the noise $\xi(t)$. The notation $\delta F\{\xi\}/\delta\xi(s)$ denotes the functional derivative; it can be viewed of as a usual derivative if we set

$$\frac{\delta F\{\xi\}}{\delta\xi(s)} = \left. \frac{dF\{\xi(t) + \lambda\delta(t - s)\}}{d\lambda} \right|_{\lambda=0} \tag{3.18}$$

assuming that both sides exist.

For a stationary *Gaussian random force* $\xi(t)$ one then finds with $C_2(t, s) \equiv C(t - s) = \langle \xi(t)\xi(s)\rangle - \langle \xi(t)\rangle\langle \xi(s)\rangle = \langle \xi(t)\xi(s)\rangle$, the useful result [104, 105]

$$\langle \xi(t)G\{\xi\}\rangle = \int_{t_0}^{t} C(t - s)\left\langle \frac{\delta G\{\xi\}}{\delta\xi(s)} \right\rangle ds \tag{3.19}$$

For two functionals $F\{\xi\}$, $G[\xi]$ one obtains for stationary Gaussian noise [102, 103]

$$\langle F\{\xi\}G\{\xi\}\rangle = \langle F\{\xi\}\rangle\langle G\{\xi\}\rangle + \sum_{n=1}^{\infty} \frac{1}{n!} \int_{t_0}^{t} \cdots \int_{t_0}^{t} \left\langle \frac{\delta^n F\{\xi\}}{\delta\xi(t_1) \cdots \delta\xi(t_n)} \right\rangle \times \left\langle \frac{\delta^n G\{\xi\}}{\delta\xi(s_1) \cdots \delta\xi(t_n)} \right\rangle \prod_{i=1}^{n} C(t_i - s_i)\, dt_i\, ds_i \tag{3.20}$$

With these results in hands we are well equipped to tackle the master equation for colored noise in Eq. (3.15) in greater detail.

D. The Colored Noise Master Equation

Here we only consider the case of stationary Gaussian noise $\xi(t)$ of vanishing mean $\langle \xi(t)\rangle = 0$ and correlation $\langle \xi(t)\xi(s)\rangle = C(t - s)$ [see Eq. (3.19)]. Then, the rate of change of the single-event probability $p_t(x) =$

$\langle \delta(x(t)-x) \rangle$ from Eqs. (3.15) and (3.19) is given by

$$\dot{p}_t(x) = -\frac{\partial}{\partial x}[f(x)p_t(x)] - \frac{\partial}{\partial x} g(x) \int_{t_0}^{t} C(t-s) \left\langle \frac{\delta[\delta(x(t)-x)]}{\delta \xi(s)} \right\rangle ds \tag{3.21}$$

With

$$\frac{\delta}{\delta \xi(s)} \delta(x(t)-x) = \left[-\frac{\partial}{\partial x} \delta(x(t)-x) \right] \frac{\delta x(t)}{\delta \xi(s)} \tag{3.22}$$

we have from the dynamical law in Eq. (3.12) for the functional derivative $\delta x(t)/\delta \xi(s)$ the integral equation [101, 102]

$$\frac{\delta x(t)}{\delta \xi(s)} = \theta(t-s) \left\{ g(x(s)) + \int_s^t du \left(\frac{\partial \dot{x}(u)}{\partial x(u)} \right) \frac{\delta x(u)}{\delta \xi(s)} \right\} \tag{3.23}$$

Here $\theta(t-s)$ is the unit step function expressing causality. Its solution is readily found to read

$$\frac{\delta x(t)}{\delta \xi(s)} = \theta(t-s) g(x(s)) \exp \int_s^t \frac{\partial \dot{x}(u)}{\partial x(u)} du \tag{3.24}$$

$$= \theta(t-s) g(x(s)) \exp \int_s^t \{ f'(x(u)) + g'(x(u)) \xi(u) \} du \tag{3.25}$$

where $h'(x)$ denotes differentiation $dh(x)/dx$. Equation (3.25) can be recast in alternative, and more appealing form [103, 106]

$$\frac{\delta x(t)}{\delta \xi(s)} = \theta(t-s) g(x(t)) \exp \int_s^t \left\{ f'(x(u)) - f(x(u)) \frac{g'(x(u))}{g(x(u))} \right\} du \tag{3.26}$$

Combining Eq. (3.22) with Eqs. (3.21) and (3.26) we then have for *Gaussian noise* $\xi(t)$ the formally exact result [101–103],

$$\dot{p}_t(x) = -\frac{\partial}{\partial x}[f(x)p_t(x)] + \frac{\partial}{\partial x} g(x) \frac{\partial}{\partial x} g(x) \cdot \int_{t_0}^{t} ds C(t-s) x \Big\langle \delta(x(t)-x)$$

$$\times \exp \int_s^t [f' - (fg'/g)] \, du \Big\rangle \tag{3.27}$$

which with different notation has been given first in [101]. At this point, the exact relation in Eq. (3.27) cannot be generally simplified further. Because of the function $\delta(x(t)-x)$, a closed expression that involves only the single-event probability $p_u(x), t \geq u \geq t_0$ only results if either $\delta x(t)/\delta\xi(s)$ does not depend on the process $x(t)$, or if it depends on $x(s)$ solely on its endpoint time t. Classes of such exact, closed colored noise master equations (e.g., all linear processes, $\dot{x} = a + bx + c\xi(t)$, nonlinear processes driven by two-state noise and/or white noise) have been discussed in [101–103]. The main result obtained in Eq. (3.27) will serve as our appropriate starting point in Section V to construct various approximation schemes.

E. Master Equation for a Linear Process Driven by Gaussian Colored Noise

We shall illustrate the result in Eq. (3.27) for a linear colored noise process. Let

$$\dot{x} = a - bx + \xi(t) \tag{3.28}$$

From Eq. (3.26) we obtain

$$\frac{\delta x(t)}{\delta \xi(s)} = \theta(t-s)\exp[-b(t-s)] \tag{3.29}$$

Thus, the master equation, Eq. (3.27), takes on a Fokker–Planck like form, that is,

$$\dot{p}_t(x) = -a\frac{\partial}{\partial x}p_t(x) + b\frac{\partial}{\partial x}[xp_t(x)] + \left\{\int_{t_0}^{t} ds C(t-s)\exp[-b(t-s)]\right\} \times \frac{\partial^2}{\partial x^2}p_t(x) \tag{3.30}$$

Note that with time-independent drift coefficients in Eq. (3.28) and stationary Gaussian noise $\xi(t)$ the effective diffusion in Eq. (3.30) is *time dependent*, and it may even take on negative values when $C(t-s)$ is, for example, an oscillatory-like function. The solution of Eq. (3.30) constitutes with $p_0(x) = \delta(x - x_0)$ a Gaussian, non-Markovian probability, which explicitly reads [96, 107]

$$p_t(x) \equiv R_t(x \mid x_0) = \frac{\exp\{-\frac{1}{2}\alpha^{-1}(t, t_0)[x - \beta(t, x_0, t_0)]^2\}}{[2\pi\alpha(t, t_0)]^{1/2}} \tag{3.31}$$

where

$$\alpha(t, t_0) = \int_{t_0}^{t} \exp[-2b(t-s)]D(s)\, ds \tag{3.32}$$

$$D(s) = 2\int_{t_0}^{s} dr C(s-r)\exp[-b(s-r)] \tag{3.33}$$

$$\beta(t, x_0, t_0) = x_0 \exp[-b(t-t_0)] + a\int_{t_0}^{t} \exp[-b(t-s)]\, ds \tag{3.34}$$

Thus, $\mathbf{R}(t\,|\,t_0)$ is a Gaussian, and with the initial probability p_0 also Gaussian, the time evolution of $p(t) = \mathbf{R}(t\,|\,t_0)p_0$ will remain a Gaussian. This clearly no longer holds for a non-Gaussian initial probability $p_0(x_0)$. We close this section with some general observations about non-Markovian master equations as exhibited by Eq. (3.27). The non-Markovian character of the process $x(t)$, Eq. (3.12), is reflected by the dependence of $\dot{p}_t(x)$ on the initial time of preparation $t = t_0$ in Eq. (3.27). Moreover, the initial rate of change of $p_t(x)$ is given by

$$\dot{p}_{t=t_0}(x) = -\frac{\partial}{\partial x}[f(x)p_{t=t_0}(x)] \tag{3.35}$$

Equation (3.35) holds true for any noise statistics with nonsingular cumulants $C_n(t_1, \ldots, t_n)$.

IV. COLORED TWO-STATE NOISE

In Section III the emphasis has been put on the equation of motion for the time evolution of the single-event probability, that is, the master equation. The environmental colored noise fluctuations are frequently based on the cumulative effect of an abundance of environmental factors. The central limit theorem implies then that the fluctuations are distributed Gaussian. As demonstrated in Section III, any Gaussian process leads to a closed equation of motion of the probability. Moreover, its solution for the conditional probability is solely determined by the vector of mean values and the covariance matrix [51]. Moreover, any process resulting from a linear transformation of Gaussian processes (Markovian or non-Markovian) is again Gaussian. Colored noise processes that are the result of a *nonlinear* transformation of a Gaussian process can also be considered to be exactly solvable. A set of criteria, which show when a process $y(t)$ can be related (via a nonlinear transformation) to a linear transformation of a Gaussian process, can be inferred from the literature [108]. The models of Hongler [109] are precisely of this form, being a nonlinear transformation of a Gaussian process [108, 110]. For Gaussian colored noise sources that result from an embedding of an n-dimensional Gauss–Markov process the statistical properties together with the spectral behavior are known explicitly [111]. Likewise, colored noise Markovian processes, which via the Darboux procedure, the Abraham–Moses procedure, the Pursey procedure, or the supersymmetry procedure, are isospectral with the quantum harmonic oscillator [112–117] can be considered as exactly solvable. Such specific examples are the models by Hongler and Zheng [118, 119] and by Razavy [120]. Other examples of noise sources, described by a Fokker–Planck process that can be related to exactly solvable 1-D Schrödinger equations [5, 7, 121], can be found in [122]. Next we shall focus on a class of exactly solvable colored noise driven nonlinear systems, whose stationary probability and mean first-passage times can be obtained (up to quadratures) in exact closed form for an arbitrarily chosen nonlineary $f(x)$ [see Eq. (2.8)].

A. Correlated Two-State Noise

A class of correlated noise that has found applications in numerous systems is *two-state noise*, that is, a noise that switches back and forth between two prescribed state values with a waiting time probability that is Poissonian. Note that within any switching process in which intradomain-of-attraction motion is filtered out can satisfactorily be modeled by such two-state noise. For the sake of simplicity we confine the discussion here to symmetric two-state noise (for asymmetric two-state noise the reader may consult chapter 9 in [6]), which switches back and forth between the state $\xi = a$ and $\xi = -a$; that is,

$$\xi(t) = a(-1)^{n(t)} \tag{4.1}$$

where $n(t) \equiv n(0, t)$ is a Poisson process with parameter λ. Put differently, $\xi(t)$ in Eq. (4.1) is a two-state Markovian process [6, 7]. Let us investigate its statistical properties. With $\xi(0) = a$, its mean reads,

$$\begin{aligned} \langle \xi(t) \rangle &= a \sum_{m=0}^{\infty} (-1)^m P[n(t) = m] \\ &= a \exp(-\lambda t) \sum_{m=0}^{\infty} \frac{(-1)^m (\lambda t)^m}{m!} \\ &= a \exp[-2\lambda t] \qquad t > 0 \end{aligned} \tag{4.2}$$

Likewise one readily evaluates the correlation as

$$\begin{aligned}\langle \xi(t)\xi(s)\rangle &= a^2\langle(-1)^{n(0,t)+n(0,s)}\rangle \qquad t<s \\ &= a^2\langle(-1)^{2n(0,s)+n(s,t)}\rangle \\ &= a^2\langle(-1)^{n(s,t)}\rangle \\ &= a^2\exp(-2\lambda|t-s|) \end{aligned} \tag{4.3}$$

In particular, note that the correlation is time homogeneous although $\xi(t)$ in Eq. (4.2) is not stationary. Let us now distribute the initial value. By use of the symmetric initial probability for the state variable $\rho_0(u) = \frac{1}{2}\{\delta(u+a)+\delta(u-a)\}$ the noise $\xi(t)$ assumes a zero mean. For the sequence of time instants $[t_1 \geq t_2 \geq \ldots \geq t_n]$ we then find for the nth correlation m_n

$$m_n = \langle \xi(t_1)\cdots\xi(t_n)\rangle = \langle \xi(t_1)\xi(t_2)\rangle\langle \xi(t_3)\cdots\xi(t_n)\rangle \tag{4.4}$$

or

$$m_n = a^2\exp[-2\lambda(t_1-t_2)]m_{n-2} \tag{4.5}$$

Here we used the fact that the statistics of nonoverlapping time intervals are independent of each other. The result in Eq. (4.5) can be generalized to yield

$$\langle \xi(t_1)\xi(t_2)G[\xi(s)]\rangle = \langle \xi(t_1)\xi(t_2)\rangle\langle G[\xi(s)]\rangle \tag{4.6}$$

where with $s \leq t_2 \leq t_1$, $G[\xi(s)]$ is a functional of the two-state noise. The analogue of the Furutsu [104] and Novitrov [105] correlation formula for the case of two-state noise has been derived by Klyatskin [83, 84]; that is, with $\theta(x)$ being the step function

$$\langle \xi(t)G[\xi]\rangle = a^2\int_0^t ds\,\exp[-2\lambda(t-s)]\cdot\left\langle \frac{\delta}{\delta\xi(s)}G[\xi(u)\theta(s-u^-)]\right\rangle \tag{4.7}$$

where the noise dependence in $G[\xi]$ is *switched-off* for times $u>s$; that is, $\xi \equiv 0$ for $u>s$.

B. Master Equation for Colored Two-State Noise Driven Nonlinear Flows

Give the nonlinear flow in Eq. (3.12), that is, $\dot{x} = f(x) + g(x)\xi(t)$, with $\xi(t)$ being *correlated two-state noise* in Eq. (4.1), the rate of change of $p_t(x)$ in

Eq. (3.15) can, by use of Eq. (4.7), be recast as

$$\dot{p}_t(x) = -\frac{\partial}{\partial x}[f(x)p_t(x)] - \frac{\partial}{\partial x} g(x)a^2 \int_0^t ds[\exp -2\lambda(t-s)]\left\langle \frac{\delta}{\delta\xi(s)} \hat{\delta}_t(x(t)-x)\right\rangle \quad (4.8)$$

where $\hat{\delta}_t = \delta_t[\xi(u)\theta(s-u)]$ and $t_0 = 0$. From the dynamical equation of motion we find that $\delta_t \equiv \delta(x(t)-x)$ satisfies

$$\dot{\delta}_t = -\frac{\partial}{\partial x}[f(x)\delta_t] - \frac{\partial}{\partial x}[g(x)\xi(t)\delta_t] \quad (4.9)$$

and $\hat{\delta}_{t=s} = \delta(x(s)-x)$. With $\xi(u)$ being switched off for times $u > s$ we thus find for $\hat{\delta}_t$ the differential equation

$$\frac{\partial}{\partial t}\hat{\delta}_t = -\frac{\partial}{\partial x}[f(x)\hat{\delta}_t] \qquad t > s \quad (4.10)$$

Therefore, its solution can be cast in operator form to read

$$\hat{\delta}_t = \exp\left\{-\frac{\partial}{\partial x} f(x)(t-s)\right\}\delta(x(s)-x) \quad (4.11)$$

Observing that $(\delta/\delta\xi_s)\delta(x(s)-x) = -\frac{\partial}{\partial x} g(x)\delta(x(s)-x)$ we end up with a closed-form master equation

$$\dot{p}_t(x) = -\frac{\partial}{\partial x}[f(x)p_t(x)] + a^2\frac{\partial}{\partial x} g(x)\int_0^t ds \exp\left\{-(t-s)\left[2\lambda + \frac{\partial}{\partial x} f(x)\right]\right\}\frac{\partial}{\partial x} g(x)p_s(x) \quad (4.12)$$

The stationary probability $p(x;\lambda)$ is obtained from Eq. (4.12) if we integrate between $s\varepsilon[0,\infty]$, and equate the probability current at zero value, that is,

$$f(x)p(x;\lambda) = a^2 g(x)\left[\frac{1}{2\lambda + (d/dx)f(x)}\right]\frac{d}{dx}\{g(x)p(x;\lambda)\} \quad (4.13)$$

After multiplication from left with $g^{-1}[2\lambda + (d/dx)f]$ one finds an ordinary first-order equation for $p(x;\lambda)$. Its solution therefore is readily found as [6, 83–85, 102]

$$p(x;\lambda) = Z^{-1} \frac{|g(x)|}{[a^2g^2(x) - f^2(x)]} \exp\left\{2\lambda \int^x dy \frac{f(y)}{[a^2g^2(y) - f^2(y)]}\right\} \tag{4.14}$$

where Z^{-1} is the normalization constant. Note that $p(x;\lambda)$ has a support on all those x values for which the term $[a^2g^2(x) - f^2(x)]$ takes on a positive value! With the correlation time $\tau \equiv (2\lambda)^{-1}$, $p(x;\lambda)$ depends exponentially on the colored noise correlation τ. We conclude this section by presenting (without proof) a few further relations that are of use in applying two-state noise in colored noise driven flows.

The curtailed characteristic functional

$$\phi_t[v] = \langle \exp i \int_0^t ds \xi(s) v(s) \rangle \tag{4.15}$$

obeys the exact second ordinary differential equation

$$\frac{d^2}{dt^2}\phi_t + \left[2\lambda - \frac{1}{v(t)}\frac{dv(t)}{dt}\right]\frac{d\phi_t}{dt} + a^2v^2(t)\phi_t = 0 \tag{4.16}$$

which with $\phi_0 = 1$, and $\dot{\phi}_{t=0} = 0$ generally is not explicitly solvable. Equivalently, Eq. (4.16) can be recast as an integrodifferential equation

$$\frac{d}{dt}\phi_t = -a^2v(t)\int_0^t ds v(s) \exp[-2\lambda(t-s)]\phi_s \tag{4.17}$$

From a nonequilibrium Brownian motion driven by correlated two-state noise $\xi(t)$, that is, with $a^2 \equiv 2kT\gamma\lambda$

$$\begin{aligned} \dot{x} &= u \\ \dot{u} &= f(x) - \gamma u + \xi(t) \end{aligned} \tag{4.18}$$

we obtain for the stationary probability $p(x, u; \lambda)$ the exact equation

$$\Gamma p(x,u;\lambda) = \frac{\partial}{\partial u}\left(\frac{2kT\gamma\lambda}{2\lambda + \Gamma}\right)\frac{\partial}{\partial u} p(x,u;\lambda) \tag{4.19}$$

where $\Gamma \equiv u(\partial/\partial x) + f(x)(\partial/\partial u) - \gamma(\partial/\partial u)u$ is the deterministic drift operator. With the parameter $\lambda \to \infty$, telegraphic noise approaches Gaussian white noise of vanishing mean and correlation $2kT\gamma\delta(t)$. With $\lambda \to \infty$, Eq. (4.19) reduces to the usual equation for Brownian motion in a force field $f(x) = -[dU(x)/dx]$. Equation (4.19) can further be recast as a partial differential equation, that is,

$$\left\{(2\lambda + \Gamma)^2\Gamma - 2kT\gamma\lambda(2\lambda + \Gamma)\frac{\partial^2}{\partial u^2} - \gamma(2\lambda + \Gamma)\Gamma + 2kT\gamma\lambda\frac{\partial^2}{\partial x\,\partial u} - \frac{df}{dx}\Gamma\right\}p(x, u; \lambda) = 0 \quad (4.20)$$

which generalizes the usual Klein–Kramers equation [45, 46, 123]. With Eq. (4.18) violating the fluctuation–dissipation relation for any finite λ, the solution of Eq. (4.20) clearly no longer exhibits the Boltzmann form, but the coordinate x and its velocity $\dot{x} = u$ now become statistically dependent variables.

C. Mean First-Passage Times

A quantity that carries valuable dynamic information is the first average of the first-passage time random variable, the so-called mean first-passage time (MFPT). The MFPT can be used to characterize relevant time scales in nonlinear dynamical problems such as they originate in chemical kinetics, decay of arbitrary metastable states, decay of unstable states, and nucleation [123], to name only a few. With a colored noise driven flow, the concept of the MFPT becomes rather nontrivial [124, 125]. For two-state noise with exponentially distributed waiting time, however, the complexity can be handled in analytical closed form.

In Section IV.B we already made extensive use of the fact that the stationary probability obeys an ordinary differential equation being of first order. Not totally surprising, this fact also holds true for the derivative of the MFPT [86].

Let $T_+(y)$ denote the MFPT for a particle, which started out at initial time $t_0 = 0$ at $x = y$, with initial velocity $\xi(0) = +a$, i.e. $\rho_0(u) = \delta(a - u)$. Here y is restricted to some a priori prescribed interval $I = [x_A, x_B]$. $T_-(y)$ is the MFPT for a particle starting out with initial velocity $\xi(0) = -a$, i.e., $\rho_0(u) = \delta(a + u)$. For the dynamical flow

$$\dot{x} = f(x) + g(x)\xi(t) \quad (4.21)$$

The coupled equation for $T_\pm(y)$ is explicitly given by [86, 90]

$$\begin{aligned} (f + ag)T_+' - \lambda T_+ + \lambda T_- &= -1 \\ (f - ag)T_-' - \lambda T_- + \lambda T_+ &= -1 \end{aligned} \quad (4.22)$$

Here the prime denotes differentiation after y, that is, $T'(y) = dT/dy$.

Upon eliminating T_+ or T_- one obtains an ordinary first-order differential equation for T'_+ or $T^{-\prime}$, respectively. The MFPT is therefore readily integrated if only the boundary conditions are known. For absorbing boundaries at x_A and x_B the exact MFPT has been obtained first in [86], and has been reobtained by use of alternative techniques in [87–89, 126, 127]. The essential difficulty in obtaining the MFPT for non-Markovian processes is the incorporation of the correct boundary conditions [86, 90, 124, 125]. For a detailed discussion of implementing the correct absorbing and/or reflecting boundary conditions we refer the interested reader to the original literature [86, 90]. For the important problem of escape from a domain of attraction, where with $x_S > x_A$, x_S being a metastable state and $x_* < x_B$, x_* denoting an unstable (barrier toplike) state, the MFPT with x_A being *reflecting* and x_B being *absorbing* yields the time scale for the escape; and its inverse yields the reaction rate, respectively. This MFPT can then be obtained in closed form, that is, from Eq. (4.14) in [90] we have with $D \equiv a^2/2\lambda$ and $p \equiv p(x; \lambda)$ given in Eq. (4.12)

$$\begin{aligned} T_+^{\text{ref}}(y) = & \int_y^{x_B} dz \frac{g}{D[g + (f/a)]^2[(g - (f/a)]p} \int_{x_A}^{z} du g^{-1}(g + f/a)p \\ & + \frac{D}{a} p(x_A; \lambda)[g^2 - (f/a)^2]_{x=x_A} g^{-1}(x_A) \\ & \cdot \int_y^{x_B} du \frac{g}{D[g + (f/a)]^2[g - (f/a)]p} \end{aligned} \tag{4.23}$$

and a similar expression holds for $T_-^{\text{ref}}(y)$ [90]. At weak noise strength $D \ll 1$, the use of the steepest descent approximation yields the reaction rate $k \equiv 1/T_+^{\text{ref}}$ as [90, 128–131]

$$k = \frac{1}{2\pi} [|f'(x_1)| f'(x_*)]^{1/2} \exp(-\Delta\phi/D) \tag{4.24}$$

with the "Arrhenius energy" given by

$$\Delta\phi = -\int_{x_1}^{x_*} du \frac{f}{[g - (f/a)][g + (f/a)]} \tag{4.25}$$

Alternatively, this reaction rate can be evaluated directly (via the method of flux over population), if one solves Eq. (4.12) for a constant, vanishing probability flux [128, 129]. The result again can be cast into a closed form involving only two quadratures in terms of the stationary probability. This exact quadrature expression thus intrinsically incorporates all the corrections to the steepest-descent expression in Eq. (4.24). These latter corrections are of relevance for finite but small effective barriers $\Delta\phi/D$.

V. COLORED NOISE THEORY: APPROXIMATION SCHEMES

Apart from the specific set of classes of systems (see Sections III.D and IV) that yield a closed-form master equation, the relation in Eq. (3.27) cannot be evaluated explicitly. Further theoretical progress must therefore invoke some form of approximation. In practice, such approximate schemes become useful only if the approximation reduces to an approximate Fokker–Planck process, or at best, a Fokker–Planck like master equation for the single event probability $p_t(x)$. The tacit assumption with such a procedure is that the resulting approximate solution in fact presents a useful estimate for the actual non-Markovian process in Eq. (3.12). Of course, such an approximation is not expected to describe *all* of the statistical information of the true non-Markovian process but only some limited statistical quantities such as its stationary probability, or its transient initial correlation function. As it will become clear below, approximation schemes that also approximate the dynamics equally well, such as the stationary two-point correlation function, the relaxation time, or its mean first-passage time, are much more difficult to obtain. Next, we shall report, extend, and interpret various novel approximation schemes for colored noise driven Langevin equations. Particular emphasis will be put, wherever possible, on a study of the regime of validity of such corresponding approximation schemes.

A. Small Correlation Time Expansion

If the noise color is "off-white", that is, close to the white noise limit, it seems appropriate to search for an effective Fokker–Planck like equation. Our starting point for this approximation is the formally exact master equation in Eq. (3.21) or (3.27). If we expand $\delta x(t)/\delta\xi(s)$ into a Taylor series around the latest time t,

$$\frac{\delta x(t)}{\delta \xi(s)} = \frac{\delta x(t)}{\delta \xi(t)} + \sum_{n=1}^{\infty} \frac{(-1)^n}{n!} (t-s)^n \left[\frac{d^n}{ds^n} \frac{\delta x(t)}{\delta \xi(s)} \right]_{s=t^-} \tag{5.1}$$

one finds from Eqs. (3.12) and (3.23)

$$\frac{\delta x(t)}{\delta \xi(s)} = \theta(t-s)\Big\{ g(x(t)) + [g'(x(t))f(x(t)) - g(x(t))f'(x(t))](t-s)$$
$$+ \Big[\Big\{ f^2\Big[f\Big(\frac{g}{f}\Big)'\Big]' - g^2\Big[g\Big(\frac{f}{g}\Big)'\Big]'\xi(t)\Big\}_{x=x(t)} (t-s)^2\Big] + \cdots \Big\} \tag{5.2}$$

Here, the prime ($'$) again denotes differentiation with respect to x. Obviously, this expansion involves the noise $\xi(t)$ already in second order. This leads to new correlations that again must be disentangled with relations such as Eq. (3.19). We will now specify the Gaussian noise $\xi(t)$ to the Ornstein–Uhlenbeck process,

$$C_2(t-s) = \frac{D}{\tau} \exp(-|t-s|/\tau) \tag{5.3}$$

If we *truncate* Eq. (5.1) at first order (i.e., $n = 1$) one finds for the master equation in Eq. (3.27) with $t_0 \equiv 0$ [65, 66]

$$\dot{p}_t(x) = -\frac{\partial}{\partial x}[f(x)p_t(x)] + D\frac{\partial}{\partial x} g(x) \frac{\partial}{\partial x}[g(x)h(x,t)p_t(x)] \tag{5.4}$$

where

$$h(x,t) = [1 - \exp(-t/\tau)] + \tau g(x)\Big(\frac{f(x)}{g(x)}\Big)' \Big\{[1 - \exp(-t/\tau)] - \frac{t}{\tau}\exp(-t/\tau)\Big\} \tag{5.5}$$

This is the time-dependent small τ approximation describing the time evolution of $p_t(x)$. Next we address the long-time limit, that is, we neglect the transients in Eq. (5.5) to obtain the *standard small τ approximation*

$$\dot{p}_t(x) = -\frac{\partial}{\partial x}[f(x)p_t(x)] + D\Big\{\frac{\partial}{\partial x} g(x) \frac{\partial}{\partial x} g(x)[1 + \tau g(x)\{f(x)/g(x)\}']\Big\} p_t(x) \tag{5.6}$$

By far this presents the most often used small correlation time approximation [55–77, 132, 133]. The stationary probability $p(x;\tau)$ is given by

$$p(x;\tau) = \frac{Z^{-1}}{|g(x)\{1 + \tau g(x)[f(x)/g(x)]'\}|} \cdot \exp \int^{x} \frac{f(y)\,dy}{Dg^{2}(y)\{1 + \tau g(y)[f(y)/g(y)]'\}} \tag{5.7}$$

where Z is the normalization constant. This very result has been repeatedly derived in the literature by a variety of methods mentioned in Section II.B below Eq. (2.11). Some authors [57, 61–63, 66, 67, 132, 133] also consider higher order corrections to Eq. (5.6) being proportional to $D\tau^{n}$. By doing so, however, one simply neglects the noise-dependent contributions of the type in Eq. (5.2), which also yield additional Fokker–Planck terms together with non-Fokker–Planck terms: As first pointed out in [134], and later reiterated in [135, 136–138], such a formal ordering of the τ expansions is fictitious, and does not improve the approximation consistently. In short, these higher order terms are of the same order as other neglected Fokker–Planck and non-Fokker–Planck terms. We next state a few properties of the approximation in Eq. (5.6).

1. For $\tau = 0$, one recovers the white noise result from both Eqs. (5.4) and (5.6); that is, the white noise Fokker–Planck equation for a white noise Langevin equation Eq. (3.12), being interpreted in the Stratonovich sense.
2. The drift and diffusion coefficients in Eq. (5.6) differ in order τ from the corresponding Markovian Fokker–Planck equation. In particular, with increasing τ the diffusion coefficient in Eq. (5.6) may take on zeros, and negative values, thereby introducing unphysical, approximation-related boundaries for the non-Markovian process.
3. With the diffusion in Eq. (5.6) generally not satisfying strict positivity, there exists no white noise Langevin equation which is stochastically equivalent with Eq. (5.6), that is, $p_t(x)$ cannot be sampled in terms of random trajectories.
4. The solution $p_t(x \mid x_0)$ of the time-dependent equation Eq. (5.4) with initial condition $p_0 = \delta(x - x_0)$ represents an approximation to the conditional probability $R(xt \mid x_0 t_0 = 0)$ of the non-Markovian process with correlation-free initial preparation (see Section III.B). Thus, Eq. (5.4) can be utilized for the calculation of *initial* correlations in the regime of small noise color τ.
5. As demonstrated below, the regime of validity of the small τ

approximation is limited to small correlation times $\tau \to 0$, with (τ/D) being a small quantity, and to regimes in state space where $\tau g(x)[f(x)/g(x)]' < 1$. In particular with (τ, D) both understood as being dimensionless, the weak noise asymptotic regime $\tau \to 0$; $\tau/D \gg 1$ is not within the regime of applicability of the small τ approximation in Eqs. (5.4) and (5.6).

Now, let us consider the contribution of the second Taylor coefficient in Eq. (5.2) in greater detail. This part contributes, with $g(x) \equiv 1$, to the master equation in Eq. (5.5) the term (see in [134])

$$D\tau^2 \frac{\partial^2}{\partial x^2} \left\{ [(f'(x))^2 - f(x)f''(x)]p_t(x) + \frac{D}{\tau} f''(x) \frac{\partial}{\partial x} \int_0^t \left\langle \delta(x(t) - x) \frac{\delta x(t)}{\delta \xi(s)} \right\rangle \exp\left(-\frac{(t-s)}{\tau}\right) ds \right\} \tag{5.8}$$

If we approximate $\delta x(t)/\delta\xi(s)$ by its first term [see Eq. (3.23)], that is, $\delta x(t)/\delta\xi(s) \simeq 1$, we find, upon neglect of transients, the following third-order non-Fokker–Planck contribution to Eq. (5.6)

$$D^2\tau^2 \frac{\partial^2}{\partial x^2} f(x) \frac{\partial}{\partial x} p_t(x) \tag{5.9}$$

By use of a nonequilibrium potential $\phi_t(x, \tau)$; that is, $p_t \propto \exp(-\phi_t(x, \tau)/D)$, we note that each Kramers–Moyal moment yields a contribution of order D^{-1}, D^0, and higher to $\dot{p}_t(x)$. If we collect the singular terms we find the following contributions to $\dot{p}_t(x)$ [134] with Eq. (5.8)

$$\dot{p}_t(x) \sim p_t \left\{ \frac{A_1^{(1)}}{D} + \frac{A_1^{(2)}}{D} + \frac{\tau}{D} [A_2^{(2)} + \tau A_3^{(2)} + \tau A_1^{(3)}] \right\} + O(D^0) \tag{5.10}$$

Here, the superscript in the functions $\{A_n^{(i)}\}$ indicates a contribution stemming from the i-th order Kramers–Moyal moment. Thus, we immediately see that it is not consistent to keep contributions of order $D\tau^n$, $n > 1$ in the Fokker–Planck like equation in Eq. (5.5), while at the same time neglecting non-Fokker–Planck terms. Moreover, with $\tau \neq 0$, the correction to the white noise limit should be small, that is, the *parameter* (τ/D) *must be small* in order for Eq. (5.6) to present a meaningful correction to the white noise limit! In recent work on small noise color $\tau \ll 1$, Fox [106, 137] attempted to patch up some of the shortcomings inherent in Eq. (5.6), such as the problem with unphysical boundaries. In this approximate treatment he obtains for the effective diffusion operator the result [106]

$$D \frac{\partial}{\partial x} g(x) \frac{\partial}{\partial x} \frac{g(x)}{[1 - \tau g(x)(f(x)/g(x))']} \tag{5.11}$$

It corresponds to formally summing up a geometric series, that is, $[1 + \tau h(x)] \rightarrow 1/[1 - \tau h(x)]$. This expression has the advantage that the small τ theory in Eq. (5.6) with the diffusion coefficient substituted by Eq. (5.11) yields the exact (Gaussian) stationary probability for a linear process. However, the diffusion in Eq. (5.11) is in general still not strictly positive for all x values. With Eq. (5.11) the corresponding stationary probability $p^{\mathrm{Fox}}(x;\tau)$ reads

$$p^{\mathrm{Fox}}(x;\tau) = Z^{-1} \frac{|[1 - \tau g(x)(f(x)/g(x))']|}{|g(x)|} \times \exp \int^x dy \frac{f(y)[1 - \tau g(y)(f(y)/g(y))']}{D g^2(y)} \tag{5.12}$$

Here we stress that the validity of the Fox approximation in Eq. (5.11) is *restricted to the very same regime of validity as the standard small τ approximation* in Eq. (5.6); that is, $\tau \rightarrow 0$ with $\tau / D \ll 1$.

We bring out further complications not present in the 1-D non-Markovian flow Eq. (3.12) by turning to the multidimensional stochastic flow in Eq. (3.3). Use of the functional methods in Section III.C for the multidimensional flow yields in terms of the functional derivative in Eq. (3.23) (we use, apart from the index i, the summation convention over equal indexes)

$$\frac{\delta x_\alpha}{\delta \xi_i(s)} = \theta(t-s) \left\{ \left(\int_s^t du \left[\frac{\partial f_\alpha}{\partial x_\beta} + \frac{\partial g_{\alpha i}}{\partial x_\beta} \right] \frac{\delta x_\beta(u)}{\delta \xi_i(s)} \right) + g_{\alpha i}(x(s)) \right\} \tag{5.13}$$

being an analogue for Eq. (3.23) for the multivariable case [139–142]. One finds that *generally there does not even exist* a consistent Fokker–Planck like structure in first order in the correlation time τ. Such a small τ multidimensional Fokker–Planck like approximation does exist, however, if the Gaussian correlations $\langle \xi_i(t) \xi_j(s) \rangle = C_{ij}(t-s) \equiv D_{ij} \gamma_{ij}(t-s)$, with correlation time τ_{ij}, are diagonal and all are of equal correlation time, $\tau_{ii} = \tau_i \equiv \tau$ for all i. It can also be obtained whenever the antisymmetric tensor $K_{\beta ij}$ vanishes, that is, if [141]

$$K_{\beta ij} \equiv g_{\alpha i}\left(\frac{\partial g_{\beta j}}{\partial x_\alpha}\right) - \left(\frac{\partial g_{\beta i}}{\partial x_\alpha}\right) g_{\alpha j} = 0 \tag{5.14}$$

Here again, a summation convention over α is implied. Moreover, if $\{g_{\alpha i}\}$ has an inverse obeying

$$\frac{\partial g^{-1}_{\alpha\mu}}{\partial x_\nu} = \frac{\partial g^{-1}_{\alpha\nu}}{\partial x_\mu} \tag{5.15}$$

one can transform multiplicative noise in Eq. (3.3) into additive noise; therefore trivially obeying Eq. (5.14). This multidimensional, small τ Fokker–Planck like approximation, whose precise form is given in [139–142], has, of course, the same regime of validity discussed above, that is,

$$\tau_i \to 0 \qquad \frac{\tau_i}{D_i} \ll 1 \qquad \frac{\tau_{ij}}{D_{ij}} \ll 1 \tag{5.16}$$

B. Decoupling Approximation

As noted in Section V.A, there is a definite need to consider approximation schemes that *do not*, *a priori restrict* the noise correlation time to small values, $\tau_n \ll \tau_s$, only. Let us go back to Eq. (3.27): On inspecting the structure in Eq. (3.27) we note that a Fokker–Planck like master equation results if we *decouple* the correlation entering the second part of Eq. (3.27), that is,

$$\left\langle \delta(x(t) - x) \exp \int_s^t [f' - (fg'/g)]\, d\tau \right\rangle \to \left\langle \exp \int_s^t [f' - (fg'/g)]\, d\tau \right\rangle p_t(x) \tag{5.17}$$

Consistent use of this decoupling procedure yields for Eq. (3.27) the approximation

$$\dot{p}_t(x) = -\frac{\partial}{\partial x}[f(x)p_t(x)] + \left(\int_{t_0}^t ds C(t-s) \exp \int_s^t [\langle f' \rangle - \langle fg'/g \rangle]\, d\tau\right) \cdot \frac{\partial}{\partial x} g(x) \frac{\partial}{\partial x} g(x) p_t(x) \tag{5.18}$$

Next, we also only consider the long-time limit of Eq. (5.18); that is, with $t \to \infty$ we neglect transients consistently and use the stationary average in all occurring averaging prescriptions. This yields the Fokker–Planck approximation for a general stationary Gaussian colored noise with correlation $C(t)$, that is,

$$\dot{p}_t(x) = -\frac{\partial}{\partial x}[f(x)p_t(x)] + \left(\int_0^\infty dt C(t) \exp\{t[\langle f' \rangle - \langle fg'/g \rangle]\}\right) \cdot \frac{\partial}{\partial x} g(x) \frac{\partial}{\partial x} g(x) p_t(x) \tag{5.19}$$

For Ornstein–Uhlenbeck noise in Eq. (5.3) this reduces to [143–145],

$$\dot{p}_t(x) = -\frac{\partial}{\partial x}[f(x)p_t(x)] + \left(\frac{D}{\{1 - \tau[\langle f' \rangle - \langle fg'/g \rangle]\}}\right) \times \frac{\partial}{\partial x} g(x) \frac{\partial}{\partial x} g(x) p_t(x) \tag{5.20}$$

This approximation thus retains the white noise Fokker–Planck form wherein the diffusion strength is substituted by the effective diffusion $D_{\text{eff}}(\tau)$

$$D \to D_{\text{eff}}(\tau) \equiv \frac{D}{\{1 - \tau[\langle f' \rangle - \langle fg'/g \rangle]\}} \tag{5.21}$$

which must be determined from Eq. (5.20) self-consistently. In practice, however, it is usually sufficient to evaluate the stationary averages in Eqs. (5.19–5.21) within the white noise approximation for the stationary probability. Note also that with the neglect of transients and the consistent replacement of averages by stationary averages, the Fokker–Planck equation in Eqs. (5.19) and (5.20) *is restricted to yield reliable information about the stationary probability* $p(x; \tau)$ only. The stationary solution of Eq. (5.20) explicitly reads

$$p(x; \tau) = \frac{Z^{-1}}{|g(x)|} \exp\left\{\frac{[1 - \tau(\langle f' \rangle - \langle fg'/g \rangle)]}{D} \int^x \frac{f(y)}{g^2(y)} dy\right\} \tag{5.22}$$

where Z denotes a normalization constant. For globally stable physical systems, that is, $\langle f' \rangle$ *is less than zero*, we find the relation $0 < D_{\text{eff}}(\tau) < D_{\text{eff}}(\tau = 0) = D$. Thus, the stationary probability in Eq. (5.22) generally [e.g., for $g(x) = const$] exhibits a *sharpening* of the probability peaks upon increasing the noise color τ. Indeed, numerical studies verify this typical colored noise effect (see Section VI). The approximation scheme in Eq. (5.17) *does not restrict the value of the noise color* τ. The decoupling ansatz in Eq. (5.17), however, neglects correlations, and thus is expected to be a valid procedure only for narrow distributions, that is, generally $D \ll 1$. Normally, the decoupling approximation is not suitable to approximate multidimensional features such as multidimensional probabilities that may exhibit color dependent correlations among the state variables [146–148]. Thus, the approximation in Eqs. (5.19–5.22) can be viewed as a weak noise approximation to the colored noise flow in Eq. (3.12), that is, the (dimensionless) noise intensity must be small, $D \ll 1$. This latter weak noise condition is fulfilled in most physical applications, see, for example, in [148, 149]. It is not straightforward to evaluate a generally valid estimate on the error induced by the decoupling ansatz. In principle, the decoupling procedure in Eq. (5.17) can be corrected to higher order if we observe the exact relation in Eq. (3.20). The approximation in Eq. (5.17) just presents the first term in Eq. (3.20). For example, we have with $g(x) = 1$ from Eq. (3.20) to second order ($n = 0$ and $n = 1$)

$$\begin{aligned}\left\langle \delta(x(t) - x) \exp \int_s^t f' \, d\tau \right\rangle &= \left\langle \exp \int_s^t f' \, d\tau \right\rangle p_t(x) \\ &\quad - \frac{\partial}{\partial x} \int_{t_0}^t du \int_{t_0}^t dv C(u - v) \left\langle \delta(x(t) - x) \exp \int_u^t f' \, d\tau \right\rangle \\ &\quad \cdot \left\langle \left(\exp \int_s^t f' \, dr \right) \int_s^t f'' \left(\exp \int_v^\tau f' \, dr \right) d\tau \right\rangle \end{aligned} \tag{5.23}$$

Thus, with a repeatedly applied decoupling procedure as outlined above we obtain the result that non-Fokker–Planck contributions already enter at second order. Such an improved approximation is thus not tractable from a practical viewpoint. Nevertheless, the decoupling approximation to lowest order in Eqs. (5.20–5.22) has successfully been applied to model moderate-to-large noise color in a dye laser [138, 149, 190], the optical bistability [135, 144], and the ring-laser gyroscope [150].

C. Unified Colored Noise Approximation

In Section V.B we belabored an approximation for weak noise $D \ll 1$ which, however, does not restrict the noise correlation time τ. The decoupling scheme, however, involves the averaging of state functions. This means that the approximation is of a global character. In other words, local effects such as colored noise induced shifts of probability extrema are likely not sensitively accounted for. With weak noise intensity $D \ll 1$, such effects are generally strongly suppressed; nevertheless, the local character can be substantially misrepresented with the decoupling ansatz in regions of small probability as it occurs with the tails of the probability, or with minima of the probability in bistable situations. Moreover, neither the small τ approximation in Section V.A, nor the decoupling approximation in Section V.B can be used to evaluate the stationary dynamics such as the stationary correlation function $\langle x(t)x(0)\rangle$, the relaxation time T,

$$T \equiv \frac{\int_0^\infty \langle x(t) - \langle x\rangle\rangle(x(0) - \langle x\rangle)\, dt}{\langle x^2\rangle - \langle x\rangle^2} \tag{5.24}$$

or other quantities of dynamical origin. The authors recently have put forward an approximation scheme that effectively overcomes most of these restrictions. We have termed it the unified colored noise approximation (UCNA) (see [151]).

1. *UCNA for Colored One-Dimensional Flows*

Let us consider Eq. (5.25),

$$\dot{x} = f(x) + g(x)\xi(t) \tag{5.25}$$

where $\xi(t)$ is an exponentially correlated Gaussian noise [see Eq. (2.10)] of vanishing mean. First, let us consider additive noise, that is,

$$\dot{x} = f(x) + \xi(t) \tag{5.26}$$

which with $\xi(t)$, an Ornstein–Uhlenbeck process, constitutes a 2-D Markovian process driven by Gaussian white noise $\zeta_w(t)$,

$$\dot{x} = f(x) + \xi \tag{5.27}$$

$$\dot{\xi} = -\frac{1}{\tau}\xi + \frac{D^{1/2}}{\tau}\zeta_w(t) \tag{5.28}$$

where $\langle \zeta_w(t)\zeta_w(s)\rangle = 2\delta(t-s)$. If we follow our original work [151] we eliminate ξ in Eq. (5.27) by use of Eq. (5.28). Then we obtain a Langevin equation for a noisy nonlinear oscillator,

$$\ddot{x} + \dot{x}[\tau^{-1} - f'(x)] - f(x)/\tau = \frac{D^{1/2}}{\tau}\zeta_w(t) \tag{5.29}$$

with a nonlinear damping function. On the new time scale $s = t\tau^{-1/2}$ this nonlinear oscillator dynamics is recast as (a dot indicates the differentiation with respect to time s)

$$\ddot{x} + \gamma(x, \tau)\dot{x} - f(x) = \frac{D^{1/2}}{\tau^{1/4}} \zeta_w(s) \tag{5.30}$$

where $\langle \zeta_w(s)\zeta_w(s')\rangle = 2\delta(s - s')$. The nonlinear damping γ explicitly reads

$$\gamma(x, \tau) = \tau^{-1/2} + \tau^{1/2}[-f'(x)] \tag{5.31}$$

With multiplicative noise, $g(x)\xi(t)$, see Eq. (5.25), the corresponding nonlinear friction would read

$$\gamma(x, \tau) = \tau^{-1/2} + \tau^{1/2}\left[-f'(x) + f(x)\frac{g'(x)}{g(x)}\right] \tag{5.32}$$

If the expression in the squared brackets in Eq. (5.31) or (5.32), is positive, the damping will become *large for both small* and large correlation times τ. The positivity condition is with $f'(x) < 0$, obeyed in regions of state space, where the noise-free flow is locally stable. The condition of large positive damping $\gamma(x, \tau) \gg 1$, allows the adiabatic elimination of $\dot{x} = v$. Setting $\dot{v} = \ddot{x} = 0$ then yields a truely Markovian approximation of the colored noise flow in Eq. (5.26),

$$\dot{x} = \frac{f(x)}{\gamma(x, \tau)} + \frac{D^{1/2}}{\tau^{1/4}\gamma(x, \tau)} \zeta_w(s) \tag{5.33}$$

Within the original time variable $t = \tau^{1/2}s$, and with multiplicative noise $g(x)\xi(t)$, the analogue of Eq. (5.33) reads [144, 151]

$$\begin{aligned}\dot{x} &= f(x)[1 - \tau(f'(x) - f(x)g'(x)/g(x))]^{-1} \\ &\quad + D^{1/2}g(x)[1 - \tau(f'(x) - f(x)g'(x)/g(x))]^{-1}\zeta_w(t)\end{aligned} \tag{5.34}$$

which is to be *interpreted in the Stratonovich sense* [3–7]. Equations (5.33) and (5.34) define a truely 1-D (Stratonovich) Fokker–Planck process, whose equation is readily written down [7]. We must emphasize the true Markovian (approximate) description in Eq. (5.34) of the original non-Markovian process. This feature has a striking advantage over the small correlation time theories outlined in Section V.A. Not only does Eq. (5.34) *become exact* both at correlation time $\tau = 0$ and $\tau \to \infty$, and hence is expected to be a useful approximation for intermediate noise color, it also provides an approximation for the time-homogeneous conditional probability of $x(t)$, that is, the stationary conditional probability $R(x, t \mid y, t_0) = R(x, t+\tau \mid y, t_0+\tau)$ obeys the very same Fokker–Planck equation. Thus, Eq. (5.34) and its corresponding Fokker–Planck equation, can be used to evaluate approximate stationary correlation functions, and so on. The approximation scheme is valid for both small and large correlation times τ, and in parts of the state space where the nonlinear damping $\gamma(x, \tau)$ is positive. In contrast to the dynamics of the small correlation time approximation, which corresponds to the correlation-free preparation (see Section III.B), the UCNA in Eqs. (5.33) and (5.34), closely models the stationary preparation class of the non-Markovian process $x(t)$ [144].

We now discuss the *regime of validity of the UCNA* in more detail. We recall that the UCNA is valid only for regimes in state space where $\gamma(x, \tau)$, Eq. (5.32), is positive. Based on the noise intensity D we form the characteristic length scale L

$$L = \frac{D^{1/2}}{\gamma(x, \tau)} \tag{5.35}$$

Then the adiabatic elimination procedure $\dot{v} = \ddot{x} = 0$ implies that Eq. (5.33) or (5.34), respectively, is a good approximation only on time scales $t > \tau^{1/2}\gamma^{-1}$, that is, with $\gamma > 0$

$$t > \tau[1 + \tau(-f' + fg'/g)]^{-1} \tag{5.36}$$

and if on the characteristic length scale L the drift force is not varying appreciably, that is, $L|f'| \ll |f|$ [144, 151]. This latter condition is the analogue of the condition for the validity of the Smoluchowski approximation in Brownian motion theory [151], wherein $L = (kTm^{-1}\gamma^{-2})^{1/2}$ denotes the thermal length scale. Let $\hat{x}$ denote a characteristic value within the regime where $\gamma > 0$. Then, we obtain for the validity of the UCNA the relation

$$\gamma(\hat{x}, \tau) \gg D^{1/2}\left|\frac{f'(\hat{x})}{f(\hat{x})}\right| \tag{5.37}$$

Thus, we deduce from Eq. (5.37) that the UCNA improves in accuracy for *increasing nonlinear damping* $\gamma \to \infty$, and *decreases in accuracy* with increasing noise intensity. Keeping the restrictions in Eqs. (5.36) and (5.37) in mind we study the solution of Eqs. (5.33) and (5.34). With the effective multiplicative noise function $g_{\mathrm{UCNA}}(x, \tau)$

$$g_{\mathrm{UCNA}}(x, \tau) = g(x)[1 - \tau(f' - fg'/g)]^{-1} \tag{5.38}$$

the (Stratonovich) Fokker–Planck equation for the UCNA in Eq. (5.34) reads[4]

$$\dot{p}_t^{\mathrm{UCNA}}(x) = -\frac{\partial}{\partial x}[f(x)g^{-1}(x)g_{\mathrm{UCNA}}(x,\tau)p_t(x)] + D\frac{\partial}{\partial x}g_{\mathrm{UCNA}}(x,\tau)\frac{\partial}{\partial x}[g_{\mathrm{UCNA}}(x,\tau)p_t(x)] \qquad (5.39)$$

Its stationary solution $p^{\mathrm{UCNA}}(x,\tau)$ reads

$$p^{\mathrm{UCNA}}(x,\tau) = \frac{Z^{-1}}{|g(x)|}\,|[1-\tau g(x)(f(x)/g(x))']| \times \exp\left\{\int^{x}\frac{f(y)[1-\tau g(y)(f(y)/g(y))']\,dy}{Dg^2(y)}\right\} \qquad (5.40)$$

being valid both for small and moderate-to-large correlation times τ. Note also that the Fokker–Planck equation in Eq. (5.39) substantially *differs* from the small τ Fox theory [138] in Eq. (5.11). Nevertheless, the stationary probability $p^{\mathrm{Fox}}(x,\tau)$ in Eq. (5.12) precisely coincides with $p^{\mathrm{UCNA}}(x,\tau)$ in Eq. (5.40). Keep in mind, however, that [in *clear contrast* to UCNA in Eq. (5.40)] the theory of Fox, that is, its dynamics, is *restricted* nevertheless to the small τ regime discussed in Section V.A. The extrema of $p^{\mathrm{UCNA}}(x,\tau)$ are located at position $\{\bar{x}\}$, which obey

$$[1-\tau g(f/g)']\{[1-\tau g(f/g)']f - Dg'g\} + Dg^2[1-\tau g(f/g)']' = 0 \qquad (5.41)$$

In particular, even in the case where the noise is additive only, one obtains a colored noise induced shift of extrema of $p^{\mathrm{UCNA}}(x,\tau)$ located at

$$-D\tau f''(\bar{x}) + [1-\tau f'(\bar{x})]^2 f(\bar{x}) = 0 \qquad (5.42)$$

By use of the Markov character in Eq. (5.39) we can also give the explicit formula for the relaxation time T in Eq. (5.24) [144, 152]

$$T(D,\tau) = \frac{1}{\langle x^2\rangle - \langle x\rangle^2}\int_0^\infty dy\left(\frac{f^2(y)}{Dg_{\mathrm{eff}}^2(y)p^{\mathrm{UCNA}}(y,\tau)}\right) \qquad (5.43)$$

where $f(y) \equiv -\int_0^y dx(x-\langle x\rangle)p^{\mathrm{UCNA}}(x)$.

[4] If in Eq. (5.25) additional Gaussian white noise $\eta(t)$ is present, that is, if $\dot{x} = f(x) + g(x)\xi(t) + \eta(t)$, with $\langle\eta(t)\eta(s)\rangle = 2T\delta(t-s)$, the UCNA can be generalized by introducing the auxiliary process $u = \xi + (f/g)\{1 + (T/Dg^2)[1-\tau g(f/g)']\}^{-1}$. An adiabatic elimination of $u(t)$; i.e. $\dot{u}(t) = 0$, then provides the UCNA with the correct behavior as $\tau \to 0$, and which with $u \to 0$ as $\tau \to \infty$ is corrected also for $\tau \to \infty$; for applications cf. Section VII.

D. Remarks on Sundry Colored Noise Approximation Schemes

The small noise color approximation reviewed in Section V.A, the decoupling theory (often also termed Hänggi–Ansatz [135, 138, 149, 190]) in Section V.B, and the UCNA in Section V.C.1, are by and large the most often employed perturbation schemes in the study of dynamical flows driven by correlated random forces. There exist, of course, other possibilities that might be preferred from time to time. For example, for flows driven by Markovian colored noise, such as the exponentially correlated Ornstein–Uhlenbeck noise, the physics can be studied in terms of an enlarged phase space, which renders the dynamics Markovian again [69, 70, 134, 153]. As pointed out in [134] some care, however, must be observed if one compares the dynamics in full space with the one in reduced space; because the correlations in the enlarged space are richer as compared to the reduced, non-Markovian dynamics. Nevertheless, the approximation schemes available for the study of higher dimensional Markovian systems, which unfortunately are rather sparse indeed, can be invoked. Usually, this reasoning has been utilized thus far only for the investigation of stationary quantities, such as the stationary probability [69, 70, 153]. The study of colored noise in the asymptotic regime of large correlation time is another topic that has attracted considerable interest in recent years [154–158]. The UCNA approach clearly does cover this regime, as demonstrated in [158]. The very asymptotic extreme large colored noise regime, can also be more directly addressed by noting that the noise with correlation time $\tau \to \infty$ becomes extremely slowly varying. This then leads to the quasistatic "switching-curve reasoning" originally put forward by Horsthemke and Lefever [159]: With $\tau \to \infty$, the variable x is in a quasi-stationary state with respect to the instantaneous value of the fluctuation force $\xi(t)$, that is, one finds with $\dot{x} \cong 0$ from Eq. (2.7)

$$f(x) + g(x)\xi(t) = 0 \tag{5.44}$$

Setting $\xi = y^{-1}(x)$, and observing the identity between the probabilities,

$$p(x)\,dx = \rho(\xi)\,d\xi \tag{5.45}$$

one therefore finds

$$p(x, \tau \to \infty) = \rho[y^{-1}(x)]\left|\frac{d\xi}{dx}\right| \tag{5.46}$$

where $\rho(\cdots)$ denotes the stationary probability for the noise ξ.

Finally, let us take another look at Eq. (3.27): With Ornstein–Uhlenbeck noise in mind [see Eq. (5.3)] the exact master equation in Eq. (3.27) reads

$$\dot{p}_t(x) = \frac{-\partial}{\partial x}[f(x)p_t(x)] + \frac{D}{\tau}\frac{\partial}{\partial x}g(x)\frac{\partial}{\partial x}g(x)\int_{t_0}^{t} ds \\ \times \exp\left(-\frac{t-s}{\tau}\right)\left\langle \delta[x(t)-x]\exp\int_s^t\left[f' - \left(\frac{fg'}{g}\right)\right]du\right\rangle \tag{5.47}$$

To obtain a workable equation for the probability we must close Eq. (5.47). First we let $t_0 \to -\infty$, so that we can safely neglect transient effects. We observe that Eq. (5.47) can be closed in a variety of different ways:

1. We recover the decoupling theory if we make the "Hanggi–Ansatz", thereby the average in Eq. (5.47) is decoupled. This yields a Fokker–Planck equation with a diffusion operator

$$(i)\ D(x,\tau) = \frac{\partial}{\partial x}g(x)\frac{\partial}{\partial x}g(x)\frac{D}{1-\tau[\langle f'\rangle - \langle fg'/g\rangle]} \tag{5.48}$$

2. If we approximate the stochastic process $x(u)$, $t_0 \le u \le t$, for all times by the final value $x(u) = x(t)$, the δ function in Eq. (5.47) implies a closure with the state-dependent diffusion operator given by

$$(ii)\ D(x,\tau) = \frac{\partial}{\partial x}g(x)\frac{\partial}{\partial x}g(x)\frac{D}{1-\tau[f'(x) - f(x)g'(x)/g(x)]} \tag{5.49}$$

Clearly, this approximation implicitly requires a small noise correlation time τ. This form of approximation actually coincides precisely with the approximation put forward by Fox [106], see in Eq. (5.11).

3. Instead of using the small τ approximation $x(u) \simeq x(t)$, we could instead follow the reasoning inherent in the decoupling theory and replace the stochastic process $f[x(u)]$ not by a stationary average, but by a time-dependent (deterministic) solution $f[x(u)] \to f[x_{\text{det}}(u)]$. A good candidate would be to use the path $x_e(t)$, which extremalizes the action of the corresponding Onsager–Machlup functional in a path integral solution of the corresponding non-Markovian process [156, 157, 160, 161].

With $x_e(t_0) = x_s$, being chosen as an attractor, so that $x_e(t) = x$ is attained only at very long times, we obtain from Eq. (3.27) upon a change of variables $dt' = dx_e / \dot{x}_e$ an x dependent, effective diffusion operator given by

$$(iii)\ D(x, \tau) = \frac{D}{\tau} \frac{\partial}{\partial x} g(x) \frac{\partial}{\partial x} g(x) \int_{x_s}^{x} \frac{dy}{\dot{y}} \exp \int_{y}^{x} \left\{ -\frac{1}{\tau} + g(z) \left(\frac{f}{g} \right)' \right\} \frac{dz}{\dot{z}} \tag{5.50}$$

Hereby we used stationary Ornstein–Uhlenbeck noise [see Eq. (5.3)] and the velocities are determined from the extremal action path. The reasoning to obtain this effective diffusion has recently been applied by Venkatesh and Patniak [162] in a study of colored noise driven bistability. An appealing feature of Eq. (5.50) must be pointed out: The effect of noise color enters the effective diffusion in Eq. (5.50) in two ways: First, there is the influence of the noise correlation $C(t) = (D/\tau)\exp(-t/\tau)$; second, there is the τ dependence in the extreme action path $x_e(t)$. Moreover, just as with UCNA, the approximation in Eq. (5.50) is not restricted solely to small noise color.

VI. COLORED NOISE DRIVEN BISTABLE SYSTEMS

Noisy bistable dynamics is an archetype phenomenon in many areas of physics, chemistry, and biology. It is therefore important to develop a detailed understanding of the fluctuation-related statistical characteristics, such as lifetimes of metastable states. Realistic modeling of noise sources, however, requires us to take into account finite correlation times. An important application of the theoretical framework of colored noise driven dynamical systems, provided in Sections II–V is therefore bistable dynamics. In this section we review key results obtained for probability densities and escape rates in colored noise driven bistable systems. Special focus is the dependence of those characteristics on the correlation time of the noise.

As a model, we are using a Ginzburg–Landau type potential and an exponentially correlated noise. The equation of motion reads [134]

$$\dot{x} = -V'(x) + \varepsilon(t) \tag{6.1}$$

with $\varepsilon(t)$ being Gaussian, exponentially correlated noise, that is,

$$\begin{aligned} \langle \varepsilon(t)\varepsilon(t') \rangle &= \frac{D}{\tau} \exp\left(-\frac{|t - t'|}{\tau} \right) \\ \langle \varepsilon(t) \rangle &= 0 \end{aligned} \tag{6.2}$$

The potential $V(x)$ is given by

$$V(x) = -\frac{a}{2}x^2 + \frac{b}{4}x^4 \tag{6.3}$$

with positive constants a and b. The potential $V(x)$ is bistable with minima at $x_{1,2} = \pm\sqrt{a/b}$, and a relative maximum at $x = 0$. Introducing scaled variables $\bar{x} = x\sqrt{b/a}$, $\bar{t} = at$, $\bar{\varepsilon} = \varepsilon\sqrt{b/a^3}$, $\bar{\tau} = a\tau$, $\bar{D} = (b/a^2)D$, the normalized Langevin equation reads

$$\dot{\bar{x}} = \bar{x} - \bar{x}^3 + \bar{\varepsilon}(t) \tag{6.4}$$

where the autocorrelation function of the noise variable $\bar{\varepsilon}$ is given by

$$\langle \varepsilon(t)\bar{\varepsilon}(t')\rangle = \frac{\bar{D}}{\bar{\tau}} \exp\left(-\frac{|t-t'|}{\bar{\tau}}\right) \tag{6.5}$$

The potential is shown in scaled variables in Fig. 6.1. In the case of a large typical system time scale $1/a$ in comparison to the correlation time of the noise τ, that is, $\bar{\tau} = \tau/(1/a) \rightarrow 0$, the correlation function $\langle \varepsilon(t)\varepsilon(t')\rangle$ approaches a δ function. The variance of the noise, $\langle \varepsilon^2 \rangle = D/\tau$, which up to a factor of two equals the total power of the noise, diverges in this *white noise* limit. In the opposite limit, $\bar{\tau} \rightarrow \infty$, the variance vanishes, that is, the total power of the noise vanishes.

From now on, we will only use the normalized variables, but drop the bar for the sake of convenience.

A. Embedding in a Two-Dimensional Markovian Process

The stochastic process Eq. (6.4) defines a non-Markovian stochastic process. The time evolution of the probability distribution is thus not given by a Fokker–Planck equation for the state variable x (see Section

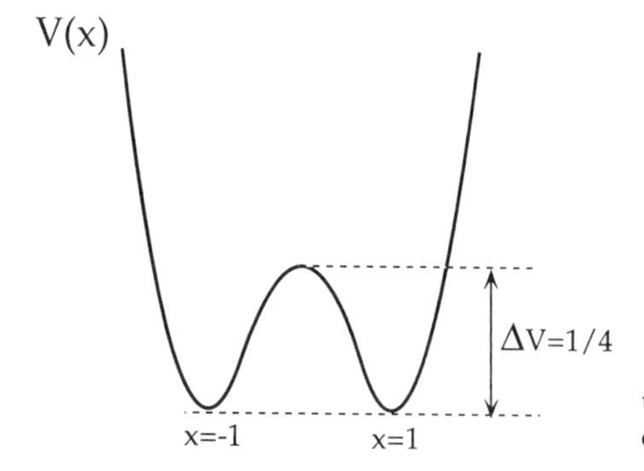

Figure 6.1. The double-well potential [Eq. (6.3)] is shown in normalized coordinates.

III). Nevertheless, one can find an equivalent 2-D pair process $(x(t), \varepsilon(t))$, with the auxiliary variable ε, obeying the linear white noise driven stochastic differential equation [134, 163]

$$\dot{\varepsilon} = -\frac{1}{\tau}\varepsilon + \frac{\sqrt{D}}{\tau}\xi(t) \tag{6.6}$$

with the Gaussian white noise

$$\begin{aligned} \langle \xi(t) \rangle &= 0 \\ \langle \xi(t)\xi(t') \rangle &= 2\delta(t - t') \end{aligned} \tag{6.7}$$

where the stationary autocorrelation function of $\varepsilon(t)$ is given by the first equation of Eq. (6.5). The pair process $(x(t), \varepsilon(t))$ is a Markovian stochastic process and the time evolution of the joint probability density, $W(x, \varepsilon, t)$, is given by the two-variable Fokker–Planck equation,

$$\begin{aligned} \frac{\partial}{\partial t} W(x, \varepsilon, t) &= \mathbf{L}_{\text{em}}(x, \varepsilon) W(x, \varepsilon, t) \\ \mathbf{L}_{\text{em}}(x, \varepsilon) &= -\frac{\partial}{\partial x}(x - x^3 + \varepsilon) + \frac{1}{\tau}\frac{\partial}{\partial \varepsilon}\varepsilon + \frac{D}{\tau^2}\frac{\partial^2}{\partial \varepsilon^2} \end{aligned} \tag{6.8}$$

In order to guarantee that the correlation function for ε is stationary for all times, we have to require that at the preparation time $t = 0$ the probability distribution in ε is stationary [134, 144], that is,

$$\int_{-\infty}^{\infty} W(x, \varepsilon, t = 0)\, dx = \frac{1}{\sqrt{2\pi\tau/D}} \exp\left(-\frac{\tau\varepsilon^2}{2D}\right) \tag{6.9}$$

1. Basic Properties of the Embedding Fokker–Planck Operator

Since the Fokker–Planck (FP) operator $\mathbf{L}_{\text{em}}$ is symmetric with respect to inversion,

$$\mathbf{L}_{\text{em}}(x, \varepsilon) = \mathbf{L}_{\text{em}}(-x, -\varepsilon) \tag{6.10}$$

the corresponding eigenfunctions

$$\mathbf{L}_{\text{em}}(x, \varepsilon)\psi_\lambda(x, \varepsilon) = -\lambda\psi_\lambda(x, \varepsilon) \tag{6.11}$$

can be classified into even and odd eigenfunctions [163],

$$\begin{aligned}\psi_\lambda^{(e)}(x, \varepsilon) &= \psi_\lambda^{(e)}(-x, -\varepsilon)\\ \psi_\lambda^{(o)}(x, \varepsilon) &= -\psi_\lambda^{(o)}(-x, -\varepsilon)\end{aligned} \tag{6.12}$$

The stationary probability, being the eigenfunction corresponding to the vanishing eigenvalue, is an even eigenfunction

$$W_{st}(x, \varepsilon) = W_{st}(-x, -\varepsilon) \tag{6.13}$$

Since the stochastic differential equation for x does not couple to the equation for ε, the eigenvalues of the Ornstein–Uhlenbeck process for ε are also eigenvalues of the stochastic pair process. This result can be seen more clearly at the adjoint eigenvalue equation

$$\begin{aligned}\mathbf{L}_{\text{em}}^{\dagger}(x, \varepsilon)\psi_\lambda^{\dagger}(x, \varepsilon) &= -\lambda\psi_\lambda^{\dagger}(x, \varepsilon)\\ \mathbf{L}_{\text{em}}^{\dagger} &= (x - x^3 + \varepsilon)\frac{\partial}{\partial x} - \frac{1}{\tau}\varepsilon\frac{\partial}{\partial\varepsilon} + \frac{D}{\tau^2}\frac{\partial^2}{\partial\varepsilon^2}\end{aligned} \tag{6.14}$$

which is solved by the adjoint eigenfunctions (H_n denotes the Hermite polynomial of order n)

$$\psi_{0n}^{\dagger}(x, \varepsilon) = H_n\left(\frac{\varepsilon}{\sqrt{2D/\tau}}\right) \tag{6.15}$$

with the corresponding eigenvalues

$$\lambda_{0n} = n\frac{1}{\tau} \tag{6.16}$$

It is important to note that for large correlation times these eigenvalues become small and that the corresponding relaxation modes can therefore influence even the large time behavior of dynamical quantities such as correlation functions!

The symmetry of the pair process $(x(t), \varepsilon(t))$ allows us to construct two isospectral Fokker–Planck systems. An equivalent pair process to Eqs. (6.4) and (6.6) is given by

$$\begin{aligned}\dot{x} &= x - x^3 - \varepsilon\\ \dot{\varepsilon} &= -\frac{1}{\tau}\varepsilon + \frac{\sqrt{D}}{\tau}\xi(t)\end{aligned} \tag{6.17}$$

The spectrum of the corresponding Fokker–Planck operator

$$\tilde{\mathbf{L}}_{\mathrm{em}}(x, \varepsilon) = -\frac{\partial}{\partial x}(x - x^3 - \varepsilon) + \frac{1}{\tau}\frac{\partial}{\partial \varepsilon}\varepsilon + \frac{D}{\tau^2}\frac{\partial^2}{\partial \varepsilon^2} \tag{6.18}$$

is therefore isospectral with the Fokker–Planck operator $\mathbf{L}_{\mathrm{em}}(x, \varepsilon)$ [164]. The eigenfunctions of $\mathbf{L}_{\mathrm{em}}(x, \varepsilon)$ follow from the eigenfunctions of $\tilde{\mathbf{L}}_{\mathrm{em}}(x, \varepsilon)$ by the substitution $\varepsilon \to -\varepsilon$. Another isospectral Fokker–Planck operator can be constructed by inverting the potential [164]. This is a general property of a colored noise driven *overdamped* system. In order to show this we convert the eigenvalue problem Eq. (6.11) [here we use a general force field $h(x)$ instead of $x - x^3$] according to

$$\psi_\lambda(x, \varepsilon) = \exp\left(\frac{H(x)}{2D} - \frac{\tau}{2D}\varepsilon^2\right)\hat{\psi}_\lambda(x, \varepsilon) \tag{6.19}$$

where the (negative of the) potential is given by

$$H(x) = \int^x h(y)\, dy \tag{6.20}$$

into the eigenvalue problem for $\hat{\psi}_\lambda(x, \varepsilon)$

$$\hat{\mathbf{L}}_{\mathrm{em}}\hat{\psi}_\lambda(x, \varepsilon) = -\lambda\hat{\psi}_\lambda(x, \varepsilon) \tag{6.21}$$

with

$$\hat{\mathbf{L}}_{\mathrm{em}} = -\hat{\mathbf{a}}_x\mathbf{a}_x - \hat{\mathbf{a}}_x^2 - \hat{\mathbf{a}}_\varepsilon\mathbf{a}_\varepsilon + \tfrac{1}{2}\hat{\mathbf{a}}_x(\mathbf{a}_\varepsilon + \hat{\mathbf{a}}_\varepsilon) \tag{6.22}$$

and the operators are defined as

$$\begin{aligned}
\hat{\mathbf{a}}_x &= -\sqrt{D}\frac{\partial}{\partial x} - \frac{h(x)}{2\sqrt{D}} \\
\mathbf{a}_x &= \sqrt{D}\frac{\partial}{\partial x} - \frac{h(x)}{2\sqrt{D}} \\
\hat{\mathbf{a}}_\varepsilon &= -\frac{\sqrt{D}}{\tau}\frac{\partial}{\partial \varepsilon} + \frac{\varepsilon}{2\sqrt{D}} \\
\mathbf{a}_\varepsilon &= \frac{\sqrt{D}}{\tau}\frac{\partial}{\partial \varepsilon} + \frac{\varepsilon}{2\sqrt{D}}
\end{aligned} \tag{6.23}$$

The eigenvalue problem with the inverted potential, that is, $H(x) \to -H(x)$ can be treated analogous, yielding the converted Fokker–Planck operator

$$\tilde{\hat{\mathbf{L}}}_{\text{em}} = -\tilde{\hat{\mathbf{a}}}_x \tilde{\mathbf{a}}_x - \tilde{\hat{\mathbf{a}}}_x^2 - \tilde{\hat{\mathbf{a}}}_\varepsilon \tilde{\mathbf{a}}_\varepsilon + \tfrac{1}{2}\tilde{\hat{\mathbf{a}}}_x(\tilde{\mathbf{a}}_\varepsilon + \tilde{\hat{\mathbf{a}}}_\varepsilon) \tag{6.24}$$

where the operators with the tilde are given by

$$\begin{aligned} \tilde{\hat{\mathbf{a}}}_x &= -\mathbf{a}_x \\ \tilde{\mathbf{a}}_x &= -\hat{\mathbf{a}}_x \\ \tilde{\hat{\mathbf{a}}}_\varepsilon &= -\mathbf{a}_x \\ \tilde{\mathbf{a}}_\varepsilon &= -\hat{\mathbf{a}}_x \end{aligned} \tag{6.25}$$

In view of the isospectral property with respect to inversion of ε, we have also performed the inversion $\varepsilon \to -\varepsilon$. From the equations above, we can establish the following operator relation

$$\mathbf{a}_x \hat{\mathbf{L}}^\dagger(x, \varepsilon) = \tilde{\hat{\mathbf{L}}}(x, -\varepsilon)\mathbf{a}_x \tag{6.26}$$

If $\hat{\psi}^\dagger(x, \varepsilon)$ is an eigenfunction of $\hat{\mathbf{L}}^\dagger(x, \varepsilon)$, that is,

$$\hat{\mathbf{L}}^\dagger(x, \varepsilon)\tilde{\psi}^\dagger(x, \varepsilon) = -\lambda\hat{\psi}^\dagger(x, \varepsilon) \tag{6.27}$$

then we find by multiplying from the left with $\mathbf{a}_x$ and by using Eq. (6.26)

$$\tilde{\hat{\mathbf{L}}}(x, -\varepsilon)\mathbf{a}_x \hat{\psi}^\dagger(x, \varepsilon) = -\lambda \mathbf{a}_x \hat{\psi}^\dagger(x, \varepsilon) \tag{6.28}$$

that is, $\mathbf{a}_x \hat{\psi}^\dagger(x, \varepsilon)$ is an eigenfunction of $\tilde{\hat{\mathbf{L}}}(x, -\varepsilon)$ with the *same eigenvalue* λ. The isospectral property is thus proven.

2. *Application of the Matrix Continued Fraction Technique*

The two-variable Fokker–Planck equation in the extended phase space can be solved by using the matrix continued fraction (MCF) technique [5]. Since our Fokker–Planck operator $\mathbf{L}_{\text{em}}(x, \varepsilon)$ has inversion symmetry, we make this technique more efficient [163] by expanding the even and odd eigenfunctions separately in complete sets of orthogonal functions with respect to both variables x and ε. Thus,

$$\psi^{(e)}(x, \varepsilon) = \rho_0(x) w_0(\varepsilon) \sum_{n,m=0}^{\infty} [c_{2n}^{2m} \varphi_{2m}(x) w_{2n}(\varepsilon) + c_{2n+1}^{2m+1} \varphi_{2m+1}(x) w_{2n+1}(\varepsilon)]$$

$$\psi^{(o)}(x, \varepsilon) = \rho_0(x) w_0(\varepsilon) \sum_{n,m=0}^{\infty} [c_{2n}^{2m+1} \varphi_{2m+1}(x) w_{2n}(\varepsilon) + c_{2n+1}^{2m} \varphi_{2m}(x) w_{2n+1}(\varepsilon)] \tag{6.29}$$

where the complete set $\{w_n(\varepsilon)\}$ is given by the eigenfunctions of the operator

$$\mathbf{L}_H = \frac{D}{\tau^2} \frac{\partial^2}{\partial \varepsilon^2} - \frac{1}{4D} \varepsilon^2 + \frac{1}{2\tau} \tag{6.30}$$

that is,

$$w_n(\varepsilon) = \frac{1}{\sqrt{2^n n! \sqrt{2\pi D/\tau}}} H_n\left(\frac{\varepsilon}{\sqrt{2D/\tau}}\right) \exp\left(-\frac{\tau\varepsilon^2}{4D}\right) \tag{6.31}$$

and the complete set in x is given by the Hermite functions

$$\varphi_n(x) = \sqrt{\frac{\alpha}{n! 2^n \sqrt{\pi}}} H_n(\alpha x) \exp\left(-\frac{1}{2}\alpha^2 x^2\right) \tag{6.32}$$

The constant α is an adjustable positive parameter to optimize the speed of the convergence. The form function $\rho_0(x)$ is assumed to be symmetric and positive and should decay to zero for $x \to \pm\infty$. Inserting those expansions into the eigenvalue equation Eq. (6.11), we find a coupled system of algebraic equations for the expansion coefficients c_n^m, which can be arranged in a tridiagonal vector recurrence relation. We find for the even eigenfunctions [163]

$$\mathbf{0} = \left(\underline{\underline{A}}_0 - \left(\frac{k}{\tau} - \lambda\right)\right)\mathbf{c}_k + \sqrt{\frac{kD}{\tau}} \underline{\underline{B}}_{0e} \mathbf{c}_{k-1} + \sqrt{\frac{(k+1)D}{\tau}} \underline{\underline{B}}_{0e} \mathbf{c}_{k+1} \tag{6.33}$$

for odd k, and

$$\mathbf{0} = \left(\underline{\underline{A}}_e - \left(\frac{k}{\tau} - \lambda\right)\right)\mathbf{c}_k + \sqrt{\frac{kD}{\tau}} \underline{\underline{B}}_{e0} \mathbf{c}_{k-1} + \sqrt{\frac{(k+1)D}{\tau}} \underline{\underline{B}}_{e0} \mathbf{c}_{k+1} \tag{6.34}$$

for even k. The components of the matrices $\underline{\underline{A}}_e$, $\underline{\underline{A}}_0$, $\underline{\underline{B}}_{e0}$, $\underline{\underline{B}}_{0e}$ are given by

$$\begin{aligned}
(\underline{\underline{A}}_0)^{i,j} &= \int_{-\infty}^{\infty} \rho_0^{-1}(x)\varphi_{2i+1}(x)\Big(-\frac{\partial}{\partial x}(x-x^3)\Big)\rho_0(x)\varphi_{2j+1}(x)\,dx = A^{2i+1,2j+1} \\
(\underline{\underline{A}}_e)^{i,j} &= \int_{-\infty}^{\infty} \rho_0^{-1}(x)\varphi_{2i}(x)\Big(-\frac{\partial}{\partial x}(x-x^3)\Big)\rho_0(x)\varphi_{2j}(x)\,dx = A^{2i,2j} \\
(\underline{\underline{B}}_{e0})^{i,j} &= \int_{-\infty}^{\infty} \rho_0^{-1}(x)\varphi_{2i}(x)\Big(-\frac{\partial}{\partial x}\Big)\rho_0(x)\varphi_{2j+1}(x)\,dx = B^{2i,2j+1} \\
(\underline{\underline{B}}_{0e})^{i,j} &= \int_{-\infty}^{\infty} \rho_0^{-1}(x)\varphi_{2i+1}(x)\Big(-\frac{\partial}{\partial x}\Big)\rho_0(x)\varphi_{2j}(x)\,dx = B^{2i+1,2j}
\end{aligned} \tag{6.35}$$

while the components of the vectors $\mathbf{c}_k$ read

$$(\mathbf{c}_k)^i = \begin{cases} c_k^{2i} & \text{for even } k \\ c_k^{2i+1} & \text{for odd } k \end{cases}$$

For odd eigenfunctions the conditions for k have to be interchanged. The tridiagonal vector recurrence relation Eqs. (6.33) and (6.34) can be solved for the eigenvalues by iterating a matrix continued fraction. For the form junction, the Gaussian, $\rho_0(x) = \exp(-cx^2)$, has been chosen, where the constant c has been adjusted to obtain a good convergence of the matrix continued fraction. The matrices $A^{m,n}$ and $B^{m,n}$ read for this Gaussian form function

$$\begin{aligned}
A^{m,n} = &-\frac{\beta^6}{2D}\{\sqrt{(m+1)(m+2)(m+3)(m+4)(m+5)(m+6)}\}\delta_{n,m+6} \\
&+\Big\{\frac{\beta^6}{2D}\sqrt{m(m-1)(m-2)(m-3)(m-4)(m-5)}\Big\}\delta_{n,m-6} \\
&+\Big\{\frac{1}{2}\beta^2 + 2c\beta^4\Big\}\sqrt{(m+1)(m+2)(m+3)(m+4)}\delta_{n,m+4} \\
&+\Big\{-\frac{1}{2}\beta^2 + 2c\beta^4\sqrt{(m(m-1)(m-2)(m-3)}\Big\}\delta_{n,m-4} \\
&+\Big\{-\frac{1}{2} + \beta^2(m+3+2c) + 4c\beta^2(2m+3)\Big\}\sqrt{(m+1)(m+2)}\delta_{n,m+2} \\
&+\Big\{\frac{1}{2} - \beta^2(m-2-2c) + 4c\beta^4(2m-3)\Big\}\delta_{n,m-2}
\end{aligned}$$

$$+\left\{-\frac{1}{2}+3\beta^2\left(m+\frac{1}{2}+2c(2m+1)\right)+12c\beta^4 m(m+1)+6c\beta^4\right\}\delta_{nm}$$
$$(6.36)$$

and

$$B^{n,m}=-\left\{\frac{1}{2\beta}+2c\beta\right\}\sqrt{m+1}\,\delta_{n,m+1}+\left\{\frac{1}{2\beta}-2c\beta\right\}\delta_{n,m-1} \quad (6.37)$$

with $\beta = \alpha/\sqrt{2}$.

3. *Stationary Probability Density in the Extended Phase Space*

For large times, the probability distribution $W(x, \varepsilon, t)$ approaches the time-independent stationary density $W_{\mathrm{st}}(x, \varepsilon)$. This is the stationary joint probability density, that is, the probability of finding x in the interval $[x, x+dx]$ and the auxiliary variable ε in the interval $[\varepsilon, \varepsilon + d\varepsilon]$. The expansion of the (even) stationary probability density $W_{\mathrm{st}}(x, \varepsilon)$ in the complete sets (see Section VI.A.2) with respect to x and ε reads

$$W_{\mathrm{st}}(x, \varepsilon) = \rho_0(x) w_0(\varepsilon) \sum_{n,m=0}^{\infty} [d_{2n}^{2m} \varphi_{2m}(x) w_{2n}(\varepsilon) + d_{2n+1}^{2m+1} \varphi_{2m+1}(x) w_{2n+1}(\varepsilon)] \quad (6.38)$$

where the expansion coefficients have been determined by using the matrix continued fraction method [163]. In Fig. 6.2, $W_{\mathrm{st}}(x, \varepsilon)$ is shown by using altitude charts. For increasing correlation times τ, the distribution exhibits a scewing [147, 163, 165], that is, the ridge of the distribution lies on a curved manifold. This manifold is approximately given by

$$\varepsilon(x) = x^3 - x \quad (6.39)$$

At a critical value of the correlation time $\tau_c(D)$, the character of the probability distribution at the origin changes from a saddle point to a minimum. One should note, however, that this crater is very shallow, cf. Fig. 6.2(d), that is, it does not significantly affect transport properties such as escape rates. One should bear in mind, however, that the whole region around the origin is very flat [166].

For small correlation times τ, one can derive an approximate expression for the stationary probability density [69]. Expanding the potential

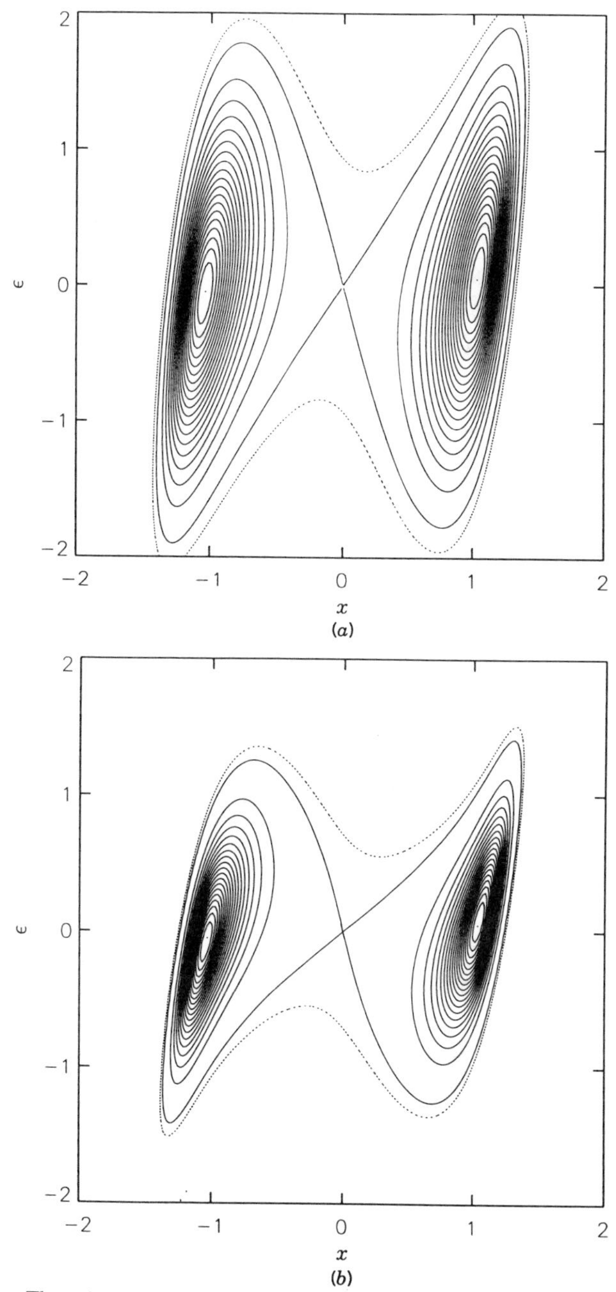

Figure 6.2. The colour lines of the stationary probability density in the extended phase space [Eq. (6.38)] are shown for $\tau = 0.2(a)$, $\tau = 0.5(b)$, $\tau = 1(c)$, and $\tau = 3.333(d)$ for $D = 0,$ 1.

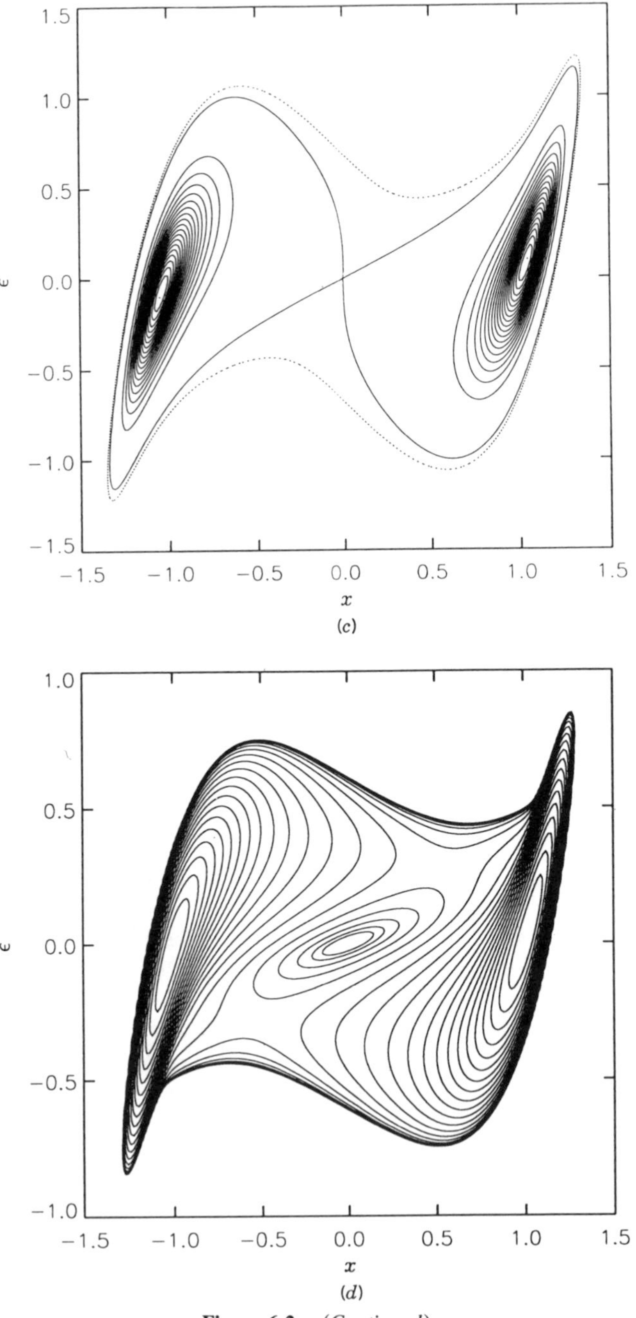

Figure 6.2. (*Continued*)

$\phi(x, \varepsilon)$, defined by

$$W_{st}(x, \varepsilon) = \exp - \left(\frac{\phi(x, \varepsilon)}{D}\right) \tag{6.40}$$

obeying the nonlinear partial differential equation with $h(x) = x - x^3$

$$[h(x) + \varepsilon]\frac{\partial \phi}{\partial x} + \frac{1}{\tau}\frac{\partial \phi}{\partial \varepsilon}\left(-\varepsilon + \frac{1}{\tau}\frac{\partial \phi}{\partial \varepsilon}\right) - D\left(h'(x) + \frac{1}{\tau^2}\frac{\partial^2 \phi}{\partial \varepsilon^2}\right) = 0 \tag{6.41}$$

in a power series in τ, that is,

$$\phi(x, \varepsilon) = \phi^{(0)}(x, \varepsilon) + \tau\phi^{(1)}(x, \varepsilon) + \tau^2\phi^{(2)}(x, \varepsilon) + \tau^3\phi^{(3)}(x, \varepsilon) + O(\tau^4) \tag{6.42}$$

we find by equating all terms of equal power in τ

$$\phi(x, \varepsilon) = \frac{\tau}{2}(1 - \tau h')(\varepsilon + h)^2 - \int^x \left\{h(y)[1 - \tau h'(y)] + \frac{1}{2}\tau^2 h(y)h''(y)\right\} dy + \frac{1}{2}\tau^3\varepsilon^2 h''\left[\frac{1}{2}h(x) + \frac{1}{3}\varepsilon\right] + \frac{3}{2}D\tau h'(x) \tag{6.43}$$

By inserting the corresponding probability distribution into the Fokker–Planck equation one confirms that the errors are only of the orders $\tau^{n\geq 3}$ and $D\tau^{n\geq 2}$. Neglecting τ^3 terms, one finds in leading order

$$\phi(x, \varepsilon) = \frac{\tau}{2}(1 - \tau h'(x))(\varepsilon + h(x))^2 - \int^x \left(h(y)(1 - \tau h'(y)) + \frac{1}{2}\tau^2 h(y)h''(y)\right) dy + \frac{3}{2}D\tau h'(x) \tag{6.44}$$

The agreement with numerical solutions is very good even for correlation times up to $\tau = 0.5$.

4. *Eigenvalues and Eigenfunctions*

The eigenvalues of the Fokker–Planck operator in the extended-phase space describe the relaxation towards the stationary state,

$$W(x, \varepsilon, t) = \sum_{n,m=0}^{\infty} c_{nm}\psi_{nm}(x, \varepsilon)\exp(-\lambda_{nm}t) \tag{6.45}$$

Most important is the smallest nonvanishing eigenvalue. It describes the relaxation on the longest time scale. The eigenvalues have been com-

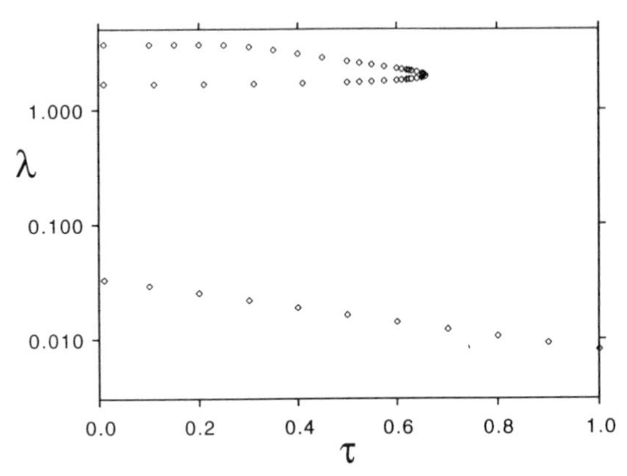

Figure 6.3. The first three branches of real valued eigenvalues, corresponding to odd eigenfunctions, are shown at $D = 0.1$ as a function of the correlation time of the noise. The intersection of the two branches of real values eigenvalues indicates the birth of a pair of complex conjugate eigenvalues.

puted in [154, 163] by applying the matrix continued fraction technique [5]. Our focus is on the dependence of the eigenvalues on the noise correlation time τ. In Fig. 6.3, this dependence is shown for several eigenvalues. The smallest nonvanishing eigenvalue decreases with increasing correlation time of the noise. It is also worthwhile to mention that at that critical value of τ, where the stationary probability distribution changes its shape from a saddle point to a minimum, none of the eigenvalues exhibits any characteristic behavior.

B. Stationary Probability Density

The stationary probability density $P_{st}(x)$ is obtained from the stationary joint probability density $P_{st}(x, \varepsilon)$ by tracing out the auxiliary variable ε,

$$P_{st}(x) = \int_{-\infty}^{\infty} P_{st}(x, \varepsilon)\, d\varepsilon \tag{6.46}$$

In terms of the expansion coefficients d_m^n in Eq. (6.38) the symmetric stationary probability density reads

$$P_{st}(x) = \rho_0(x) \sum_{m=0}^{\infty} d_0^{2m} \varphi_{2m}(x) = P_{st}(-x) \tag{6.47}$$

In Fig. 6.4, $P_{st}(x)$ is plotted at $D = 0.1(a)$ and $D = 0.05(b)$. We observe [163],

1. That for increasing correlation times, the peaks become higher and

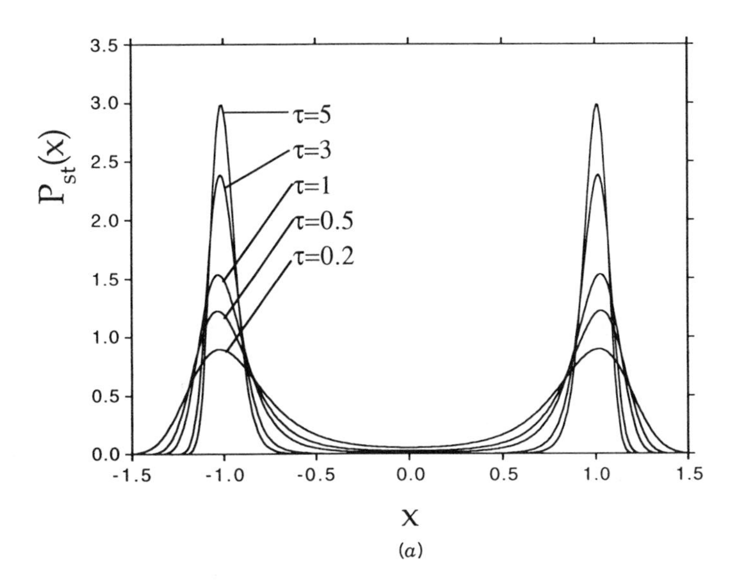

(a)

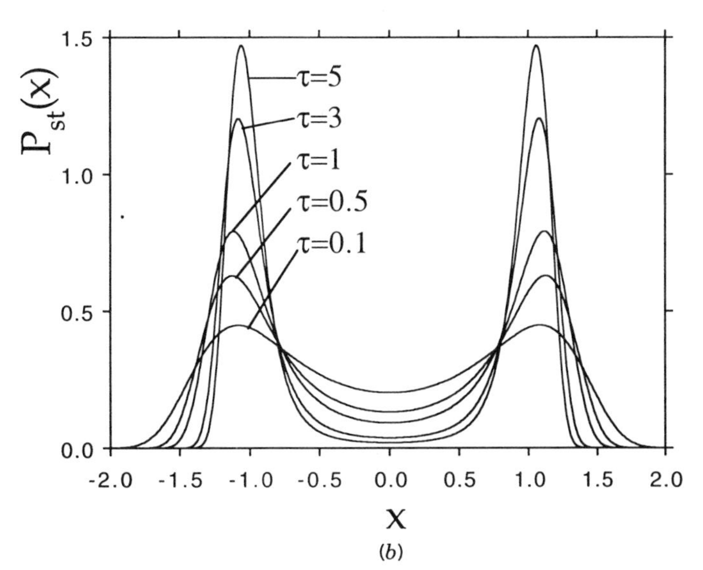

(b)

Figure 6.4. The numerically evaluated stationary probability density [Eq. (6.47)] is shown at $D = 0.1(a)$ and $D = 0.5(b)$ for various values of the correlation time τ of the noise.

more narrow and the probability density at $x = 0$ becomes smaller. This is in accordance with the reasoning put forward in Eq. (5.21) with the decoupling theory.

2. That starting at $\tau = 0$, the maxima for increasing τ shift to larger values of $|x|$ and shift back towards $|x| = 1$ for further increasing correlation times. The maximum shift increases with decreasing noise strength D; being in agreement with the UCNA prediction in Eq. (5.42).

The first observation can be explained by the decrease of the variance of the noise with increasing correlation times τ of the noise. The second observation can be explained qualitatively by changes of the stability of the oscillator equation in the variables $(x, v = \dot{x} = x - x^3 + \varepsilon)$.

Using the approximation schemes (introduced in Section V) we can derive approximate expressions for the stationary probability density. Within the small correlation time approximation, that is, for $\tau \to 0$, $\tau/D \to 0$, one finds with $V_0(x) = x^4/4 - x^2/2$ for the stationary probability density

$$P_{\mathrm{st}}(x) = \frac{1}{Z}\left(1 - \tau|1 - 3x^2| - \frac{\tau}{2D}(x - x^3)^2\right) \exp\left\{-\frac{1}{D} V_0(x)\right\} \tag{6.48}$$

defined in the finite region of support

$$|x| < \sqrt{\frac{1+\tau}{3\tau}} \tag{6.49}$$

Using the unified colored noise approximation we find

$$P_{\mathrm{st}}(x) = \frac{1}{Z}|1 - \tau(1 - 3x^2)| \exp\left\{-\frac{1}{D} V_0(x) - \frac{\tau}{2D}(x - x^3)^2\right\} \tag{6.50}$$

valid in the region of support, given by

$$1 - \tau(1 - 3x^2) > 0 \tag{6.51}$$

The decoupling approximation yields

$$P_{\mathrm{st}}(x) = \frac{1}{Z} \exp\left\{\frac{-1}{\bar{D}} V_0(x)\right\} \tag{6.52}$$

with

$$\bar{D} = \frac{D}{1 + \tau(3\langle x^2 \rangle - 1)} \tag{6.53}$$

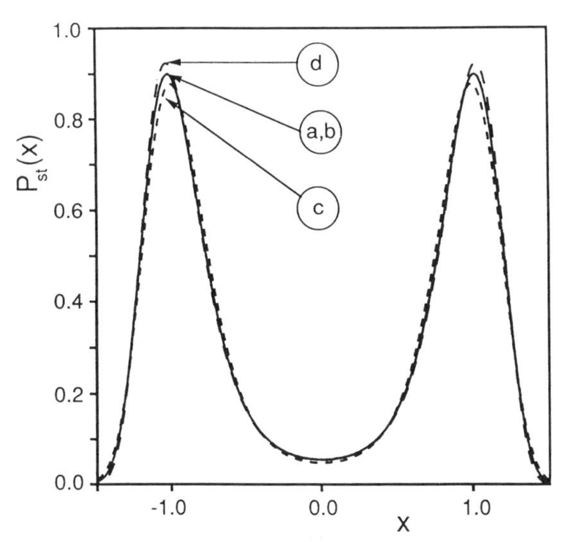

Figure 6.5. The approximate expressions, Eq. (6.48) (small correlation time approximation) (d), Eq. (6.50) (unified colored noise approximation) (b), and Eq. (6.52) (decoupling approximation) (c) are compared with the numerical solutions (a) at $\tau = 0.1$ and $D = 0, 1$.

In all expressions, Z is the respective normalization factor. In Fig. 6.5, the approximate expressions for the stationary probability densities are compared with numerical results obtained from the full numerical solutions at $D = 0.1$ and $\tau = 0.1$. The agreement is good for all approximations. In Fig. 6.6 we have compared the numerical result against the stationary densities obtained by using unified colored noise approximation and by using the decoupling ansatz for $\tau = 1$. The unified colored noise approximation breaks down locally at $x = 0$. The overall agreement, however, is still good. The decoupling approximation yields a distribution with infinite support, but the overall agreement is not very good.

C. Colored Noise Induced Escape Rates and Mean First-Passage Times

A central, but in recent years also very controversial problem [134–138, 144, 154–158, 161–176, 193] is the dependence of the escape rates and the mean first-passage times on the noise correlation time τ. To describe a decay process out of a region Ω in phase space by a *escape rate*, the decay of the population in this region has to be *exponential* on its longest time scale, that is, Ω has to be a basin of attraction. The system has escaped when it has crossed the basin boundary of the basin of attraction. To uniquely identify a basin of attraction and a basin boundary we have to consider the stochastic dynamics in the extended-phase space. In 1-D x

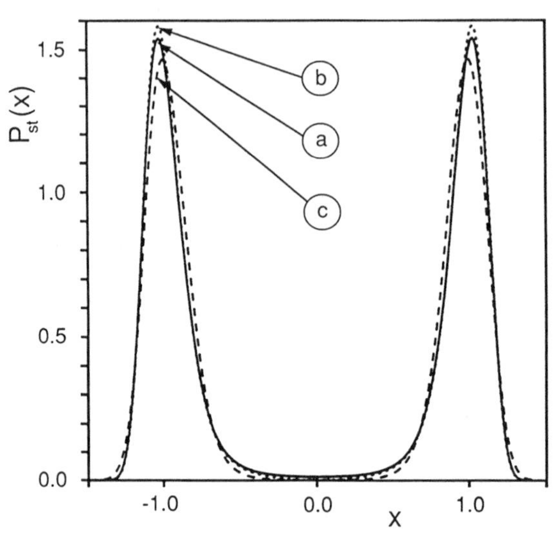

Figure 6.6. The approximate expressions, Eq. (6.50) (unified colored noise approximation) (*b*) and Eq. (6.52) (decoupling approximation) (*c*) are compared with the numerical stationary probabilities (*a*) at $\tau = 1$ and $D = 0, 1$.

space, a certain value of x cannot be considered to be in one or the other basin of attraction, since due to the memory of the noise, the time evolution depends on the prehistory of the process. The attractors of the 2-D system of equations

$$\begin{aligned} \dot{x} &= x - x^3 + \varepsilon \\ \dot{\varepsilon} &= -\frac{1}{\tau}\varepsilon \end{aligned} \tag{6.54}$$

are given by the points ($x_1 = -1$, $\varepsilon_1 = 0$) (stable node), ($x_2 = 1$, $\varepsilon_2 = 0$) (stable node), and ($x_3 = 0$, $\varepsilon_3 = 0$) (saddle point). The separatrix is obtained as the solution of the differential equation [158]

$$\frac{d\varepsilon}{dx} = -\frac{1}{\tau}\frac{\varepsilon}{x - x^3 + \varepsilon} \tag{6.55}$$

with the initial condition

$$\varepsilon(0) = 0 \tag{6.56}$$

Full analytical solutions of Eq. (6.55) are not known to the authors. Near the saddle point ($x = 0$, $\varepsilon = 0$), the solution is found to be [158]

$$\varepsilon(x) = -\left(1 + \frac{1}{\tau}\right)x \tag{6.57}$$

while for large correlation times τ the solution is

$$\varepsilon(x) = \begin{cases} x^3 - x & \text{for } |x| < 1/\sqrt{3} \\ \frac{2}{3}\frac{1}{\sqrt{3}} & \text{for } x \leq -1/\sqrt{3} \\ -\frac{2}{3}\frac{1}{\sqrt{3}} & \text{for } x \geq 1/\sqrt{3} \end{cases} \tag{6.58}$$

In Fig. 6.7, we show some numerically obtained trajectories, cf. Eq. (6.54)], and the separatrix for $\tau = 0.1$, $\tau = 1$, $\tau = 10$, and $\tau = 50$. We note that the asymptotic separatrix for large τ is approached only at extremely large values of τ ($\tau = 50$ is certainly not sufficient). The noise induced escape across the separatrix from the left to the right well takes place at positive values of ε. For small correlation times, the actual escape takes place at large values of $|\varepsilon|$, since the noise acts only in the ε direction. For increasing τ the separatrix bends over and the escape takes place across the separatrix at smaller values of $|\varepsilon|$. For large τ, the trajectories avoid a region around the origin. This unstable region is also responsible for the formation of the crater of the stationary joint probability density $W_{\text{st}}(x, \varepsilon)$ in the extended phase space.

For weak noise, that is, $D \to 0$, the mean first-passage time to the separatrix T_s is related to the escape rate r_s by [123, 167]

$$r_s = \frac{1}{2T_s} \tag{6.59}$$

The smallest nonvanishing eigenvalue $\lambda_{\min}$ is related to the escape rate r_s by

$$\lambda_{\min} = 2r_s = \frac{1}{T_s} \tag{6.60}$$

The smallest nonvanishing eigenvalues obtained in [154], have been plotted as a function of the correlation time of the noise in Fig. 6.3. In Fig. 6.8, we depict the color induced rate suppression

$$\eta(\tau, D) = \frac{r_s(\tau)}{r_s(\tau = 0)} \tag{6.61}$$

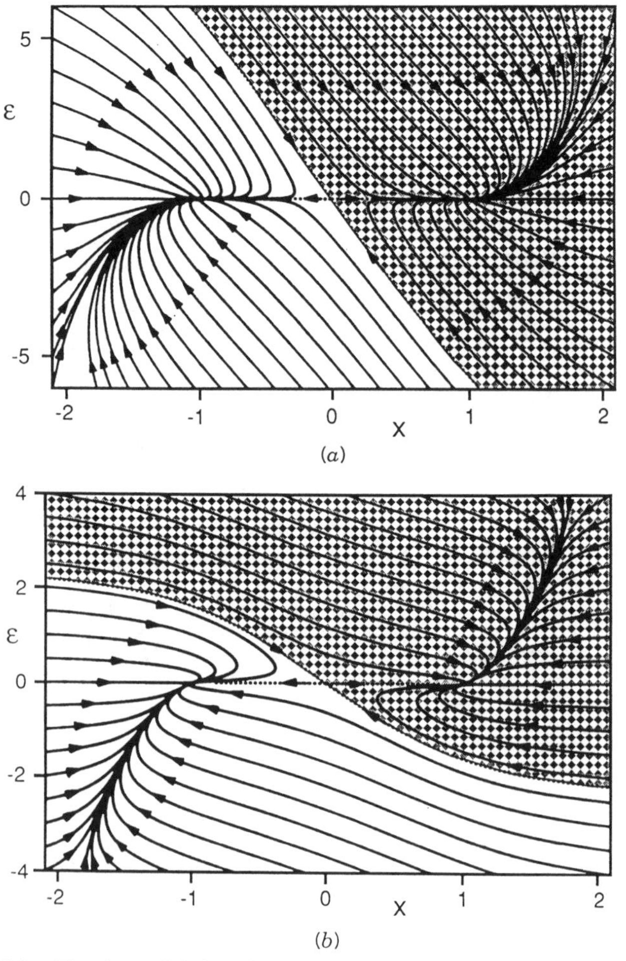

Figure 6.7. The deterministic trajectories, Eq. (6.54), are shown for $\tau = 0.1(a)$, $\tau = 1(b)$, $\tau = 10(c)$, and $\tau = 50(d)$. The dotted lines indicate the separatrix $\varepsilon(x)$ in (c) and (d) while in (a) and (b) the separatrix is the border between the hatched and nonhatched regions—the basins of attractions. The limiting result in Eq. (6.58) for $\varepsilon(x)$ is indicated by the dash-dotted line in (d).

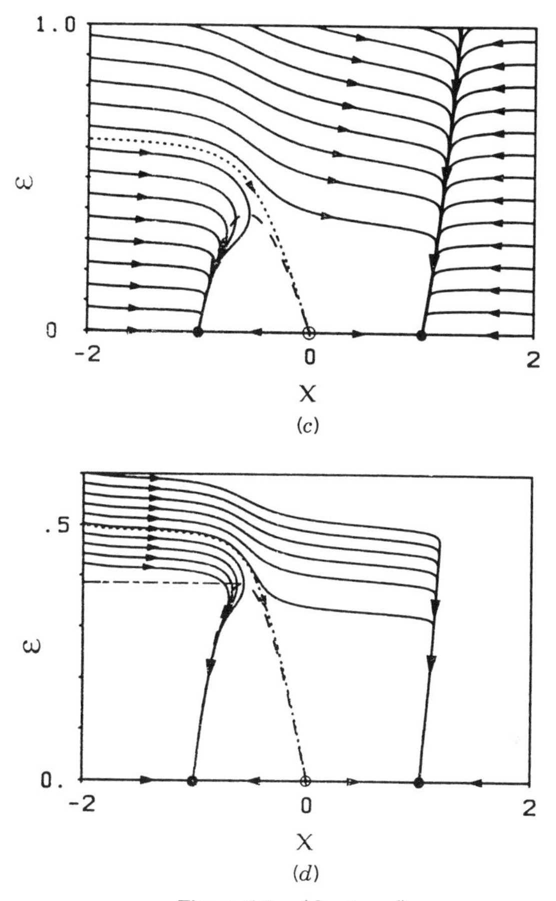

(c)

(d)

Figure 6.7. (*Continued*)

1. *Escape Rates for Weakly Colored Noise*

For small correlation times τ, the small correlation time approximation, in Eq. (5.6) valid on a large system time scale (i.e. small a), yields the 1-D eigenvalue equation

$$\mathbf{L}_{\mathrm{SRTA}}\psi_{\min} = -\lambda_{\mathrm{SRTA}}\psi_{\min}$$
$$\mathbf{L}_{\mathrm{SRTA}} = -\frac{\partial}{\partial x}(x - x^3) + \frac{\partial^2}{\partial x^2}[1 + \tau(1 - 3x^2)] \tag{6.62}$$

Accordingly, the approximation by Fox [106, 137] leads to the 1-D

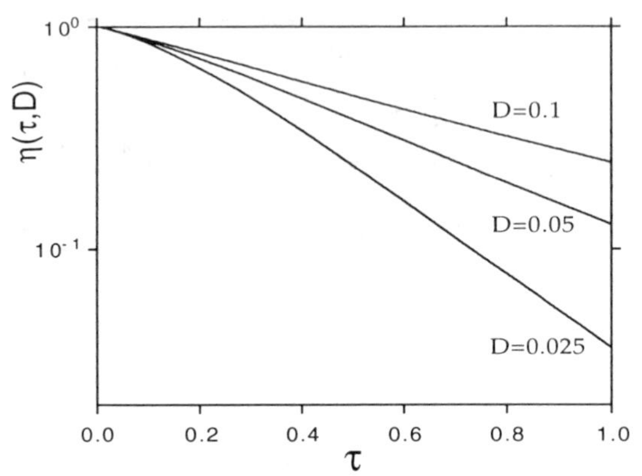

Figure 6.8. The noise color induced suppression of the escape rate $\eta(\tau, D)$ [Eq. (6.61)] is shown for $D = 0.1$, $D = 0.05$, and $D = 0.025$.

eigenvalue equation

$$\begin{aligned} \mathbf{L}_{\text{Fox}}\psi_{\text{min}} &= -\lambda_{\text{Fox}}\psi_{\text{min}} \\ \mathbf{L}_{\text{Fox}} &= -\frac{\partial}{\partial x}(x - x^3) + D\frac{\partial^2}{\partial x^2}\frac{1}{1 - \tau(1 - 3x^2)} \end{aligned} \tag{6.63}$$

The results of the numerical solutions of Eqs. (6.62) and (6.63) are compared in Fig. 6.9 with the smallest nonvanishing eigenvalue, obtained from the full 2-D Fokker–Planck equation.

Using the method of the effective small τ potential [168], one can find, within the small correlation time theory, an expression for the correlation time corrections of the escape rate valid for weakly colored noise. In the new variable

$$y = x - \tfrac{1}{2}\tau(x - x^3) \tag{6.64}$$

the diffusion coefficient of the Fokker–Planck operator in Eq. (6.62) is a constant, and the drift term corresponds to the potential

$$V_{\text{eff}}(y) = \tfrac{1}{4}x^4 - \tfrac{1}{2}(1 + 3D\tau)x^2 \tag{6.65}$$

which is again a potential of the quartic type, but with renormalized coefficients. Applying Kramers' formula for the escape rate valid for

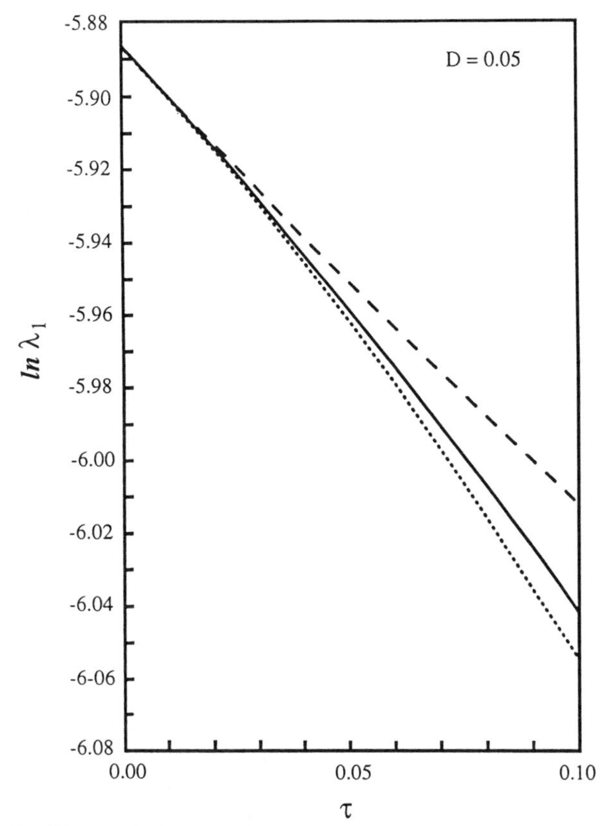

Figure 6.9. The smallest nonvanishing eigenvalues, obtained numerically from the approximate Fokker–Planck equations (6.62) (dotted) and (6.63) (dashed), are compared with the eigenvalues obtained from the full 2-D eigenvalue equation (6.11) (solid) at $D = 0.1$.

weak noise (for a recent review, see [123])

$$r_s = \frac{1}{2\pi} \sqrt{|V''_{\text{eff}}(x_{\text{min}}) V''_{\text{eff}}(x_{\text{saddle}})|} \exp\left(-\frac{\Delta V}{D}\right) \tag{6.66}$$

where ΔV is the barrier height, x_{min} is the position of the local potential minimum, and x_{saddle} is the position of the barrier top. Inserting the expression for the effective potential, we find

$$r_s(\tau) = r_s(\tau = 0)[1 - \beta(D)\tau)] \tag{6.67}$$

where

$$\beta(D) = \tfrac{3}{2} - 3D \tag{6.68}$$

This expression is asymptotically for $\tau \to 0$ exact. The dilemma, however, is that within the small correlation time theory for $D \to 0$, we also have to reduce τ according to the condition $\tau/D \to 0$.

In the limit $D \to 0$ and *small* but finite τ, corrections to the rate have been obtained by use of a variety of methods. The result reads [123, 134, 154, 156, 161, 169–171]

$$r_s(\tau) = r_s(\tau = 0)\{1 - \beta(D)\tau\} \exp - \left[\frac{\tau^2}{8D} + \frac{3\tau^4}{10D} + O\frac{\tau^6}{D}\right] \tag{6.69}$$

2. *Escape Rates for Strongly Colored Noise*

For large correlation times, corrections to the exponential part of the escape rates have been determined by using path integral techniques, adiabatic arguments, or by using the unified colored noise approximation. The result in leading order reads [154, 157]

$$r_s \propto \exp\left(-\frac{2\tau}{27D}\right) \tag{6.70}$$

Comparison with numerical solutions for the escape rate at finite values of τ, however, shows that this result is actually very asymptotic [154, 158, 166]. The dependence of the exponential part of the numerically evaluated rate in fact shows a dependence of the type in Eq. (6.70) but with a factor other than 2/27, that is $2/27 = 0.074\ldots \to \sim 0.1$ [154]. Luciani and Verga [156] derived a bridging formula, connecting the approximations at small τ and large τ, given for $\Delta V = \frac{1}{4}$

$$r_s = \frac{1}{\sqrt{2}\pi}(1 + 3\tau)^{-1/2} \exp\left[-\frac{1}{4D}\left\{\frac{1 + \frac{27}{16}\tau + \frac{1}{2}\tau^2}{1 + \frac{27}{16}\tau}\right\}\right] \tag{6.71}$$

3. *Mean First-Passage Times for Other Boundary Conditions*

So far, we have discussed escape rates and its connection to mean first-passage times to leave the basin of attraction. Since the concept of mean first-passage times is valid for more general regions in phase space, one can also ask for the mean first-passage time to leave the right or left infinite one-half plane, $x < 0$ or $x > 0$. This problem has been studied by Doering et al. [172] for weakly colored noise. They solved the Fokker–Planck equation in the extended-phase space $(x, z = \varepsilon/\langle\varepsilon^2\rangle)$ with a

Gaussian source term $\delta(x-1)\rho_0(z)$ and an absorbing boundary at $x=0$ for $z>0$. Those boundary conditions yield a current-carrying stationary solution from which they can derive an expression for the mean first-passage time. In order to obtain a solution for weakly colored noise, they expand the stationary current carrying solution in functions of the orders of $\sqrt{\tau}$. As a result one finds

$$T_{x=0}=\frac{\pi}{\sqrt{2}}\exp\left(\frac{1}{4D}\right)\left[1+\sqrt{\frac{2}{\pi}}\lambda_M\sqrt{\tau}+\frac{3}{2}\tau\right] \tag{6.72}$$

with the Milne extrapolation length λ_M, given in terms of the Riemannian ζ function, by $\lambda_M=-\zeta(1/2)=1.460354$. This $\sqrt{\tau}$ correction has been established by simulating the Langevin equations $\dot{x}=x-x^3+\varepsilon$, $\dot{\varepsilon}=-(1/\tau)\varepsilon+\sqrt{D}/\tau\xi(t)$. Those results have been contrasted with the escape rates over the separatrix in detail in [173–176].

D. Colored Noise Driven Systems with Inertia

Up to now, we have neglected the inertial effect completely, that is, we have assumed that the velocity relaxation takes place on a very fast time scale in comparison with other time scales. For finite inertia, we introduce another finite time scale $\tau_r=1/\gamma$ into the system. The normalized Langevin equations read

$$\begin{aligned}\dot{x}&=v\\ \dot{v}&=-\gamma v+f(x)+\varepsilon\\ \dot{\varepsilon}&=-\frac{1}{\tau}\varepsilon+\frac{\sqrt{D}}{\tau}\xi(t)\end{aligned} \tag{6.73}$$

with Gaussian white noise $\xi(t)$,

$$\begin{aligned}\langle\xi(t)\rangle&=0\\ \langle\xi(t)\xi(t')\rangle&=2\delta(t-t')\end{aligned} \tag{6.74}$$

The corresponding Fokker–Planck equation

$$\frac{\partial}{\partial t}P(x,v,\varepsilon,t)=\left[-\frac{\partial}{\partial x}v+\gamma\frac{\partial}{\partial v}(v-f(x)-\varepsilon)+\frac{1}{\tau}\frac{\partial}{\partial\varepsilon}\varepsilon+\frac{D}{\tau^2}\frac{\partial^2}{\partial\varepsilon^2}\right]\times P(x,v,\varepsilon,t) \tag{6.75}$$

can be solved analytically for a quadratic potential, $f(x)=-\omega_0^2x$, only.

For the stationary distribution one then obtains

$$P_{\mathrm{st}}(x, v, \varepsilon) = \frac{1}{Z} \exp\left(-\frac{\phi(x, v, \varepsilon)}{D}\right) \tag{6.76}$$

with the potential

$$\begin{aligned} \phi(x, v, \varepsilon) = &\tfrac{1}{2}\gamma\omega_0^2(1 + \tau^2\omega_0^2)x^2 + \gamma^2\tau^2\omega_0^2 xv + \tfrac{1}{2}\gamma[(1+\gamma\tau)^2 + \tau^2\omega_0^2]v^2 \\ &+ \tfrac{1}{2}\tau(1+\gamma\tau)\varepsilon^2 - \gamma\tau(1+\gamma\tau)\varepsilon v - \gamma\tau^2\omega_0^2 x\varepsilon \end{aligned} \tag{6.77}$$

The quadratic form for the potential can be diagonalized by introducing the new variable q instead of ε [111]

$$q = -\gamma v + \varepsilon - \frac{\gamma\tau\omega_0^2 x}{1+\gamma\tau} \tag{6.78}$$

The potential then factorizes, that is,

$$\phi(x, v, q) = \tfrac{1}{2}\tau(1+\gamma\tau)q^2 + \tfrac{1}{2}(1 + \gamma\tau + \tau^2\omega_0^2)\gamma v^2 + \tfrac{1}{2}\gamma\omega_0^2\left(1 + \frac{\tau^2\omega_0^2}{1+\gamma\tau}\right)x^2 \tag{6.79}$$

This factorization for the parabolic potential implies for a general force field $f(x)$ the introduction of the new variable [111]

$$q = -\gamma v + \varepsilon - \frac{\gamma\tau f(x)}{1+\gamma\tau} \tag{6.80}$$

The system of Langevin equations then reads [111]

$$\begin{aligned} \dot{x} &= v \\ \dot{v} &= q + \frac{f(x)}{1+\gamma\tau} \\ \dot{q} &= -\left(\gamma + \frac{1}{\tau}\right)q + \frac{\gamma}{\tau}\left[\frac{\tau^2}{1+\gamma\tau} f'(x) - 1\right]v + \frac{\sqrt{D}}{\tau}\,\xi(t) \end{aligned} \tag{6.81}$$

1. *Small Correlation Time Approximation*

Starting from the Fokker–Planck equation in Eq. (6.75),

$$\frac{\partial}{\partial t} P = (\mathbf{A} + \varepsilon\mathbf{B} + \mathbf{L}_\varepsilon)P \tag{6.82}$$

where

$$\mathbf{A} = -\frac{\partial}{\partial x} v + \gamma \frac{\partial}{\partial v} v - f(x) \frac{\partial}{\partial v}$$

$$\mathbf{B} = -\frac{\partial}{\partial v} \tag{6.83}$$

$$\mathbf{L}_\varepsilon = \frac{1}{\tau} \frac{\partial}{\partial \varepsilon} \varepsilon + \frac{D}{\tau^2} \frac{\partial^2}{\partial \varepsilon^2}$$

we apply the same small τ-expansion technique as presented in Appendix A1 of [5] (generalized to a three dimensional (3-D) Fokker–Planck equation). As the result one finds for large times and $\gamma \ll 1/\tau$ up to order $O(\tau^2)$ the 2-D Fokker–Planck approach

$$\frac{\partial}{\partial t} P(x, v, t) = \mathbf{L}_{\text{SCTA}} P(x, v, t) \tag{6.84}$$

where

$$\mathbf{L}_{\text{SCTA}} = \mathbf{A} + D\mathbf{B}^2 + \tau D \mathbf{B}[\mathbf{A}, \mathbf{B}] + D\tau^2 \mathbf{B}[\mathbf{A}, [\mathbf{A}, \mathbf{B}]]$$
$$+ D^2\tau^2(\mathbf{B}[[\mathbf{B}, \mathbf{A}], \mathbf{B}]\mathbf{B} + \tfrac{1}{2}\mathbf{B}^2[[\mathbf{B}, \mathbf{A}], \mathbf{B}]) \tag{6.85}$$

with $[\mathbf{A}, \mathbf{B}] = \mathbf{AB} - \mathbf{BA}$. The same result can also be obtained by extending the functional technique to higher dimensions [146, 177]. Inserting the expressions for the operators **A** and **B**, we obtain the Fokker–Planck type operator (see Appendix of [111])

$$\mathbf{L}_{\text{SCTA}} = -\frac{\partial}{\partial x} v + \gamma \frac{\partial}{\partial v} v - f(x) \frac{\partial}{\partial v} + D(1 - \gamma\tau + \tau^2 f'(x) + \tau^2\gamma^2) \frac{\partial^2}{\partial v^2}$$
$$+ \tau D(1 - \gamma\tau) \frac{\partial^2}{\partial x\, \partial v} + O(\tau^3) \tag{6.86}$$

The stationary distribution can be obtained analytically up to order τ and reads

$$P_{\text{st}}^{(1)}(x, v) = \frac{1}{Z} \exp\left[-\frac{\gamma}{D} U(x) - \frac{\gamma}{2D(1 - \gamma\tau)} v^2\right] \tag{6.87}$$

where

$$U(x) = -\int^x f(y)\, dy \tag{6.88}$$

The distribution in x and v factorize within this approximation. The shape of the distribution function in x only is in first order of τ, independent of the correlation time of the noise. For a harmonic potential $U(x) = \frac{1}{2}\omega_0^2 x^2$, the stationary probability density can be computed and obtained up to second order τ^2, yielding for the variance $\sigma_{xx} = \langle x^2 \rangle_{\rm st} - \langle x \rangle_{\rm st}^2 = \langle x^2 \rangle_{\rm st}$

$$\sigma_{xx} = \frac{D}{\omega_0^2 \gamma}(1 - \omega_0^2 \tau^2) \tag{6.89}$$

which agrees with the exact solution up to order τ^2.

2. *Unified Colored Noise Approximation*

The starting point for the application of the unified colored noise approximation is the set of equations in Eq. (6.81). Performing a time scale transformation $t \to t/\sqrt{\tau}$, we observe that we can adiabatically eliminate q for [111]

$$\frac{\gamma\sqrt{D\tau}}{(\sqrt{1+\gamma\tau})^3} \ll 1 \tag{6.90}$$

and

$$t > \frac{\tau}{1+\gamma\tau} \tag{6.91}$$

The Fokker–Planck equation for this approach reads [111]

$$\frac{\partial}{\partial t} P(x, v, t) = \mathbf{L}_{\rm UCNA} P(x, v, t) \tag{6.92}$$

where

$$\begin{aligned} \mathbf{L}_{\rm UCNA} = &- \frac{\partial}{\partial x} v + \left[\frac{\gamma}{1+\gamma\tau} \left(1 - \frac{\tau^2}{1+\gamma\tau} f(x) \right) \right] \frac{\partial}{\partial v} v \\ &- \frac{f(x)}{1+\gamma\tau} \frac{\partial}{\partial v} + \frac{D}{(1+\gamma\tau)^2} \frac{\partial^2}{\partial v^2} \end{aligned} \tag{6.93}$$

The stationary probability density can only be obtained analytically for the harmonic potential $U(x) = \frac{1}{2}\omega_0^2 x^2$. In this case, it precisely agrees with the exact stationary probability density Eqs. (6.76) and (6.77). The unified colored noise approximation—as a truly Markovian approximation—also correctly describes the dynamical quantities, such as correlation functions within its range of validity Eqs. (6.90) and (6.91). In contrast to the small correlation time approximation, the high friction limit can be carried through by another adiabatic approximation, yielding the (Stratonovich) Langevin equation

$$\dot{x} = \frac{f(x)}{\gamma - \tau f'(x)} + \frac{\sqrt{D}}{\gamma - \tau f'(x)} \xi(t) \tag{6.94}$$

which is up to a time scale $(1/\gamma)$ identical with the standard unified colored noise approximation for overdamped systems, discussed in Section V.C. This higher dimensional UCNA in Eqs. (6.92) and (6.93) has recently been applied to study colored noise driven bistability by Schimansky-Geier and Zülicke [178], and for modeling the dynamics in dye lasers by Cao et al. [179].

3. *Decoupling Approximation*

The probability density $P(x, v, t)$ for the non-Markovian stochastic process, described by the Langevin equation, Eq. (6.73), obeys the integrodifferential equation [146]

$$\begin{aligned}
\frac{\partial}{\partial t} P(x, v, t) &= \frac{\partial}{\partial t} \langle \delta(x(t) - x)\delta(v(t) - x) \rangle \\
&= -v \frac{\partial}{\partial x} P(x, v, t) - f(x) \frac{\partial}{\partial v} P(x, v, t) + \gamma \frac{\partial}{\partial v} (vP(x, v, t)) \\
&\quad + \frac{\partial^2}{\partial x \, \partial v} \left[\frac{D}{\tau} \int_0^t \exp -\left(\frac{t-s}{\tau} \right) \left\langle \delta(x(t) - x)\delta(v(t) - v) \frac{\delta x(t)}{\delta \xi(s)} \right\rangle \right] \\
&\quad + \frac{\partial^2}{\partial v^2} \left[\frac{D}{\tau} \int_0^t \exp -\left(\frac{t-s}{\tau} \right) \left\langle \delta(x(t) - x)\delta(v(t) - v) \frac{\delta v(t)}{\delta \xi(s)} \right\rangle \right]
\end{aligned} \tag{6.95}$$

where the functional derivatives $\delta(x(t))/\delta(\xi(s))$ and $\delta(v(t))/\delta(\xi(s))$ obey the coupled integrodifferential equations [146]

$$\begin{aligned}
\frac{\delta x(t)}{\delta \xi(s)} &= \theta(t - s) \left[\int_s^t \frac{\delta v(r)}{\delta \xi(s)} \, dr \right] \\
\frac{\delta v(t)}{\delta \xi(s)} &= \theta(t - s) \left[1 - \int_s^t f(x(r)) \frac{\delta x(r)}{\delta \xi(s)} \, dr - \gamma \int_s^t \frac{\delta v(r)}{\delta \xi(s)} \, dr \right]
\end{aligned} \tag{6.96}$$

Factorization of the probability density and the functional derivatives yields a closed equation of the Fokker–Planck type for the probability density $P(x, v, t)$, that is [146],

$$\frac{\partial}{\partial t} P(x, v, t) = \mathbf{L}_{\mathrm{Dec}} P(x, v, t)$$
$$\mathbf{L}_{\mathrm{Dec}} = -v \frac{\partial}{\partial x} + \frac{\partial}{\partial v} [\gamma v - f(x)] + \frac{D}{1 + \gamma\tau - \tau^2 \langle f'(x) \rangle} \frac{\partial^2}{\partial v^2} + \frac{D\tau}{1 + \gamma\tau - \tau^2 \langle f'(x) \rangle} \frac{\partial^2}{\partial x\, \partial v} \tag{6.97}$$

The stationary probability density factorizes in x and v, that is,

$$P_{\mathrm{st}}(x, v) = \frac{1}{Z} \exp\left(-\frac{v^2}{2\sigma_{vv}}\right) \exp\left(-\frac{U(x)}{\sigma_{xx}}\right) \tag{6.98}$$

with the potential $U(x)$ defined in Eq. (6.88) and the covariances

$$\sigma_{xx} = \frac{D}{\gamma} \frac{1}{1 - \tau^2 [\langle f'(x) \rangle / (1 + \gamma\tau)]}$$
$$\sigma_{vv} = \frac{D}{\gamma} \frac{1}{1 + \gamma\tau - \tau^2 \langle f'(x) \rangle} \tag{6.99}$$

Tracing out the velocity, we obtain the equation of motion for the probability density in the position x only [146]

$$\frac{\partial}{\partial t} P(x, t) = -\frac{\partial}{\partial x} \tilde{f}(x) P(x, t) + \frac{\tilde{D}(1 + (1/\gamma\tau))}{1 + (1/\gamma\tau) - \tau \langle \tilde{f}'(x) \rangle} \frac{\partial^2}{\partial x^2} P(x, t) \tag{6.100}$$

where

$$\tilde{f}(x) = \frac{f(x)}{\gamma}$$
$$\tilde{D} = \frac{D}{\gamma^2} \tag{6.101}$$

The prediction of the decoupling theory has been tested by using analogue simulations [147]. Although the agreement of the stationary joint probability density $P_{\mathrm{st}}(x, v)$ is not very good (incorrect symmetry due to factorization), the agreement of the reduced density with the analogue simulation result is excellent. The mean values, occurring in the

Fokker–Planck approach have to be determined self-consistently. Since the decoupling theory is not restricted to small correlation times, one can also perform the overdamped limit $\gamma\tau \gg 1$,

$$\frac{\partial}{\partial t} P(x, t) = -\frac{\partial}{\partial x} \tilde{f}(x) P(x, t) + \frac{\tilde{D}}{1 - \tau\langle \tilde{f}'(x) \rangle} \frac{\partial^2}{\partial x^2} P(x, t) \quad (6.102)$$

This equation agrees precisely with the standard decoupling theory for overdamped systems.

VII. MULTIPLICATIVE COLORED NOISE AND PHOTON STATISTICS OF DYE LASERS

The study of dynamical systems in optical sciences is attracting rapidly growing interest. In particular, the fields of optical bistability and chaos in optical systems have become the main focus of interest for many researchers. Here, we restrict ourselves to the influence of noise in nonlinear optical system. Experiments with dye lasers strongly emphasize the role played by noise sources with a finite correlation time. For dye lasers strongly correlated noise enters via the pumping mechanism and it crucially impacts their photon statistics.

Some time ago it was found that the behavior of a single mode dye laser was not very well described by the usual single mode laser theory [180–182]. Short and co-workers [10, 182] suggested taking into account fluctuations of the pump parameter, to describe the dye laser close to threshold. Graham and co-workers [183, 184] and Schenzle and Brand [185] developed a stochastic model for the field of the single mode laser, which incorporates δ correlated pump noise. Some intensity correlation functions compared very nicely with the results of this one-fit parameter model. It turned out, however, that it is not possible to explain experimental data at different working points of the laser with one value of the fit parameter [10]. Short et al. [10] concluded from their experiments that the pump noise should be slower than the fluctuations of the intensity. Dixit and Sahni [11] and Schenzle and Graham [12] first discussed the impact of colored pump fluctuations for the photon statistics of dye lasers. While the stochastic equations have been simulated in [10], Schenzle and Graham [12] used a small correlation time expansion. Numerical studies of the dye laser model for arbitrary correlation times have been put forward in [186, 187].

To describe the behavior of the laser close to the threshold correctly, one has to take into account both pump and quantum fluctuations [188]. For adiabatically eliminated inversion and polarization, the equation of motion for the complex field amplitude E reads [149, 188, 189]

$$\dot{E} = a_0 E - AE|E|^2 + p(t)E + q(t) \tag{7.1}$$

where

$$\begin{aligned} E &= E_1 + iE_2 \\ p &= p_1 + ip_2 \\ q &= q_1 + iq_2 \end{aligned} \tag{7.2}$$

In Eq. (7.1) a_0 denotes the pump parameter; A denotes the saturation parameter, which limits the stationary field amplitude to a finite value; q denotes the quantum fluctuations due to spontaneous emission processes, being important at low-field intensities; and $p(t)$ denotes the pump fluctuations. Both, quantum and pump fluctuations are assumed to be Gaussian distributed with zero mean. The correlation functions are given by

$$\begin{aligned} \langle q_i(t) q_j(t') \rangle &= 2\tilde{q}\delta(t - t')\delta_{ij} \\ \langle p_i(t) p_j(t') \rangle &= \frac{\tilde{D}}{\tilde{\tau}} \exp\left[-\frac{1}{\tilde{\tau}} \left| t - t' \right| \right] \delta_{ij} \end{aligned} \tag{7.3}$$

Some typical sets of parameters $\tilde{D}$, $\tilde{q}$ have been obtained by Zhou et al. [13] and Roy et al. [190] by comparing simulations of Eqs. (7.2) and (7.3) with experimental data. A typical set of parameters is $a_0 = 0.7 \times 10^6 \,\mathrm{s}^{-1}$, $A = 0.114 \times 10^6 \,\mathrm{s}^{-1}$, $\tilde{q} = 2 \times 10^{-3} \,\mathrm{s}^{-1}$, $\tilde{D} = 4.9 \times 10^3 \,\mathrm{s}^{-1}$, and $\tilde{\tau} = 5.0 \times 10^{-7} \,\mathrm{s}$.

Above threshold (i.e. $a_0 > 0$) we transform the equation of motion for the complex field amplitude into a set of equations for intensity I and phase ϕ. The equation of motion for the intensity reads in suitable normalized form, that is, with $t \rightarrow a_0 t$, $I \rightarrow A/a_0 I$

$$\begin{aligned} \dot{I} &= 2(I - I^2) + 2I\varepsilon + \frac{Q}{2} + \sqrt{QI}q(t) \\ \dot{\varepsilon} &= -\frac{1}{\tau}\varepsilon + \frac{\sqrt{D}}{\tau}\xi(t) \end{aligned} \tag{7.4}$$

where

$$\tau = a_0 \tilde{\tau}$$
$$D = \frac{\tilde{D}}{a_0} \tag{7.5}$$
$$Q = 4 \frac{\tilde{q}}{a_0^2} A$$

Equation (7.4) already takes advantage of the embedding property introduced in the last section. Inserting the experimental values for the parameters in Eq. (7.5), we obtain the values for the normalized parameters $D = 7 \times 10^{-3}$, $Q = 1.86 \times 10^{-9}$, $\tau = 0.35$. It is interesting to note that the quantum fluctuations are six orders of magnitude smaller than the pump fluctuations, and that the noise correlation time τ is not much smaller than the typical system time scale, which is in our normalized units of $\tau_s = 1$. Other experimental sets yield even much larger values of τ in the order 10. This makes it clear that in order to understand the dye laser one needs theories for dynamical systems driven by colored noise that are *nonperturbative* in the correlation time of the noise. If we neglect quantum fluctuations that can be safely done if we are interested only in the stationary properties of the laser light intensity and those properties that are not too close to threshold, the equation of motion for the field intensity reads

$$\dot{I} = 2(a - I)I + 2I\varepsilon$$
$$\dot{\varepsilon} = -\frac{1}{\tau}\varepsilon + \frac{\sqrt{D}}{\tau}\xi(t) \tag{7.6}$$

with the δ correlated Gaussian stochastic process $\xi(t)$. In Eq. (7.6) we have made use of a different scaling as compared to Eq. (7.4) by invoking the dimensionless pump parameter a, i.e. $a_0 \rightarrow a \cdot a_0$. This gives the possibility to vary the working point of the laser. For $a < 0$, that is, below threshold, the laser is off. In Eq. (7.6) this is reflected in a vanishing stationary mean value of the intensity, that is, $\langle I_{st} \rangle = 0$. For $a > 0$, that is, above threshold, the laser is on. Accordingly, the stationary mean value of the intensity is given by $\langle I_{st} \rangle = a$. Note that the mean value agrees identically with the behavior of the noiseless system $\dot{I} = 2(a - I)I$; the noise has not shifted the bifurcation point. Some quantities of interest are the stationary probability distribution $P_{st}(I)$, since it directly relates to the photon counting statistics [191, 192], and the correlation function

$$\phi_I(t) = \frac{\langle (I(t) - \langle I \rangle_{st})(I(0) - \langle I \rangle_{st}) \rangle}{\langle I^2 \rangle_{st} - \langle I \rangle_{st}^2} \tag{7.7}$$

which relates, via the Wiener-Khintchine theorem, to the fluctuation induced line width. The line width is characterized by the effective eigenvalue λ_{eff}, or equivalently by the inverse of the relaxation time

$$T = \int_0^\infty \phi_I(t)\, dt = \frac{1}{\lambda_{eff}} \tag{7.8}$$

A. The White Noise Limit

In the white noise limit, the (Stratonovich) Fokker–Planck equation for the intensity reads

$$\frac{\partial}{\partial t} P(I, t) = \mathbf{L}_0 P(I, t) \tag{7.9}$$

where

$$\mathbf{L}_0 = -2 \frac{\partial}{\partial I}(a - I)I + 4D \frac{\partial}{\partial I} I \frac{\partial}{\partial I} I \tag{7.10}$$

Above threshold, the stationary distribution function is given by [6]

$$P_{st}(I) = P_{st}^{\tau=0}(I) = \frac{1}{Z}(4DI)^{\kappa_0} \exp\left(-\frac{I}{2D}\right) \tag{7.11}$$

with the normalization constant (Γ denotes the Gamma function)

$$Z = \frac{1}{4D}(8D^2)^{\kappa_0 + 1}\Gamma(\kappa_0 + 1) \tag{7.12}$$

and

$$\kappa_0 = \frac{a}{2D} - 1 \tag{7.13}$$

Below threshold, the stationary probability is given by the right-sided δ function $\delta_+(I)$. The stationary distribution function above threshold shows a noise induced transition at $D = a/2$ [6]. For $D < a/2$, the stationary distribution vanishes at $I = 0$, while for $D > a/2$, $P_{st}(I)$ has an integrable singularity at $I = 0$. This transition, however, does not show up in the stationary mean values

$$\langle I^n \rangle_{st} = \int_0^\infty I^n P_{st}(I) dI = (2D)^n \Gamma(\kappa_0 + n + 1)$$
$$= (2D)^n \Gamma\left(\frac{a}{2D} + n\right) \tag{7.14}$$

The effective eigenvalue, λ_{eff} for white noise can be obtained from Equation (7.15) [152], that is,

$$\lambda_{eff}^{-1} = \frac{1}{\langle I^2 \rangle - \langle I \rangle^2} \int_0^\infty \frac{f^2(x)}{D(x) P_{st}(x)} dx \tag{7.15}$$

with $D(x) = 4Dx^2$

$$f(x) = -\int_0^x (x' - \langle I \rangle_{st}) P_{st}(x') \, dx' \tag{7.16}$$

The integrals can be evaluated exactly, yielding the simple result [152]

$$\lambda_{eff} = 2a \tag{7.17}$$

that is, the bandwidth of the laser does not depend on the noise strength.

B. The Stationary Probability with Colored Noise

The two-variable Fokker–Planck equation

$$\frac{\partial}{\partial t} W(I, \varepsilon, t) = \mathbf{L}_{em} W(I, \varepsilon, t) \tag{7.18}$$

with the embedding Fokker–Planck operator

$$\begin{aligned} \mathbf{L}_{em} &= \mathbf{A} + \varepsilon \mathbf{B} + \mathbf{L}_\varepsilon \\ \mathbf{A} &= -2 \frac{\partial}{\partial I} (a - I) I \\ \mathbf{B} &= -2 \frac{\partial}{\partial I} I \\ \mathbf{L}_\varepsilon &= \frac{1}{\tau} \frac{\partial}{\partial \varepsilon} \varepsilon + \frac{D}{\tau^2} \frac{\partial^2}{\partial \varepsilon^2} \end{aligned} \tag{7.19}$$

Expanding the stationary solution of the Fokker–Planck equation, Eq. (7.18), into complete sets of orthogonal functions with respect to both variables I and ε,

$$W_{\mathrm{st}}(I, \varepsilon) = \frac{\alpha}{\Gamma(1+\nu)} \psi_0(\varepsilon) \sum_{n,m=0}^{\infty} d_n^m (\alpha I)^\nu L_m^\nu(\alpha I) \exp(-\alpha I) \psi_n(\varepsilon) \tag{7.20}$$

with $L_m^\nu(x)$ being generalized Laguerre polynomials, α being a positive arbitrary scaling parameter, and $\psi_n(\varepsilon)$ being Hermite functions, that is,

$$\psi_n(\varepsilon) = \frac{1}{\sqrt{2^n n! \sqrt{2\pi D/\tau}}} H_n\left(\frac{\varepsilon}{\sqrt{2D/\tau}}\right) \exp\left(-\frac{\tau\varepsilon^2}{4D}\right) \tag{7.21}$$

Arranging the expansion coefficients in vectors,

$$\mathbf{c}_n = (c_n^0, c_n^1, c_n^3, \ldots) \tag{7.22}$$

we obtain the tridiagonal vector recurrence relation [186]

$$\sqrt{n\frac{D}{\tau}}\,\underline{\underline{B}}\mathbf{c}_{n-1} + \left(\underline{\underline{A}} - n\frac{1}{\tau}\underline{\underline{1}}\right)\mathbf{c}_n + \sqrt{(n+1)\frac{D}{\tau}}\,\underline{\underline{B}}\mathbf{c}_{n+1} = \mathbf{0} \tag{7.23}$$

which can be solved in terms of matrix continued fractions [186]. The matrices $\underline{\underline{A}}$ and $\underline{\underline{B}}$ are matrix representations of the operators $\mathbf{A}$ and $\mathbf{B}$, respectively. They are given by

$$\begin{aligned} A^{mn} &= 2m\left(1 - \frac{3m+2\nu+1}{\alpha}\right)\delta_{m,n} + 2m\frac{m+1+\nu}{\alpha}\delta_{m,n-1} \\ &\quad + 2m\left(\frac{3m+\nu-1}{\alpha}\right)\delta_{m,n+1} - 2m\frac{m-1}{\alpha}\delta_{m,n+2} \\ B^{mn} &= 2m(\delta_{m,n} - \delta_{m,n+1}) \end{aligned} \tag{7.24}$$

The parameter ν has been chosen as $a/(2D) - 1$ [186] to match the dependency on I at low intensities. In Fig. 7.1, the numerically obtained stationary probabilities are shown for $D = 0.25(a)$ and $D = 1(b)$ for increasing correlation times of the noise. We observe crucial changes in the distribution function for increasing correlation times. The consequences for the photon-counting statistics are evident.

Within the small correlation time approximation, one obtains the

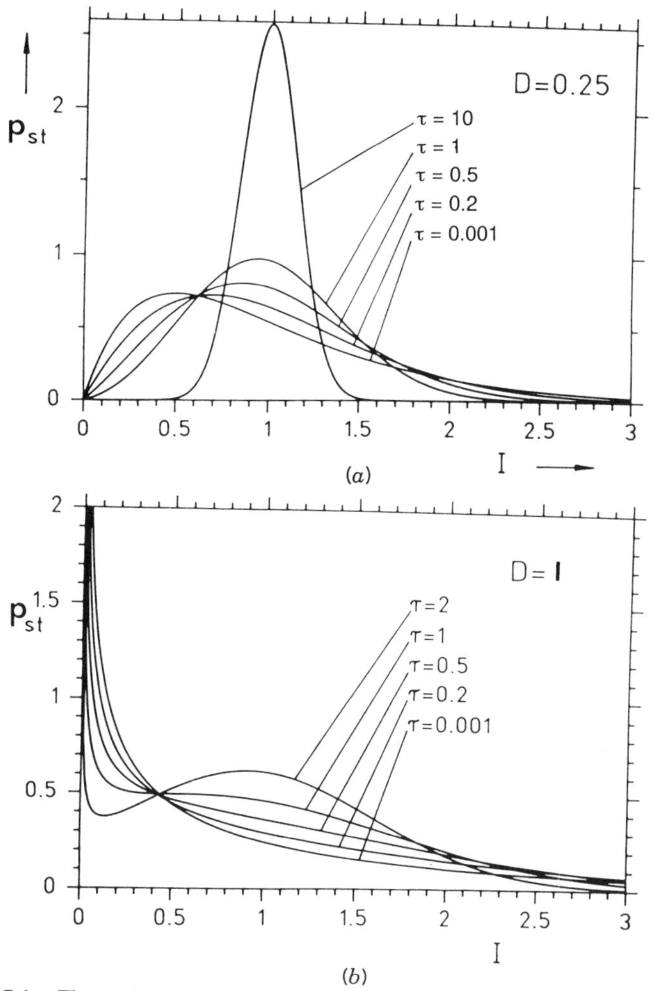

Figure 7.1. The stationary probability density [Eq. (7.20)] is shown at $D = 0.25(a)$ and $D = 1(b)$ for various values of the correlation time τ of the noise, for $a = 1$. Here, the UCNA result in Eq. (7.29) coincides within line thickness with the numerical (MCF) result in Eq. (7.20).

Fokker–Planck type equation [186]

$$\frac{\partial}{\partial t} P(I, t) = \left(-2 \frac{\partial}{\partial I} I(a - I) + 4D \frac{\partial}{\partial I} I \frac{\partial}{\partial I} I\right) P(I, t) - 8D\tau \frac{\partial}{\partial I} I \frac{\partial}{\partial I} I^2 P(I, t) \tag{7.25}$$

The stationary probability density within small correlation time approximation is obtained from Eq. (7.25) without ad hoc exponentiation as

$$P_{st}(I) = P_{st}^{\tau=0}(I)\left\{1 - \frac{\tau}{2D}[(2D + a)(a - 2I) + I^2]\right\} \tag{7.26}$$

where $P_{st}^{\tau=0}$ is the stationary probability in the white noise limit Eq. (7.11). The agreement of Eq. (7.26) with the numerical solution is obviously nonuniform in the intensity. In order that the correction term for finite correlation term remains a *small* term, the ratio τ/D has to be small. How small it has to be is determined by the value of the intensity itself. In other words, the small correlation time approximation is only valid in the finite region of support where the resulting stationary probability density is positive. Using the unified colored noise approximation [144, 151, 179], we find a Fokker–Planck equation with a strictly positive diffusion coefficient

$$\frac{\partial}{\partial t} P(I, t) = -\frac{\partial}{\partial I}\left(\frac{2(a - I)I}{1 + 2\tau I} + \frac{4DI}{(1 + 2\tau I)^3}\right)P(I, t)$$
$$+ \frac{\partial^2}{\partial I^2}\frac{4DI^2}{(1 + 2\tau I)^2} P(I, t) \tag{7.27}$$

The condition of validity is given for this particular model by

$$\gamma(I, \tau) = \frac{1}{\sqrt{\tau}} + 2I\sqrt{\tau} \gg 1 \tag{7.28}$$

which is fulfilled for small and large correlation times of the noise on the whole intensity axis, except for $I = 0$. The stationary probability density is obtained within this approximation by

$$P_{st}(I) = \frac{1}{Z}(1 + 2\tau I)\exp\left[\frac{\tau}{2D}(2a - I)I\right]P_{st}^{\tau=0}(I) \tag{7.29}$$

where Z is a normalization constant and $P_{st}^{\tau=0}(I)$ is the stationary probability in the white noise limit. In Fig. 7.2, the numerically evaluated stationary probabilities are compared with those obtained by using the unified colored noise approximation at $D = 0.5$. The agreement is good for large-to-large correlation times of the noise. The largest deviations are in agreement with what one expects from Eq. (7.29) at small intensities I.

The fluctuational line width of the dye laser (i.e., the effective eigenvalue) is given by the inverse of the integral of the normalized intensity autocorrelation function over the complete time axis Eqs. (7.7) and (7.8). In the white noise limit, this quantity could be obtained exactly, yielding the simple expression $\lambda_{eff} = 2a$. Since the unified colored

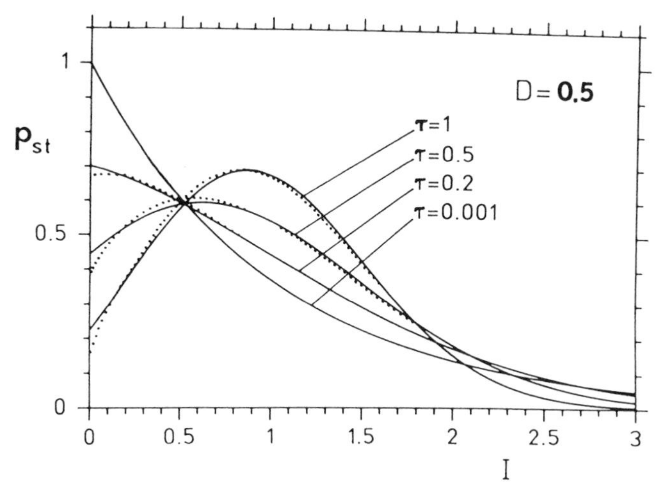

Figure 7.2. The approximate (solid line) expression [Eq. (7.29)] for the stationary probability density is compared with the full numerical (dotted line) result at $D = 0.5$, and $a = 1$.

noise approximation represents a true Markovian approximation of a non-Markovian process, we can use the closed expressions for the effective eigenvalue Eqs. (7.15) and (7.16) with $D(x) \rightarrow 4D_x^2/(1 + 2\tau x)^2$. The integrals, however, have to be evaluated numerically. The result is compared in Fig. 7.3 with the numerical results obtained in [144, 187].

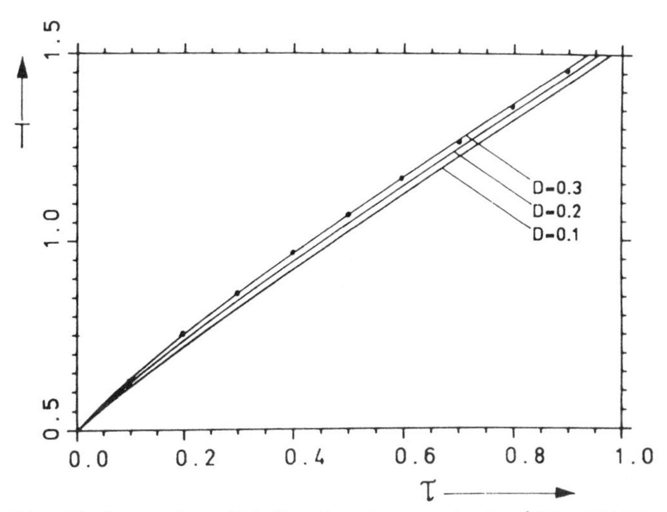

Figure 7.3. The inverse line width (i.e. the relaxation time) $\lambda_{\text{eff}}^{-1}$ [Eq. (7.15)] is depicted versus noise correlation time τ for different noise strengths D, and $a = 1$. The solid lines are the numerical exact results. The UCNA result [Eqs. (7.15, 7.27, 7.29)] for $D = 0.3$ is indicated by the full circles, being very close to the exact result.

The comparison is very favorable for the unified colored noise approximation. We should mention, also, that for the particular dye laser model we are using here, a weak noise expansion yields the analytical expression [187]

$$\lambda_{\text{eff}}^{-1} = \frac{1}{2a} + \tau + \frac{2D\tau(3a\tau + 1)}{(1 + a\tau)(1 + 2a\tau)(1 + 4a\tau)} \tag{7.30}$$

which agrees also well with numerical results.

VIII. SUMMARY AND OUTLOOK

We have reviewed current theories and applications of colored noise driven dynamical systems. Exact solutions are available for colored dichotomous noise driven dynamical systems (see Section IV) and for colored noise driven linear systems (see Section III.E, and Appendix A in [111]). For colored Gaussian noise driven nonlinear systems, one has to apply approximation schemes and/or to resort to numerical techniques (see Sections VI and VII).

For *small* correlation times τ of the noise, that is, when the noise correlation time is smaller than all system time scales, numerous approximate Fokker–Planck type equations have been derived in the literature (see Section V.A). Triggered by the need to describe experiments (photon statistics of dye lasers, electrohydrodynamic instabilities, analogue experiments, or turbulent transitions in liquid He II, see the various contributions in [194]), new theories valid for *intermediate-to-large* correlation times (decoupling approximation and unified colored noise approximation) have been developed (see Sections V.B and V.C, respectively).

In Section VI, the theoretical concepts introduced in Section V are applied to the problem of colored noise driven bistability. We focused on the noise color dependence of the escape rates and stationary correlation functions. The answers of the approximation schemes have been compared with precise numerical results. In Section VII, the impact of multiplicative colored noise for the photon-counting statistics and the fluctuational line width has been investigated.

The reader has certainly noted that throughout this chapter the emphasis has been placed on nonequilibrium systems described by stochastic differential equations driven by general multiplicative colored noise. Thus, we did not address the huge area of colored noise fluctuations within thermal equilibrium systems, being described by the epoch-making *generalized Langevin equation* [195–197] and obeying the fluctuation–dissipation theorem of the second kind with memory friction.

The vast research obtained for this thermal equilibrium colored noise has been extensively addressed in the literature, see, for example, [100, 123, 195–205].

Before we close our survey we would also like to mention other topics, related to colored noise driven dynamical systems that we did not cover in detail. One such topic is the influence of colored noise for transport quantities in periodic potentials [206–208], or the work on the relaxation times, see Eq. (5.24) [209, 210]. Another area, widely studied in recent years, refers to the *decay from unstable states* [211] when triggered by colored noise [212–218], where the decay time T obeys a characteristic scaling law (Suzuki's scaling [211]) of the form $T \propto \log(D) + B(\tau) + C$, where D denotes the noise strength. The influence of the noise color is accounted for by the function $B(\tau)$. Much of the progress on colored noise approximation schemes has originated from digital simulations. Likewise, the method of colored noise analogue simulations [194, 219, 220] has played a pioneering role in guiding the theoretical practitioners to improve upon their theoretical schemes.

Finally, throughout our survey we restricted the time evolution to continuous time. If on the other hand the dynamics is recorded by stroboscopic methods—commonly used in the study of chaotic dynamics—the dynamical flow is not governed by a stochastic differential equation but rather by a noisy map. Very recently, the study of correlated noise has been initiated for such discrete-time dynamical flows [221–223].

Undoubtedly, we shall witness more research work in future years aimed at completing, extending, and interpreting the present state of the art of colored noise driven systems in chemistry, physics, biology, and the engineering sciences.

REFERENCES

1. A. Einstein, *Investigations on the Theory of Brownian Movement*, Methuen and Co., London, 1926; republication by Dover, New York, 1956.
2. N. Wax, Ed., *Selected Papers on Noise and Stochastic Processes*, Dover Publications, New York, 1954.
3. R. L. Stratonovich, Gordon and Breach, New York, 1963, Vols. I and II.
4. N. G. van Kampen, *Stochastic Processes in Physics and Chemistry*, North-Holland, Amsterdam, 1992, revised and enlarged edition.
5. H. Risken, *The Fokker–Planck Equation: Methods of Solution and Applications*, Vol. 18 of *Springer Series in Synergetics*. Springer-Verlag, Berlin, 1984.
6. W. Horsthemke and R. Lefever, *Noise-Induced Transitions: Theory and Applications in Physics, Chemistry and Biology*, Springer-Verlag, Berlin, 1984.
7. P. Hänggi and H. Thomas, *Phys. Rep.*, **88**, 207 (1982).

8. R. Fox, *Phys. Rep.*, **48**, 179 (1978).
9. R. Kubo, A stochastic theory of line-shape and relaxation, in D. Ter Haar, Ed., *Fluctuation, Relaxation and Resonance in Magnetic Systems*, Oliver and Boyd, Edinburgh, Scotland, 1962, pp. 23–68.
10. R. Short, R. Mandel, and R. Roy, *Phys. Rev. Lett.*, **49**, 647 (1982).
11. S. N. Dixit and P. S. Sahni, *Phys. Rev. Lett.*, **50**, 1273 (1983).
12. A. Schenzle and R. Graham, *Phys. Lett. A*, **98**, 319 (1983).
13. S. Zhou, A. W. Yu, and R. Roy, *Phys. Rev. A*, **34**, 4333 (1986).
14. M. O. Scully, *Phys. Rev. Lett.*, **55**, 2902 (1985).
15. H. R. Brand, S. Kai, and S. Wakabayashi, *Phys. Rev. Lett.*, **54**, 555 (1985).
16. H. Fukunaga and H. Brand, *J. Stat. Phys.*, **54**, 1133 (1989).
17. R. Brown, *Philos. Mag.*, **4**, 161 (1828).
18. M. Gouy, *J. Phys.*, **7**, 561 (1888).
19. P. Langevin, *C. R. Acad. Sci. (Paris)*, **146**, 530 (1908).
20. I. M. Gel'fand and N. J. Wilenkin, *Verallgemeinerte Funktionen (Distributionen)*, Vol. 4. VEB Deutscher Verlag der Wissenschaften, Berlin (1964).
21. N. Wiener, *J. Math. Phys.*, **2**, 131 (1923).
22. N. Wiener, *Proc. London Math. Soc.*, **22**, 454 (1924).
23. A. Kolmogorov, *Math. Ann.*, **104**, 415 (1931).
24. A. Kolmogorov, *Math. Ann.* **108**, 149 (1933).
25. W. Feller, *Math. Ann.*, **113**, 113 (1936).
26. W. Feller, *Trans. Am. Math. Soc.*, **48**, 488 (1940).
27. P. Hänggi, *Z. Phys. B*, **36**, 271 (1980).
28. P. Hänggi, *Z. Phys. B*, **43**, 269 (1981).
29. P. Hänggi, K. E. Shuler, and I. Oppenheim, *Physica A*, **107**, 143 (1981).
30. L. S. Ornstein, *Versl. Acad. Amst.*, **26**, 1005 (1917).
31. L. S. Ornstein, *Proc. Acad. Amst.*, **21**, 96 (1919).
32. G. E. Uhlenbeck and L. S. Ornstein, *Phys. Rev.*, **36**, 823 (1930).
33. M. C. Wang and G. E. Uhlenbeck, *Rev. Mod. Phys.*, **17**, 323 (1945).
34. S. Chandrasekhar, *Rev. Mod. Phys.*, **15**, 1 (1943).
35. A. Einstein, *Ann. Phys.*, **17**, 549 (1905).
36. A. D. Fokker, Ph.D. Thesis, Leiden, 1913.
37. A. D. Fokker, *Ann. Phys.*, **43**, 812 (1914).
38. M. V. Smoluchowski, *Krakauer Ber.*, 418 (1913).
39. M. Planck, *Sitzungsberichte der Preussischen Akademie der Wissenschaften*, p. 324, 1917.
40. Lord Rayleigh, *Philos. Mag.*, **32**, 424 (1891).
41. Lord Rayleigh, *Scientific Papers of Lord Rayleigh*, Vol. III, Dover Publications, New York, 1964, p. 471.
42. M. V. Smoluchowski, *Ann. Phys.*, **48**, 1103 (1915).
43. M. V. Smoluchowski, *Phys. Z.* **17**, 557 (1916).
44. R. Fürth, *Ann. Phys.*, **53**, 177 (1917).
45. O. Klein, *Ark. Mat. Astron. Fys.*, **16**, 5 (1922).

46. H. A. Kramers, *Physica* (*Utrecht*), **7**, 284 (1940).
47. E. Schrödinger, *Phys. Z.*, **16**, 189 (1915).
48. L. Pontryagin, A. Andronov, and A. Vitt, *Zh. Eksp. Teor. Fiz*, **3**, 165 (1933).
49. L. Pontryagin, A. Andronov, and A. Vitt, On the statistical treatment of dynamical systems, in F. Moss and P. V. E. McClintock, Eds., *Noise in Nonlinear Dynamical Systems*, Vol. 1, Cambridge University Press, Cambridge, 1989, pp. 329–348.
50. J. L. Doob, *Ann. Math.*, **43**, 351 (1942).
51. P. Hänggi and H. Thomas, *Phys. Rep.*, **88**, 207, 1982, especially Eqs. (1.3.50–1.3.52) on p. 223.
52. J. L. Doob, *Ann. Am. Stat.*, **15**, 229 (1944).
53. B. J. Berne, J. P. Boon, and S. A. Rice, *J. Chem. Phys.*, **45**, 1086 (1966).
54. P. I. Kuznetsov, R. L. Stratonovich, and V. I. Tikhonov, *Nonlinear Transformation of Stochastic Processes*, Pergamon, London, 1965. See especially the reprinted articles on pp. 157, 259, 269, and 298.
55. R. L. Stratonovich, *Theory of Random Noise*, Vol. 1, Gordon and Breach, New York, 1963. See Eq. (4.180), on p. 98 wherein the third cumulant $\kappa_3 = 0$, and also note the supplement to Chapter 4, pp. 126–129.
56. M. Lax, *Rev. Mod. Phys.*, **38**, 341 (1966).
57. N. G. Van Kampen, *Phys. Rep.*, **24**, 171 (1971).
58. N. G. Van Kampen, *Physica*, **74**, 215 (1974).
59. N. G. Van Kampen, *Physica*, **74**, 239 (1974).
60. R. Fox, *Phys. Rev. A*, **37**, 911 (1988).
61. K. Lindenberg and B. West, *Physica A*, **119**, 485 (1983).
62. K. Lindenberg and B. West, *Physica A*, **128**, 25 (1984).
63. R. Der, *Phys. Lett. A*, **15**, 4464 (1987).
64. S. E. Pritvranov and V. M. Chetverikov, *Theor. Math. Phys.*, **35**, 211 (1978).
65. J. M. Sancho and M. San Miguel, *Z. Phys. B*, **36**, 357 (1980).
66. J. M. Sancho, M. San Miguel, S. L. Katz, and J. D. Gunton, *Phys. Rev. A*, **26**, 1589 (1982).
67. W. Horsthemke and R. Lefever, *Z. Phys. B*, **40**, 241 (1980).
68. H. Malchow and L. Schimansky-Geier, Noise and Diffusion in Bistable Nonequilibrium Systems, *Teubner Texte Physik*, Vol. 5, Teubner Verlagsgesellschaft, Leipzig, 1985.
69. L. Schimansky-Geier, *Phys. Lett. A*, **126**, 455 (1988).
70. L. Schimansky-Geier, *Phys. Lett. A*, **129**, 481 (1988).
71. M. Dygas, B. J. Matkowsky, and Z. Schuss, *SIAM J. Appl. Math.*, **48**, 425 (1988).
72. A. Schenzle and T. Tel, *Phys. Rev. A*, **32**, 596 (1985).
73. H. Risken, H. D. Vollmer, and M. Morsch, *Z. Phys. B*, **40**, 343 (1980).
74. K. Kaneko, *Prog. Theor. Phys.*, **66**, 129 (1981).
75. F. Marchesoni and P. Grigolini, *Physica A*, **121**, 269 (1983).
76. F. Marchesoni and P. Grigolini, *Adv. Chem. Phys.*, **62**, 29 (1985).
77. L. Hannibal, *Phys. Lett. A*, **145**, 220 (1990).
78. R. Kubo, *J. Math. Phys.*, **4**, 174 (1962).
79. C. J. Gorter and J. H. Van Vleck, *Phys. Rev.*, **72**, 1128 (1947).

80. P. W. Anderson and P. J. Weiss, *Rev. Mod. Phys.*, **25**, 269 (1953).
81. P. W. Anderson, *J. Phys. Soc. J.*, **9**, 316 (1954).
82. R. Kubo, *J. Phys. Soc. J.*, **9**, 935 (1954).
83. V. I. Klyatskin, *Radiophys. Quantum Electron.*, **20**, 382 (1978).
84. V. I. Klyatskin, *Radiofizika*, **20**, 562 (1977).
85. R. Lefever, W. Horsthemke, K. Kitahara, and I. Inaba, *Prog. Theor. Phys.*, **64**, 1233 (1980).
86. P. Hänggi and P. Talkner, *Phys. Rev. A*, **32**, 1934 (1985).
87. M. A. Rodriguez and L. Pesquera, *Phys. Rev. A*, **34**, 4532 (1986).
88. E. Hernandez-Garcia, L. Pesquera, M. A. Rodriguez, and M. San Miguel, *Phys. Rev. A*, **36**, 5774 (1987).
89. J. Masoliver, K. Lindenberg, and B. J. West, *Phys. Rev. A*, **34**, 1481 (1986).
90. V. Balakrishnan, C. Van den Broeck, and P. Hänggi, *Phys. Rev. A*, **38**, 3556 (1988).
91. A. Brissaud and U. Frisch, *J. Quant. Spectrosc. Radiat. Transfer*, **11**, 1767 (1971).
92. A. Brissaud and U. Frisch, *J. Math. Phys.*, **15**, 524 (1974).
93. I. I. Gihman and A. V. Skorohod, *The Theory of Stochastic Processes*, Vol. 1, Springer, Berlin, Heidelberg, New York, 1974, p. 159 ff.
94. P. Hanggi, *Phys. Rev. A*, **25**, 1130 (1982).
95. P. Hänggi and H. Thomas, *Phys. Rep.*, **88**, 207, 1982. See Sections 1.1, 1.3, and 2.6 therein.
96. P. Hänggi and H. Thomas, *Z. Phys. B*, **26**, 85 (1977).
97. H. Grabert, P. Talkner, and P. Hänggi, *Z. Phys. B*, **26**, 389 (1977).
98. H. Grabert, P. Talkner, P. Hänggi, and H. Thomas, *Z. Phys. B*, **29**, 273 (1978).
99. P. Hänggi, H. Thomas, H. Grabert, and P. Talkner, *J. Stat. Phys.*, **18**, 155 (1978).
100. H. Grabert, P. Hänggi, and P. Talkner, *J. Stat. Phys.*, **22**, 537 (1980).
101. P. Hänggi, *Z. Phys. B*, **31**, 407 (1978).
102. P. Hänggi, The functional derivative and its use in the description of noisy dynamical systems, in *Stochastic Processes Applied to Physics*, World Scientific, Singapore and Philadelphia, 1985, pp. 69–95.
103. P. Hänggi, Colored noise in dynamical systems: A functional calculus approach, in F. Moss and P. V. E. McClintock, Eds., *Noise in Nonlinear Dynamical Systems*, Vol. 1, Chapter 9, Cambridge University Press, Cambridge, 1989, pp. 307–328.
104. K. Furutsu, *J. Res. Natl. Bur. Standards*, **67D**, 303 (1963).
105. E. A. Novikov, *Sov. Phys. JETP*, **20**, 1290 (1965).
106. R. F. Fox, *Phys. Rev. A*, **34**, 4525 (1988).
107. P. Hänggi and H. Thomas, *Z. Phys. B*, **22**, 295, 1975. See Section 4.
108. P. Hänggi, *Z. Phys. B*, **30**, 85 (1978). See Eqs. (2.36), (2.37), and (2.39) therein.
109. M. O. Hongler, *Helv. Phys. Acta*, **52**, 280 (1979).
110. W. Horsthemke and R. Lefever, *Noise-Induced Transitions: Theory and Applications in Physics, Chemistry and Biology*, Springer Series in Synergetics. Springer-Verlag, Berlin, 1984. See pp. 204 and 210.
111. L. H'walisz, P. Jung, P. Hänggi, P. Talkner, and L. Schimansky-Geier, *Z. Phys. B*, **77**, 471 (1989).
112. M. Luban and D. L. Pursey, *Phys. Rev. D*, **33**, 431 (1986).
113. D. L. Pursey, *Phys. Rev. D*, **33**, 2267 (1986).

114. G. Darboux, *C.R. Acad. Sci.* (*Paris*), **94**, 1456 (1882).
115. P. B. Abraham and H. E. Moses, *Phys. Rev. A*, **22**, 1333 (1980).
116. R. Dutt, A. Khane, and U. P. Sakhatme, *Am.*, *J. Phys.*, **56**, 163 (1988).
117. C. A. Singh and T. H. Devi, *Phys. Lett.*, **A171**, 249 (1992).
118. M. O. Hongler and W. M. Zeng, *J. Stat. Phys.*, **29**, 317 (1982).
119. M. O. Hongler and W. M. Zeng, *Physica A*, **122**, 431 (1983).
120. M. Razavy, *Phys. Lett. A*, **72**, 89 (1979).
121. L. F. Favella, *Ann. Inst. Henri Poincare* (*Section A*), **7**, 77 (1967).
122. M. Mörsch, H. Risken, and H. D. Vollmer, *Z. Phys. B*, **32**, 245 (1979).
123. P. Hänggi, P. Talkner, and M. Borkovec, *Rev. Mod. Phys.*, **62**, 251 (1990).
124. P. Hänggi and P. Talkner, *Phys. Rev. Lett.*, **51**, 2242 (1983).
125. P. Hänggi and P. Talkner, *Z. Phys. B*, **45**, 79 (1981).
126. C. R. Doering, *Phys. Rev. A* **35**, 3166 (1987).
127. M. Kus, E. Wajnryb, and K. Wodkiewicz, *Phys. Rev. A*, **43**, 4167 (1991).
128. P. Hänggi and P. Riseborough, *Phys. Rev. A*, **27**, 3379 (1983).
129. C. Van den Broek and P. Hänggi, *Phys. Rev. A*, **30**, 2730 (1984).
130. I. L'Heureux and R. Kapral, *J. Chem. Phys.*, **88**, 7468 (1988).
131. I. L'Heureux and R. Kapral, *J. Chem. Phys.*, **90**, 2453 (1988).
132. J. Luczka, *Acta Phys. Polonica*, **A77**, 427 (1990).
133. A. M. Cetto, L. de Pena, and R. M. Velasco, *Phys. Rev. A*, **39**, 2747 (1989).
134. P. Hänggi, F. Marchesoni, and P. Grigolini, *Z. Phys. B*, **56**, 333 (1984).
135. F. Marchesoni, *Phys. Rev. A*, **36**, 4050 (1987).
136. N. G. Van Kampen, *J. Stat. Phys.*, **54**, 1284 (1989).
137. R. F. Fox, *Phys. Rev. A*, **33**, 467 (1986).
138. L. Cao, D. Wu, and X. Luo, *Z. Phys. B*, **93**, 251 (1994).
139. M. San Miguel and J. M. Sancho, *Phys. Lett. A*, **76**, 97 (1980).
140. A. Hernandez-Machado, J. M. Sancho, M. San Miguel, and L. Pesquera, *Z. Phys. B*, **52**, 335 (1983).
141. H. Dekker, *Phys. Lett. A*, **90**, 26 (1980).
142. R. F. Fox, *Phys. Lett. A*, **94**, 281 (1983).
143. P. Hänggi, T. J. Mroczkowski, F. Moss, and P. V. E. McClintock, *Phys. Rev. A*, **32**, 965 (1978).
144. P. Jung and P. Hänggi, *J. Opt. Soc. Am. B*, **5**, 979 (1988).
145. P. Hänggi, *Noise in Nonlinear Dynamical Systems*, Vol. 1, in F. Moss and P. V. E. McClintock, Eds., Chapter 9, Cambridge University Press, Cambridge, 1989, pp. 322–326.
146. L. Fronzoni, P. Grigolini, P. Hänggi, F. Moss, R. Mannella, and P. V. E. McClintock, *Phys. Rev. A*, **33**, 3320 (1986).
147. F. Moss, P. Hänggi, R. Manella, and P. V. E. McClintock, *Phys. Rev. A*, **33**, 4459 (1986).
148. W. Ebeling and L. Schimansky-Geier, Transition phenomena in multidimensional systems—models of evolution, in F. Moss and P. V. E. McClintock, Eds., *Noise in Nonlinear Dynamical Systems*, Vol. 1, Chapter 8, Cambridge University Press, Cambridge, 1989, pp. 279–306.

149. R. F. Fox and R. Roy, *Phys. Rev. A*, **35**, 1838 (1987).
150. K. Vogel, Th. Leiber, H. Risken, P. Hänggi, and W. Schleich, *Phys. Rev. A*, **35**, 4882 (1987).
151. P. Jung and P. Hänggi, *Phys. Rev. A*, **35**, 4464 (1987).
152. P. Jung and H. Risken, *Z. Phys. B*, **59**, 469 (1985).
153. Hu Gang and L. Zhi-Leng, *Phys. Rev. A*, **44**, 8027 (1991).
154. P. Jung and P. Hänggi, *Phys. Rev. Lett.*, **61**, 11 (1988).
155. G. P. Tsironis and P. Grigolini, *Phys. Rev. Lett.*, **61**, 7 (1988).
156. J. F. Luciani and A. D. Verga, *J. Stat. Phys.*, **50**, 567 (1988).
157. P. Hänggi, *Z. Phys. B*, **75**, 275 (1989).
158. P. Hänggi, P. Jung, and F. Marchesoni, *J. Stat. Phys.*, **54**, 1367 (1989).
159. W. Horsthemke and R. Lefever, *Noise-Induced Transitions: Theory and Applications in Physics, Chemistry and Biology*, Springer-Verlag, Berlin, 1984. See Section 8.5.
160. H. S. Wio, P. Colet, M. San Miguel, L. Pesquera, and M. A. Rodrigues, *Phys. Rev. A*, **40**, 7312 (1989).
161. A. J. Bray, A. J. McKane, and T. J. Newman, *Phys. Rev. A*, **41**, 657 (1990).
162. T. G. Venkatesh and L. M. Patnaik, *Phys. Rev. E*, **48**, 2402 (1993).
163. P. Jung and H. Risken, *Z. Phys.*, **B61**, 367 (1985).
164. Th. Leiber, F. Marchesoni, and H. Risken, *Phys. Rev. Lett.*, **59**, 1381 (1987).
165. F. Moss and P. V. E. McClintock, *Z. Phys. B*, **61**, 381 (1985).
166. P. Jung, F. Marchesoni, and P. Hänggi, *Phys. Rev. A*, **40**, 5447 (1989).
167. P. Talkner, *Z. Phys. B*, **68**, 201 (1987).
168. Th. Leiber and H. Risken, *Phys. Rev. A*, **38**, 3789 (1988).
169. M. Dygas, B. J. Matkowsky, and Z. Schuss, *SIAM J. Appl. Math.*, **48**, 425 (1988).
170. I. Klik, *J. Stat. Phys.*, **63**, 389 (1990).
171. R. Mannella, V. Palleschi, and P. Grigolini, *Phys. Rev. A*, **42**, 5946 (1990).
172. C. R. Doering, P. S. Hagan, and C. D. Levermore, *Phys. Rev. Lett.*, **59**, 2129 (1987).
173. P. Hänggi, P. Jung, and P. Talkner, *Phys. Rev. Lett.*, **60**, 2804 (1988).
174. C. R. Doering, R. J. Bagley, P. S. Hagan, and C. D. Levermore, *Phys. Rev. Lett.*, **60**, 2805 (1988).
175. R. F. Fox, *Phys. Rev. Lett.*, **62**, 1205 (1989).
176. P. Jung and P. Hänggi, *Phys. Rev. Lett.*, **62**, 1206 (1989).
177. L. Ramirez-Piscina and J. M. Sancho, *Phys. Rev. A*, **37**, 4469 (1988).
178. L. Schimansky-Geier and Ch. Zülicke, *Z. Phys. B*, **79**, 451 (1990).
179. Li Cao, D. Wu, and X. Luo, *Phys. Rev. A*, **47**, 57 (1993).
180. J. A. Abate, H. J. Kimble, and L. Mandel, *Phys. Rev. A*, **14**, 788 (1976).
181. R. Short, R. Roy, and L. Mandel, *Appl. Phys. Lett.*, **37**, 973 (1980).
182. K. Kaminishi, R. Roy, R. Short, and L. Mandel, *Phys. Rev. A*, **24**, 370 (1981).
183. R. Graham, M. Höhnerbach, and A. Schenzle, *Phys. Rev. A*, **48**, 1396 (1982).
184. R. Graham and A. Schenzle, *Phys. Rev. A*, **25**, 1731 (1982).
185. A. Schenzle and H. Brand, *Phys. Rev. A*, **20**, 1628 (1979).
186. P. Jung and H. Risken, *Phys. Lett. A*, **103**, 38 (1984).
187. Th. Leiber, P. Jung, and H. Risken, *Z. Phys. B*, **68**, 123 (1987).

188. P. Lett, R. Short, and L. Mandel, *Phys. Rev. Lett.*, **52**, 341 (1984).
189. R. Roy, A. W. Yu, and S. Zhu, *Phys. Rev. Lett.*, **55**, 2794 (1985).
190. R. Roy, A. W. Yu, and S. Zhou, Colored in dye laser fluctuations, in F. Moss and P. V. E. McClintock, Eds., *Noise in Nonlinear Dynamical Systems*, Vol. 3, Chapter 4, Cambridge University Press, Cambridge, 1989, pp. 90–118.
191. L. Mandel, *Proc. Phys. Soc.*, **72**, 1037 (1958).
192. L. Mandel, in *Progress in Optics*, Vol. 2, North-Holland, Amsterdam, 1963.
193. J. Müller and J. Schnakenberg, *Ann. Phys.* (8-*th Ser.*), **2**, 92 (1993).
194. F. Moss and P. V. E. McClintock, Eds., *Noise in Nonlinear Dynamical Systems*, Vol. 3, Cambridge University Press, Cambridge, 1989.
195. H. Mori, *Progr. Theor. Phys.*, **33**, 423 (1965).
196. H. Mori, *Progr. Theor. Phys.*, **34**, 399 (1965).
197. R. Kubo, *Progr. Theor. Phys.*, **29**, 255 (1966).
198. B. J. Berne and G. D. Harp, *Adv. Chem. Phys.*, **17**, 63 (1970).
199. B. J. Berne, Time correlation functions in condensed media, in H. Eyring et al., Eds., *Physical Chemistry: An Advanced Treatise*, Vol. VIIIB, Academic, New York, 1971, p. 539.
200. J. T. Hynes and J. M. Deutsch, *Physical Chemistry: An Advanced Treatise*, Vol. XIB, in D. Henderson, Ed., Academic, New York, 1975.
201. K. Kawasaki, *J. Phys. A*, **6**, 1289 (1973).
202. H. Grabert, *Springer Tracts in Modern Physics*, Vol. 95, Springer, New York, Berlin, 1980.
203. A. M. Levine, M. Shapiro, and E. Pollak, *J. Chem. Phys.*, **88**, 1959 (1988).
204. S. A. Adelmann, *Adv. Chem. Phys.*, **53**, 61 (1983).
205. M. Tuckerman and B. J. Berne, *J. Chem. Phys.*, **98**, 7301 (1993).
206. Th. Leiber, F. Marchesoni, and H. Risken, *Phys. Rev. A*, **38**, 983 (1988).
207. Th. Leiber, F. Marchesoni, and H. Risken, *Phys. Rev. A*, **40**, 6107 (1989).
208. G. Debnath, F. Moss, Th. Leiber, H. Risken, and F. Marchesoni, *Phys. Rev. A*, **42**, 703 (1990).
209. J. Casademunt and J. M. Sancho, *Phys. Rev. A*, **39**, 4915 (1989).
210. J. Casademunt, R. Manella, P. V. E. McClintock, F. E. Moss, and J. M. Sancho, *Phys. Rev. A*, **35**, 5183 (1987).
211. M. Suzuki, *Adv. Chem. Phys.*, **46**, 193 (1981).
212. M. C. Valsakumar, *J. Stat. Phys.*, **39**, 347 (1985).
213. A. K. Dhara and S. V. G. Menon, *J. Stat. Phys.*, **46**, 743 (1987).
214. M. Suzuki, Y. Lia, and T. Tsano, *Physica A*, **138**, 433 (1986).
215. M. James, F. Moss, P. Hänggi,, and Ch. Van den Broek, *Phys. Rev. A*, **38**, 4690 (1988).
216. J. M. Sancho and M. San Miguel, *Phys. Rev. A*, **39**, 2722 (1989).
217. J. Casademunt, J. I. Jimenez-Aquiro, and J. M. Sancho, *Phys. Rev. A*, **40**, 5905 (1989).
218. J. Casademunt, J. I. J. Aquino, J. M. Sancho, C. J. Lambert, R. Mannella, P. Martano, P. V. E. McClintock, and N. G. Stocks, *Phys. Rev. A*, **40**, 5913 (1989).
219. J. M. Sancho, M. San Miguel, H. Yamazaki, and T. Kawakubo, *Physica A*, **116**, 560 (1982).

220. L. Fronzoni, Analogue simulations of stochastic processes by means of minimum component electronic devices, in F. Moss and P. V. E. McClintock, Eds., *Noise in Nonlinear Dynamical Systems*, Vol. 3, Chapter 8, Cambridge University Press, Cambridge, 1989, pp. 222–242.

221. D. Fiel, *J. Phys. A*, **20**, 3209 (1987).

222. P. Reimann and P. Talkner, *Helv. Phys. Acta*, **65**, 882 (1992).

223. P. Reimann, PhD. Thesis, University of Basel, 1992.

FORMULATION OF OSCILLATORY REACTION MECHANISMS BY DEDUCTION FROM EXPERIMENTS

JANET D. STEMWEDEL AND JOHN ROSS

Department of Chemistry, Stanford University, Stanford, CA 94305, USA

IGOR SCHREIBER

Department of Chemical Engineering, Prague Institute of Chemical Technology, Technicka 5, 166 28 Prague 6, Czech Republic

CONTENTS

Advances in Chemical Physics, Volume LXXXIX, Edited by I. Prigogine and Stuart A. Rice.
ISBN 0-471-05157-8

I. INTRODUCTION

The experimental and theoretical study of kinetics (rates) and mechanisms of chemical reactions is a science at least 100 years old. There are thousands of research articles, vast compilations of rate data, treatises, and texts. In spite of the availability of a large literature, there are very few prescriptions, outlines of methods, and strategies for formulations of reaction mechanisms by means of deduction from experiments. There is no vademecum for this difficult task for either experts or the beginning student. In 1968 Edwards et al. [1] referred to this issue and, in an attempt to improve the situation, listed several rules and hints about the formulation of relatively simple reaction mechanisms derivable from the stoichiometry and empirical rate law. As far we we know, there is only one textbook on physical chemistry [2] in which the issue of how to construct a reaction mechanism is addressed, and this is done by discussing the rules in [1].

The construction of a reaction mechanism has been and is in large part an art. Elementary steps in a reaction mechanism are hypothesized on the basis of prior experience and intuition. There may be important evidence from experiments, such as isotope studies, that indicate whether and how a given atom participates in an hypothesized elementary step. Individual elementary steps in the mechanism may be known from prior work or can perhaps be studied separately. Reaction mechanisms are invented. Then the proposed mechanism is tested: The derived rate equation is checked against the empirical overall rate equation, the overall stoichiometry, and other experimental details. Furthermore, the proposed mechanism may suggest experiments. Many reaction mechanisms have thus been constructed; of these reactions many are generally accepted, but many are tentative and contested. The kineticist's "warning label" is appropriate: A reaction mechanism checked as described is consistent with these measurements but not unique [3] in that consistency.

Much research on oscillatory reactions has been concerned with the complex reaction mechanisms, with many variables and many elementary steps, which give rise to temporal oscillations of concentrations of species in the mechanism. Hypothesized reaction mechanisms have been tested by comparing calculated periods (frequencies), amplitudes, and the shape of temporal variations of oscillating species with experimentally observed behavior. This is an important but frequently not stringent test; also such comparisons are usually not suggestive of improvements in a proposed reaction mechanism. In the last few years, a series of theoretical studies and suggestions of new types of experiments have provided more stringent tests of a proposed mechanism and have led the way to a strategy for formulating reaction mechanisms, based on operational prescriptions, that is, a set of experiments. In this chapter we outline these developments. In Section II, we describe several necessary theoretical concepts and constructs of oscillatory reactions. In Section III, we list and briefly describe a series of experiments showing what information about the reaction mechanism can be deduced from each type of experiment. We also give examples of the application of these techniques to various experimental systems. In Section IV, we illustrate the strategy, for one proposed model of a reaction mechanism of the horseradish peroxidase reaction, by numerical analyses and simulations of the experiments described in Section III. Applications to experimental systems are in progress.

A difficulty in applying the strategy outlined here is connected with the analytical problem of measuring the temporal variations of the concentrations of a sufficient number of species. Nonetheless, the strategy can be helpful even if one some of the species can be measured.

We focus here on strategies of formulating reaction mechanisms of oscillatory chemical reactions in open and closed (oscillating for a limited time) systems. We believe that the strategies may be extendable to nonoscillatory systems.

II. CONCEPTS AND THEORETICAL CONSTRUCTS FOR OSCILLATORY REACTIONS

A homogeneous oscillatory chemical reaction in a closed system is described by a set of ordinary differential equations

$$\frac{dX_i}{dt} = F_i(X_1, \ldots, X_n) \qquad i = 1, \ldots, n \tag{2.1}$$

one for each of the species. The equations are obtained from a reaction mechanism consisting of elementary steps by use of mass action kinetics. In an open system [a continuous-flow stirred tank reactor (CSTR)] terms are added for inflows and outflows, giving

$$F_i(X_1, \ldots, X_n) = R_i(X_1, \ldots, X_n) + k_0(X_i^0 - X_i) \tag{2.2}$$

where $R_i(X_1, \ldots, X_n)$ is the set of reaction terms for the ith species and $k_0(X_i^0 - X_i)$ is the term due to inflow and outflow of the ith species; k_0 is the reciprocal residence time and X_i^0 is the input concentration of the ith species.

A. Jacobian Matrix Elements

The dynamics of the system to first order (i.e., to a linearized approximation) are given by

$$\frac{d\delta X_i}{dt} = \sum_{j=1}^{n} \left.\frac{\partial F_i}{\partial X_j}\right|_{X^\gamma} \delta X_j \qquad i = 1, \ldots, n \tag{2.3}$$

the Taylor expansion of Eq. (2.1) about the reference solution $X^\gamma = X_1^\gamma(t), \ldots, X_n^\gamma(t)$, where $\delta X_i = X_i - X_i^\gamma$. The Jacobian matrix elements (JMEs) are given by the partial derivative $\partial F_i / \partial X_j = \partial \dot{X}_i / \partial X_j = J_{ij}$. When the reference solution chosen is a stationary state, that is,

$$F_i(X^\gamma) = 0 \qquad i = 1, \ldots, n \tag{2.4}$$

the JMEs give useful information about both the stability of the stationary state and the connectivity among the species in the reaction mechanism of the system.

An alternative expression for δX_i is given by

$$\delta X_i(t) = \sum_{j=1}^{n} a_j \nu_i^j e^{\lambda_j t} \qquad i = 1, \ldots, n \tag{2.5}$$

where $\nu^1, \ldots, \nu^n$ are the eigenvectors, $\lambda_1, \ldots, \lambda_n$ are the eigenvalues of the Jacobian matrix $\mathbf{J}$, and $a_1, \ldots, a_n$ are expansion coefficients. If we let $w_i^j = a_j \nu_i^j$, $\mathbf{J}$ can be computed directly from

$$\mathbf{J} = \mathbf{W}\Lambda\mathbf{W}^{-1} \tag{2.6}$$

where $\mathbf{W} = \{w_i^j\}$ and Λ is a diagonal matrix containing $\lambda_1, \ldots, \lambda_n$.

In 1975, Tyson [4] classified destabilizing processes in chemical reaction systems according to mathematical relations among the JMEs. He distinguished direct autocatalysis, which includes product activation and substrate inhibition; indirect autocatalysis, as seen in competition, symbiosis, and positive feedback loops; and negative feedback loops. Luo and Epstein [5] extend Tyson's classifications, emphasizing the important interplay between positive and negative feedback, and the time delay between them, in oscillating chemical systems. They distinguished three distinct types of negative feedback: coproduct autocontrol, double autocatalysis, and flow control. Recently, Chevalier et al. [6] presented new experimental strategies for measuring JMEs at a stationary state of a complex reaction (which will be discussed in Section III) and a discussion of how JMEs may be used to construct a reaction mechanism.

B. Bifurcation Analysis

A given reaction mechanism may have various dynamic regimes in different regions of constraint, variable, and parameter space. Constraints are imposed, controllable conditions, such as influx of reactants into an open reaction system; the variables are the quantities X_i in Eq. (2.1) such as concentrations of intermediates and products, and other quantities such as temperature in nonisothermal systems; examples of parameters are rate coefficients. The dynamic regimes are often displayed in a bifurcation diagram. An example is given in Fig. 2.1, obtained by calculation [7] from a model of the chlorite–iodide reaction [8]; the space of the diagram is chosen to be the ratio of input concentrations, $[ClO_2^-]_0/[I^-]_0$, versus the logarithm of the reciprocal residence time, $\log k_0$. Several dynamic regions are shown: two different stable stationary states, which are nodes; a region of oscillations; a small region of multistability and hysteresis; and a region of excitability. The various regions are separated by boundaries corresponding to different types of transitions, or bifurcations: supercritical Hopf bifurcations, indicated by dashed lines; saddle–node infinite period bifurcations, indicated by solid lines; and saddle-node bifurcations of periodic orbits, indicated by dotted–dashed lines. For definitions and descriptions of the regions and bifurcations see

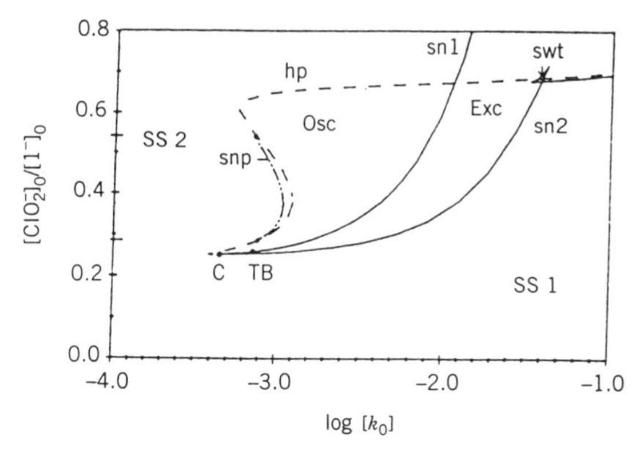

Figure 2.1. Two-dimensional (2-D) bifurcation diagram [8] calculated by continuation from the Citri–Epstein mechanism for the chlorite–iodide system [24]. The logarithm of the reciprocal residence time k_0 is plotted on the abscissa, the ratio of $[ClO_2^-]_0$ to that of $[I^-]_0$ on the ordinate. The terms SS1, SS2 = region of high $[I^-]$ and low $[I^-]$ steady states; Osc = region of periodic orbits; Exc = region of excitability; sn1, sn2 = curves of saddle–node bifurcations; hp = curve of Hopf bifurcations; snp = curve of saddle–node bifurcations of periodic orbits; swt = swallow tail (small area of tristability); C = cusp point; TB = Takens–Bogdanov point.

[8]. In Section III, we show experiments corresponding to a number of the regions and bifurcations.

One-dimensional (1-D) bifurcation diagrams contain less information than 2-D bifurcation diagrams but are still informative. Figure 2.2 shows a bifurcation diagram in 1-D calculated from the same model used to generate the 2-D bifurcation diagram given in Fig. 2.1 [8]. The bifurcation parameter here is $\log k_0$; $[ClO_2^-]_0$ and $[I^-]_0$ are fixed. The diagram indicates stable steady states as solid lines, unstable steady states as dashed lines, and branches of periodic orbits as dotted–dashed lines. From this, ranges of $\log k_0$ can be identified over which either of two steady states are seen, over which oscillations are seen (where periodic orbits and unstable steady states coexist), and where the system is excitable (where stable and unstable steady states coexist). The unstable features cannot usually be seen directly but are only deduced; this may lead to some ambiguity in identifying such features from experiments [9–14].

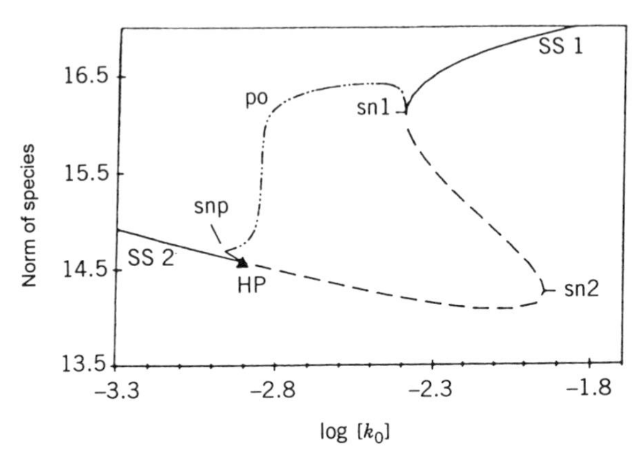

Figure 2.2. One-dimensional bifurcation diagram [8] calculated by continuation from the Citri–Epstein mechanism for the chlorite–iodide system [24]: plot of the norm of species concentration [8, 17], that is,

$$\|u\| = \left[\sum_{i=1}^{6} u_i^2\right]^{1/2} \qquad \text{for steady states}$$

$$\|u\| = \left[\frac{1}{T}\int_0^T \sum_{i=1}^{6} u_i^2(t)\, dt\right]^{1/2} \qquad \text{for periodic solutions}$$

versus $\log k_0$. Solid lines denote the stable high iodide (SS1) and stable low iodide (SS2) steady states; dashed lines denote unstable steady states; dotted–dashed lines and the label p_0 denote branches of periodic orbits; sn1, sn2 = saddle–node bifurcation point; snp = saddle–node bifurcation point of periodic orbits; HP = Hopf bifurcation point. (Reproduced from [8].)

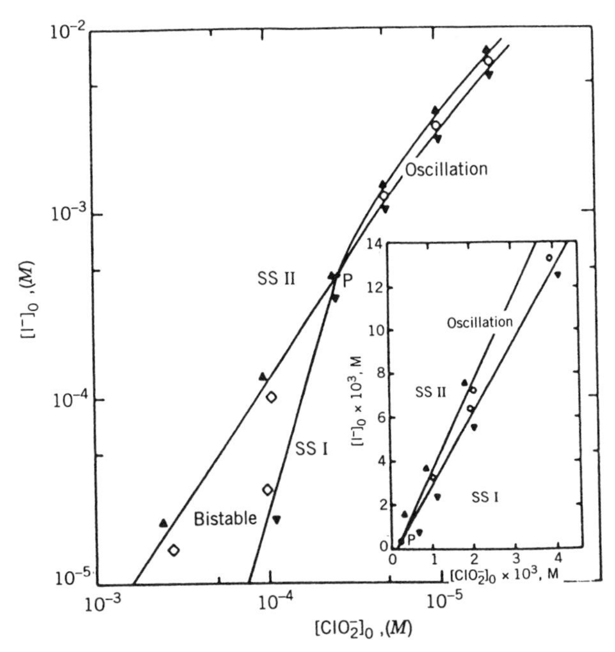

Figure 2.3. Calculated [23] section of the phase diagram of the chlorite–iodide reaction in the $[ClO_2^-]_0$–$[I^-]_0$ plane with pH 2 and $k_0 = 1.1 \times 10^{-3}\ s^{-1}$. Filled triangles indicate the high iodide steady state, filled inverted triangles indicate the low iodide steady state, open diamonds indicate bistability, and open circles indicate oscillation. (Reproduced from [16].)

A bifurcation diagram seen in many, but not all, oscillatory reactions is the so-called cross-shaped phase diagram (XPD) [15]; Fig. 2.3 shows an example calculated for the Citri–Epstein model of the chlorite–iodide reaction [16]. Four dynamic regions are separated by cross-shaped bifurcation lines: In Fig. 2.3 there are two regions of different stationary states, an oscillatory region and a region of bistability of two stationary states.

Predictions of proposed reaction mechanisms are tested against experiments. The details of bifurcation diagrams are useful in seeing the complexity of dynamic regimes of chemical reactions and in providing a series of predictions on dynamic regions and on types of bifurcations to be tested against experiments. This topic is discussed further in connection with illustrative experiments, including the analysis of Ringland [17], in Section III.L. Recently, Clarke and Jiang [18] gave a method for deriving approximate equations for Hopf and saddle–node bifurcation surfaces of chemical reaction networks, using the methods of stoichiometric network analysis (SNA) to assess the effects of adding or deleting particular reactions on the character of the bifurcation of the network. The methods of SNA will be mentioned again in Section II.C.

Some features of bifurcation diagrams are universal, in the sense that if they occur in reaction mechanisms these features do not distinguish one mechanism from another; other features are specific to a particular reaction mechanism. For example, the increase in amplitude of oscillations of concentrations near a supercritical Hopf bifurcation with some measure of distance from the bifurcation (say a measure of a constraint) is a universal function. On the other hand, the phase difference between an external periodic perturbation and the periodic response of an oscillatory reaction at the edge of an entrainment band is specific to a given oscillatory reaction mechanism. There is an inherent appeal in universality, particularly for mathematicians and physicists. For chemists and their concern for reaction mechanisms universal features give some general information, such as the presence of a feedback in the reaction mechanism with a Hopf bifurcation where oscillations occur; however, specific features may be necessary to distinguish among competing suggested mechanisms.

C. Categorization of Oscillatory Reactions

Eiswirth et al. [19] proposed a categorization of chemical oscillators. First, a distinction is made between nonessential and essential species. If the concentration of a nonessential species is held constant, oscillations continue. In contrast, if the concentration of an essential species is fixed, oscillations cease. In analyzing the core mechanism of a chemical oscillator, nonessential species may be neglected or considered as parameters, which simplifies the problem under consideration. The categorization is based on experiments, detailed later, that lead to identification of the roles of essential chemical species in producing the instability leading to oscillatory behavior, and to the *connectivity* of these species in the elementary steps of the mechanism.

Three types of nonessential species are distinguished. Nonessential species of type A are those that react to produce essential species; these can be omitted and modeled as inflows of the essential species they produce. Nonessential species of type B are inert products of the reactions, since, in the limiting case of totally irreversible reactions, they do not act as reactants anywhere in the network; they can thus be neglected without affecting any other species. Nonessential species of type C react with at least one essential species and may take part in several reactions. In the limit, the concentration of type C species may be large and constant without suppressing oscillations (e.g., a buffering species).

Next, we relate the distinction between essential and nonessential species to the theoretical description of a Hopf bifurcation. A supercritical Hopf bifurcation separates a stable steady state from an unstable steady state which coexists with a stable limit cycle. As a control parameter is varied to bring the system from the steady state through the bifurcation point to the limit cycle, a pair of complex conjugate eigenvalues of the Jacobian matrix change from negative real parts (at the steady state) to zero (at the bifurcation point) to positive real parts (on the limit cycle). In an n variable system near a supercritical Hopf bifurcation, the nearly sinusoidal small amplitude oscillations that are seen can be described with two variables. These two variables define the plane in which the limit cycle lies, the unstable center manifold. The other $(n-2)$ variables are used to describe the stable manifold. The center eigenspace, which approximates the 2-D unstable center manifold for the linearized regime of a system with three or more essential species, gets a large contribution from species with high amplitudes. However, both the amplitudes, and hence the contributions, go to zero if the concentration axis of the species is perpendicular to the center eigenspace. This corresponds to a nonoscillating species in the system.

To see how the different types of nonessential species effect the eigenvalue problem, we consider the $n \times n$ Jacobian of an n-variable system and fix the concentration of the ith species. If the system does not cease to oscillate, then the ith row and ith column of the Jacobian may be deleted, leaving an $(n-1) \times (n-1)$ matrix with $(n-1)$ eigenvalues identical to those of the original Jacobian. Such a limiting case (driving the elementary reaction forward in one direction) manifests itself in the original Jacobian in one of three ways: off-diagonal elements in the ith row vanish, and the concentration of the ith species is not affected by the concentration of any other species (case A); off-diagonal elements in the ith column vanish, and the concentration of the ith species does not influence the concentration of any other species (case B); off-diagonal elements in the ith species neither affects nor is affected by the concentration of any other species (case C). The corresponding geometrical interpretations in concentration space are shown in Fig. 2.4: the ith coordinate axis is perpendicular to the center eigenspace (i.e., the oscillations are completely described by basis vectors linearly independent from the ith coordinate axis) and, hence, the ith species does not oscillate (case A); the ith coordinate axis is parallel to the stable eigenspace (i.e., the ith coordinate axis is *not* linearly independent from the basis vectors describing the oscillations in the center eigenspace) so the ith species oscillates, but it is impossible to push the system from the limit cycle to the stable eigenspace (see discussion of quenching in Section III.I) along

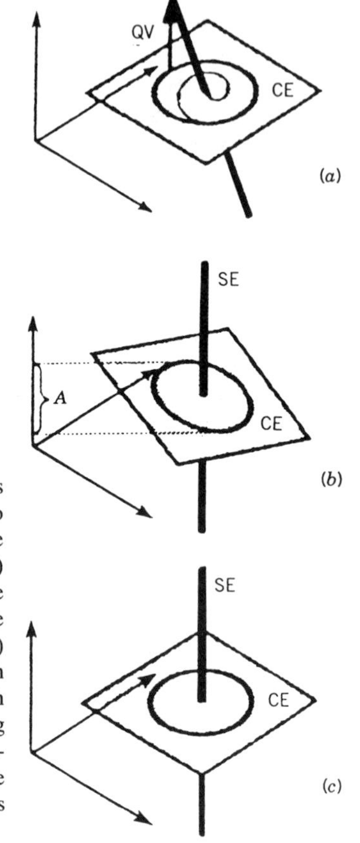

Figure 2.4. Position of the ith coordinate axis (corresponding to a nonessential species) relative to the center eigenspace (CE) and the stable eigenspace (SE) at a Hopf bifurcation. (Reproduced from [19].) (*a*) Limiting case, nonessential species of type A. The ith coordinate axis is perpendicular to the CE, and the quench vector (QV) is finite (see Section III.I). (*b*) Limiting case, nonessential species of type B. The ith coordinate axis is parallel to the SE, and the ith species oscillates with finite amplitude A_i. (*c*) Limiting case, nonessential species of type C. The ith coordinate axis is perpendicular to the CE and parallel to the SE. The QV approaches infinity and A_i approaches zero.

the ith coordinate axis (case B); the ith coordinate axis is perpendicular to the center eigenspace *and* parallel to the stable eigenspace, so the species "oscillates with zero amplitude" (i.e., does not oscillate, as with type A nonessential species) and cannot be pushed from the limit cycle to the stable eigenspace along the ith coordinate axis (case C).

The roles of essential species in mechanisms of chemical oscillators are defined as: autocatalytic or cycle species (those involved in a reaction in which an increase in the concentration of a species increases the rate of production of that species), denoted X; exit species (those reacting with cycle species to produce inert products or outflows), denoted Y; feedback species (those providing negative feedback on the autocatalytic cycle), denoted Z; and recovery species (those creating reactions allowing the autocatalysis to recover), denoted W. Not all of these roles are necessary in every category of oscillator.

In order to discuss the categorization of oscillators set forth by Eiswirth et al. [19], we must first briefly review some of the concepts set out by Clarke [20–22] in his development of SNA. *Currents* are the stationary-state solutions of a reaction *network*; the network is defined as the set of stoichiometric relations among species from elementary mechanistic steps, and the kinetic exponents of each species, without fixing rate constants (except to specify that they be nonnegative). The fundamental reaction pathways of the network are called *extreme currents* (EC) and comprise a basis set for all the reaction currents. Any given current can be decomposed to determine the dominant EC. It is generally the case that the current being decomposed cannot be uniquely expressed as a linear combination of the whole set of EC (i.e., for any given current some EC may not contribute any character at all); rather, it is necessary to choose the $(r - n)$ linearly independent EC closest to (or topologically most similar to) the current being decomposed in order to make the decomposition unique, where r is the total number of elementary reactions and pseudoreactions and n is the number of dynamical variables. Further analysis allows a determination of the percent contribution of each EC to the reaction current that has been decomposed [8]. The dominant EC will be largely responsible for the stability properties of the stationary state given by the decomposed current; Clarke outlined methods by which we can analyze the stability properties of EC using either a diagrammatic approach or an evaluation of matrix subdeterminants [20].

The categories of oscillators discussed contain the *connectivity* between the species, that is, the relations among the species in a system, as products and reactants in elementary mechanistic steps, which can produce oscillatory behavior. Two main categories of oscillators are defined on the basis of the type of autocatalytic process, or current cycle in the language of SNA, in the extreme currents of the reaction mechanism. Category 1 is the group of oscillators with critical current cycles (those in which the reaction order of the exit process is equal to the cycle order) and destabilizing exit reactions. Category 2 is the group of oscillators with strong current cycles (those in which the reaction order of the exit process is lower than the cycle order); these oscillators are unstable on their own, and hence do not require destabilizing exit reactions.

Subcategories within this classification scheme are distinguished according to the type of destabilizing feedback present; Tyson [4] and Luo and Epstein [5] describe the identification of these feedbacks from the signs of the JMEs. Category 1 oscillators can be divided into two subcategories, 1B and 1C. In subcategory 1B, exit species Y is formed in a chain of reactions from the critical current cycle via at least one intermediate Z. This negative feedback manifests itself as a relation among Jacobian matrix elements J_{ij}, which is $J_{zx} \cdot J_{xy} \cdot J_{yz} < 0$. Such feedback may be due to all three JMEs being less than zero, for which an increase in [X] leads to a decrease in [Z], an increase in [Y] leads to a decrease in [X], and an increase in [Z] leads to a decrease in [Y]. Similarly, two of the JMEs may be positive while the third is negative; for example, an increase in [X] leads to an increase in [Z], an increase in [Z] leads to a decrease in [Y], and a decrease in [Y] leads to a decrease in [X]. Oscillators in this category may not require inflows, since the exit species Y is produced internally; thus, this category includes (but is not limited to) systems that show oscillations when run experimentally under batch (closed system) conditions. In subcategory 1C, the negative feedback arises when species Z is consumed by the critical current cycle. This can be seen in the Jacobian matrix elements J_{ij}, in the relation $J_{xz} \cdot J_{zx} < 0$. One of the JMEs must be positive while the other is negative. For example, an increase in [X] leads to a decrease in [Z], while an increase in [Z] leads to an increase in [X], feedback consistent with the elementary step

$$X + Z \rightarrow 2X$$

(A more detailed description of the connection between elementary mechanistic steps and JMEs is given in Section III.G.) An inflow of Y is required in subcategory 1C oscillators. Subcategory 1C can be divided further into the subcategories 1CW and 1CX. In 1CW, species W is formed from the critical current cycle and reacts with species Y in the unstable current to allow the autocatalysis to recover. In 1CX, a finite inflow concentration of X is required for the recovery of the autocatalysis.

Category 2 oscillators, which have strong current cycles, need no additional reaction to create an instability. However, a species Z must be consumed in the autocatalysis to maintain mass balance. This creates a negative feedback that takes on the form $J_{xz} \cdot J_{zx} < 0$ in the JMEs, as was the case in subcategory 1C oscillators. The autocatalytic "species" in Category 2 oscillators need not be chemical species, but may include vacant surface sites in heterogeneous catalysis, or the temperature and change in enthalpy of the system in an exothermic thermokinetic oscillator.

Network diagrams for the prototypes of these categories are shown in

Fig. 2.5, along with the elementary mechanistic steps represented by each diagram.

III. EXPERIMENTS

In this section we briefly describe, with examples, a series of experiments from which we can deduce the essential and nonessential species in a given oscillatory system, the connectivity of the essential species producing the oscillations, and the category of the oscillatory system; these steps lead to a formulation of the core of the reaction mechanism. We also describe a variety of experiments for testing a proposed mechanism.

A. Characterization of Oscillations of Chemical Species

Experiments on the temporal variation of oscillatory species allow a comparison of the amplitude, period, and shape of concentration oscillations observed experimentally with those predicted by a proposed model. Refinements are made in the model to give better qualitative and quantitative agreement.

An example of such a comparison is seen in the modeling of the oscillating chlorite–iodide reaction. The model initially proposed by Epstein and Kustin [23] showed only fair agreement with the experimentally observed I^- evolution, and worse agreement with the experimentally observed I_2 evolution, as seen in Fig. 3.1(*a*). A revised mechanism proposed by Citri and Epstein [24] predicts oscillations quite similar in shape to the experimentally observed I^- and I_2 oscillations (Fig. 3.1(*b*)).

In many oscillatory systems the temporal variation of only a few species (essential or nonessential) can be measured. The comparison of an experimental time series with a prediction of a proposed mechanism can be made with regard to the period of the oscillations, but becomes subjective with regard to the shape of the variation. The comparisons do not easily lead to suggestions for improvement of the proposed reaction mechanism.

B. Amplitude Relations

The measurement of the amplitudes of as many oscillating species as possible leads to useful results. An examination of the relative amplitudes allows the identification of essential and nonessential species. Figure 3.2 shows traces of the oscillations of NADH and oxygen (O_2) in the horseradish peroxidase (HRP) reaction [25]. The abscissa is scaled in relative concentration units. The amplitude of the oscillations in O_2 is seen to be an order of magnitude greater than the amplitude of the oscillations in NADH.

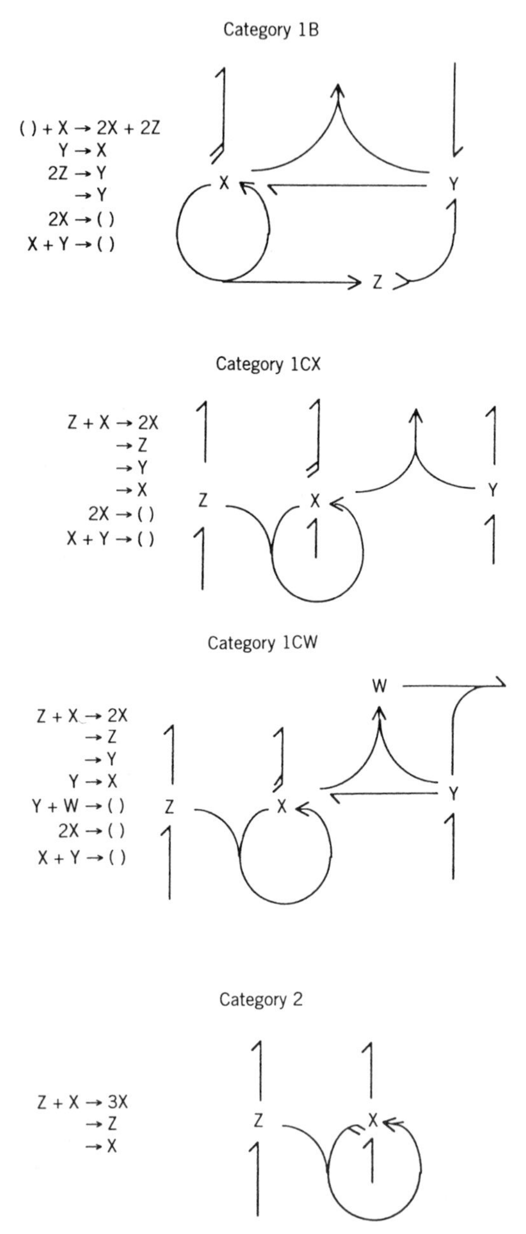

Figure 2.5. Elementary mechanistic steps and network diagrams for the prototypes of the categories of oscillators defined in [19]. In the network diagrams, the number of feathers at a species corresponds to the number of molecules of the species consumed in the reaction (no feathers shown when stoichiometric coefficient is 1); the number of barbs at a species corresponds to the number of molecules of the species produced in the reaction.

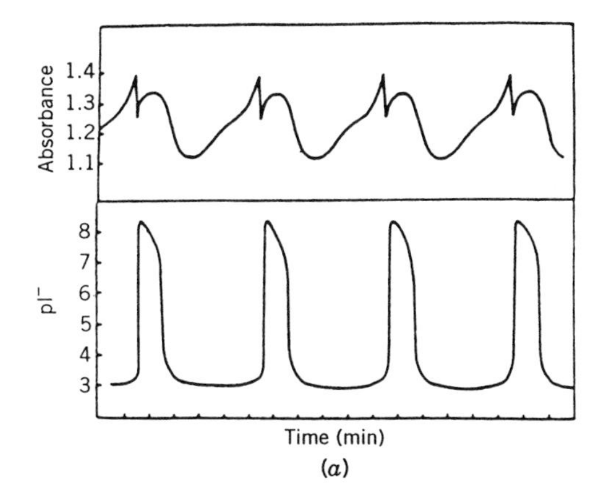

(*a*)

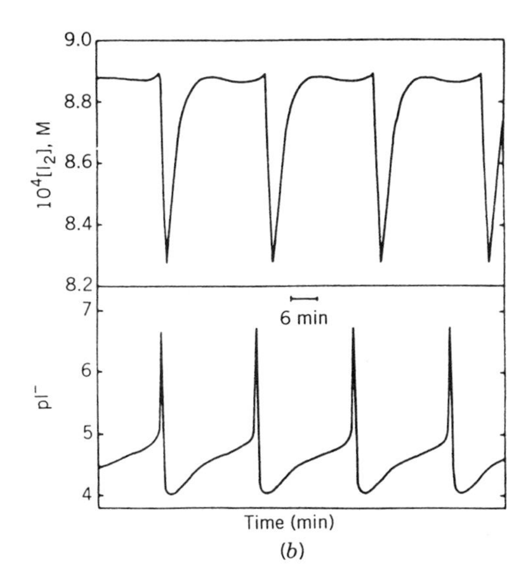

(*b*)

Figure 3.1. Comparison of experimental and theoretical oscillatory traces of I_2 absorbance and I^- potential for the chlorite–iodide reaction. (*a*) Experimental traces [16]. (*b*) Theoretical traces from the Epstein–Kustin mechanism [23] show fair qualitative agreement, but wave shapes are substantially different from experimental observations. (*c*) Theoretical traces from the Citri–Epstein mechanism [24] show marked improvement in wave shape agreement with experiments.

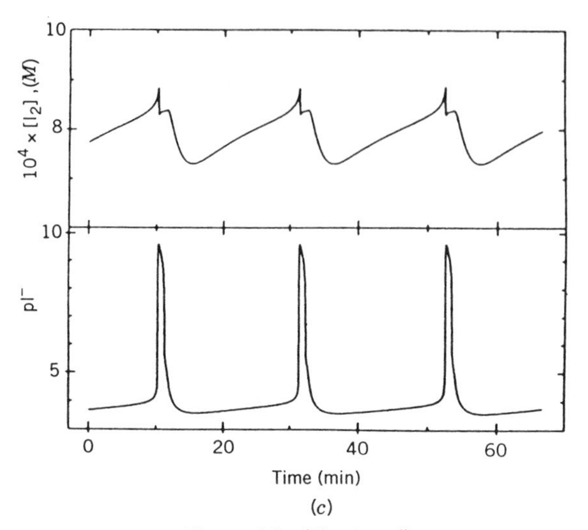

(*c*)

Figure 3.1 (*Continued*)

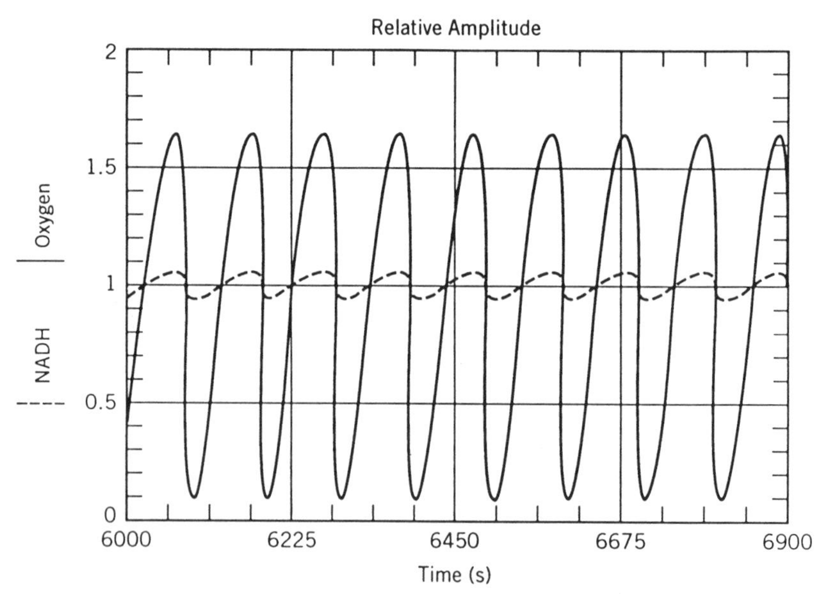

Figure 3.2. Experimentally determined relative amplitudes of O_2 and NADH in the HRP system [25]. The solid line denotes O_2 concentration, while the dashed line indicates NADH concentration. Concentrations are in arbitrary relative units.

A small relative amplitude distinguishes nonessential species of types A and C from essential species and nonessential species of type B. The data presented above suggest that O_2 is essential to oscillations of the horseradish peroxidase system, or is nonessential type B (an inert product); however, since O_2 is flowed into the system at a constant rate (i.e., O_2 is a reactant), we conclude that O_2 is essential. In contrast, the small relative amplitude of NADH oscillations indicate that NADH is a nonessential species of type A or C in this reaction.

Such a comparison is only useful for analyzing experimental data if oscillations in more than one species can be measured simultaneously or correlated. A comparison of relative amplitudes predicted by a proposed model is probably the most straightforward way to predict nonessential species, which needs to be confirmed experimentally by other feasible tests (see Section III.F) if multiple species measurements are not possible.

C. Phase Relations

From measurements of the temporal variations of chemical species comparison is made of the phases of as many essential oscillating species as possible. The phases of the species are assigned as in-phase (I), antiphase (A), advanced (+), or delayed (−) with respect to the oscillations of a reference species. This results in a sign-symbolic phase shift matrix. The sign-symbolic phase shift matrix $\Delta\Phi_{symb}$ for the prototype of each category is given in Table I. Figure 3.3 illustrates possible concurrent time series oscillations for the species in a Category 1CW oscillator.

Since we were unable to find an experimental example in which the phase relations of essential species in an autonomous oscillatory system are determined, we instead illustrate this technique with the results of simulated measurements. In simulations of the horseradish peroxidase (HRP) reaction based on the abstract model proposed by Degn [26] (the DOP model), Chevalier et al. [6] found the sign-symbolic phase shift matrix given in Table II. The entries of this matrix indicate that the oscillations in species A and B are in-phase with each other and that the oscillations in species X and Y are in-phase with each other. (Note that a species is always in-phase to itself, so all diagonal entries of $\Delta\Phi_{symb}$ are I.) Furthermore, X and Y are advanced by A and B, while A and B are delayed by X and Y.

From examination of Table III, essential species that are I with each other play the same role (except possibly X and W, which cannot be distinguished experimentally by this method). Species that are A to each other must be X and Y, since the only category with antiphase species is

TABLE I
Sign-Symbolic Phase Shift Matrices $\Delta\Phi_{symb}$ for the Essential Species in the Prototype of Each Category of Oscillatory Reactions

Category 1B

Phase relation of species . . .

. . . *with respect to*		X	Y	Z
	X	I	−	−
	Y	+	I	+
	Z	+	−	I

Category 1CX

Phase relation of species . . .

. . . *with respect to*		X	Y	Z
	X	I	A	+
	Y	A	I	−
	Z	−	+	I

Category 1CW

Phase relation of species . . .

. . . *with respect to*		X	Y	Z	W
	X	I	A	+	I
	Y	A	I	−	A
	Z	−	+	I	−
	W	I	A	+	I

Category 2

Phase relation of species . . .

. . . *with respect to*		X	Z
	X	I	+
	Z	−	I

1C; the species with peaks between those of Y and X (i.e., + with respect to Y and − with respect to X) are of type Z. The absence of species that are A to each other and the presence of only one (set of) essential species phase shifted from another indicates a Category 2 oscillator; X is + with respect to Z and Z is − with regard to X. Finally, the absence of species

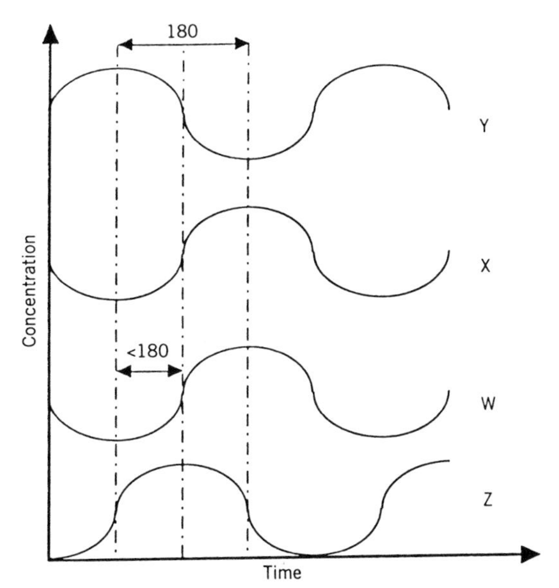

Figure 3.3. Possible concurrent time series oscillations for a Category 1CW oscillator. The peaks of species X and W are in-phase (I), those of X and Y are antiphase (A). The oscillatory peaks for species Z fall between those of Y and X (i.e., Z is − to Y and + to X), so that the order of appearance of peaks is Y, Z, X.

TABLE II
Sign-Symbolic Phase Shift Matrix $\Delta\Phi_{symb}$ Obtained from the DOP Model of the HRP Reaction

		Phase relation of species . . .			
		A	B	X	Y
. . . with respect to	A	I	I	−	−
	B	I	I	−	−
	X	+	+	I	I
	Y	+	+	I	I

that are A to each other and the presence of more than one set of phase shifts indicates a Category 1B oscillator; X is − with respect to Y and Z, Y is − with respect to X and + with respect to Z, and Z is + with respect to X and − with respect to Y. An examination of the interconnection of species in the network diagrams shown in Fig. 2.4 aids in rationalizing these observations.

TABLE III
Sign-Symbolic Concentration Shift Matrices ΔX_{symb} for the Essential Species in the Prototype of Each Category

Category 1B

Measured species \ *Perturbing species*	X	Y	Z
X	+	−	−
Y	+	+	+
Z	+	−	−

Category 1CX

Measured species \ *Perturbing species*	X	Y	Z
X	+	−	+
Y	−	+	−
Z	−	+	−

Category 1CW

Measured species \ *Perturbing species*	X	Y	Z	W
X	+	−	+	+
Y	−	+	−	−
Z	−	+	−	−
W	+	−	+	+

Category 2

Measured species \ *Perturbing species*	X	Z
X	+	+
Z	−	−

In the example of the DOP model given above, Category 1C is immediately ruled out, since no antiphase entries appear in $\Delta\Phi_{symb}$. From phase relations data alone, A and B seem to play the same roles, as X and Y. Finally, as there exists no species in $\Delta\Phi_{symb}$ that shows both advances and delays relative to other species in the system, we conclude that the system is a Category 2 oscillator, with A and B as species of type Z, and X and Y as species of type X.

The interpretation of the information in $\Delta\Phi_{symb}$ alone requires some caveats. First, phase relations are not useful in distinguishing essential and nonessential species. Chevalier et al. [6] used amplitude relations to determine that A was in fact a nonessential species for particular sets of inflow parameters, B nonessential for other sets of inflow parameters. In addition, a definitive assignment may be impossible unless the phase relations of all the essential species can be measured.

D. Concentration Shift Regulation

In this experiment, a constant inflow of one species at a time is added to the system at steady state near a supercritical Hopf bifurcation. This inflow additional to the inflow terms in Eq. 2.2 should not be large enough to shift the system from one dynamic region to another, for example, from a stationary state to an oscillatory state. The response of the concentrations of as many species as possible should be determined after the addition of each species and compared to the steady-state concentrations of the unperturbed system. These measurements allow the construction of an experimental shift matrix, which is directly related to the Jacobian matrix (see Section II.A). If we approximate the change in steady-state concentration X_i^s with respect to an inflow perturbation in X_j as dX_i^s/dX_j^0, the shift matrix ΔX is then given by

$$\Delta X \equiv dX^s/dX^0 = -k_0(dF/dX)^{-1}$$

where, as in Eq. 2.3, $dF/dX = \mathbf{J}$. Thus, the numerical shift matrix is the inverse of the Jacobian matrix times the opposite of the reciprocal residence time.

If precise numerical data cannot be obtained, qualitative assignment of the inflows as increasing (+) or decreasing (−) the steady-state concentrations of each species suffice to give the signs of the elements of $\mathbf{J}^{-1}$; from this information it is possible to distinguish roles of essential species and to assign the category of oscillator. The sign-symbolic concentration shift matrices for the essential species in the prototype of each category are given in Table III.

In measurements of the combustion of acetaldehyde, Skrumeda and Ross [27] used mass spectroscopy (MS) to measure time-dependent concentrations of eight of the reactive species in response to constant perturbations in CH_3CO and O_2. These measurements yield the partial sign-symbolic concentration shift matrix given in Table IV. These data indicate that an increased inflow of O_2 shifts the steady-state con-

TABLE IV
Partial Sign-Symbolic Shift Matrix ΔX_{symb} for the Acetaldehyde Combustion Reaction

		Perturbing species	
		O_2	CH_3CHO
Measured species	O_2	+	−
	CH_3CHO	−	+
	H_2O_2	+	
	CH_3OH	−	+
	CHOH	+	+
	CO_2	+	−
	H_2O	+	−
	CH_4	+	+

centrations of O_2, H_2O_2, CHOH, CO_2, H_2O, and CH_4 to higher values and the steady-state concentrations of CH_3CHO and CH_3OH to lower values. In addition, an increased inflow of CH_3CHO shifts the steady-state concentrations of CH_3CHO, CH_3OH, CHOH, and CH_4 to higher values and the steady-state concentrations of O_2, CO_2, and H_2O to lower values.

In the partial sign-symbolic concentration shift matrix for the acetaldehyde combustion reaction both O_2 and CH_3CHO display normal self-regulation, from which it can be concluded that neither species is an essential species of type Z. No other species were perturbed, so no information about the self-regulation of the other species is available. Without an identification of the type Z species, no assignment of the roles of the essential species or the category of the oscillator can be made. However, the experimentally observed concentration shifts can be compared to the predictions of proposed models.

For all categories of oscillatory systems, species Z may be distinguished from species X, Y, and W by inspection of the diagonal elements of the sign-symbolic concentration shift matrix ΔX_{symb} since it displays inverse (−) self-regulation [28] while the other species display normal (+) self-regulation. Species X, Y, and W, and the category of oscillator, may then be distinguished by examining the rows of the sign-symbolic concentration shift matrix. If each row has only elements of the same sign, the oscillator is in Category 2; species X shows + regulation with respect to all inflows and species Z shows − regulation with respect to all inflows. If at least one row has only positive elements, the oscillator is in Category 1B; X shows + self-regulation and − regulation with respect to other inflows, Y shows + regulation with respect to all inflows. If all rows have entries of both signs, the oscillator is in Category 1C; species X (and W) show − regulation with respect to inflow of Y, and species Y shows + self-regulation but − regulation with respect to all other inflows. Identification of the inversely self-regulated species Z is central to these assignments.

E. Concentration Shift Destabilization

A constant inflow of species is added to a system near a supercritical Hopf bifurcation (on either the steady state or the oscillatory side). It is noted whether or not a transition occurs across the bifurcation, and if so, whether the transition is from oscillations to a steady state or vice versa. The procedure is continued on both sides of the bifurcation with increasing inflow concentrations of perturbant until either a shift away from the bifurcation or a shift toward the bifurcation is observed. The shift in stability is assigned as stabilizing (s) for oscillatory to steady state and destabilizing (d) for steady state to oscillatory. This test may also be performed on either side of a saddle–node infinite period bifurcation.

Concentration shift destabilization experiments were performed [29] on the chlorite–iodide system near the supercritical Hopf bifurcation (where the iodide concentration is relatively low); the system was monitored with an iodide specific electrode. Upon addition of constant perturbing inflows of ClO_2^-, NaOCl, and I_2, either to the stable steady-state side of the bifurcation (i.e., the MI state), or to the oscillatory side of the bifurcation, the system relaxes to a stable steady state (s). Depending on the size of the perturbation, the system relaxes to either the MI steady state or to the LI steady state, even further from the supercritical Hopf bifurcation. Addition of a constant perturbing inflow of I^- to either the steady state or the oscillatory side of the bifurcation point causes the system to relax to stable oscillations (d). For relatively large I^- perturbation strength, the amplitude of the oscillations increases, indicating that the system has been shifted away from the supercritical Hopf bifurcation and toward the center of the oscillatory region. Figure 3.4(*a*) shows the experimentally observed shift in stability from oscillatory to steady state (s) due to addition of NaOCl. Figure 3.4(*b*) shows the experimentally observed shift in stability from steady state to oscillatory (d) due to addition of I^-.

Eiswirth et al. [19] presented a way of distinguishing roles of essential species and subcategories of Category 1 oscillators using s/d assignment

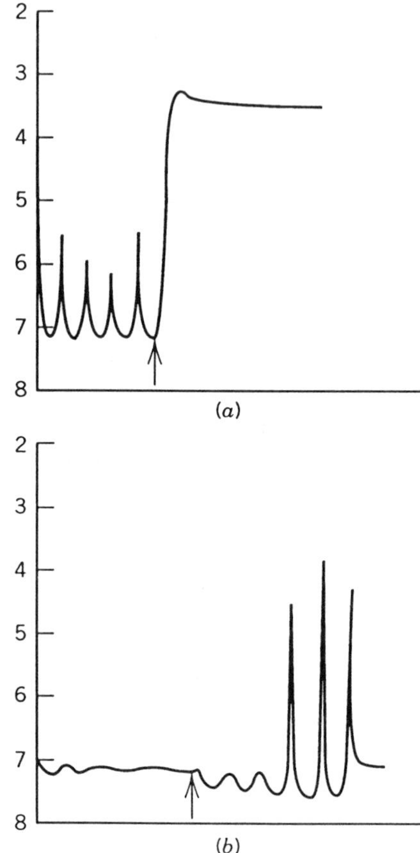

Figure 3.4. Concentration shift destabilization experiments performed on the chlorite–iodide system [50]. Traces of iodide potential versus time are shown. (*a*) System starts on the oscillatory side of supercritical Hopf bifurcation. When an inflow of NaOCl is added, the system shifts to the low iodide steady state. (*b*) System starts on the steady-state side of the Hopf bifurcation. When an additional inflow of I^- is added, the system shifts to the oscillatory regime.

and self-regulation information from concentration shift regulation experiment (Table III in [19]). The results in that table are based on the assumption that at the Hopf bifurcation near which experiments are performed, the steady-state concentration of the autocatalytic species X is high, for example, that the system is at point B on the phase diagram shown in Fig. 3.5. However, it may also be the case for a given system that the steady-state concentration of X at the Hopf bifurcation is instead low (e.g., at point A in Fig. 3.5), which reverses the s and d assignments given, but leaves the self-regulation unchanged. Thus, assignments of the roles of species and of the category of oscillator purely from concentration shift destabilization data depend on prior knowledge of the identity of the autocatalytic X species. Certain points in phase space, in particular those where the slope of the bifurcation boundaries change

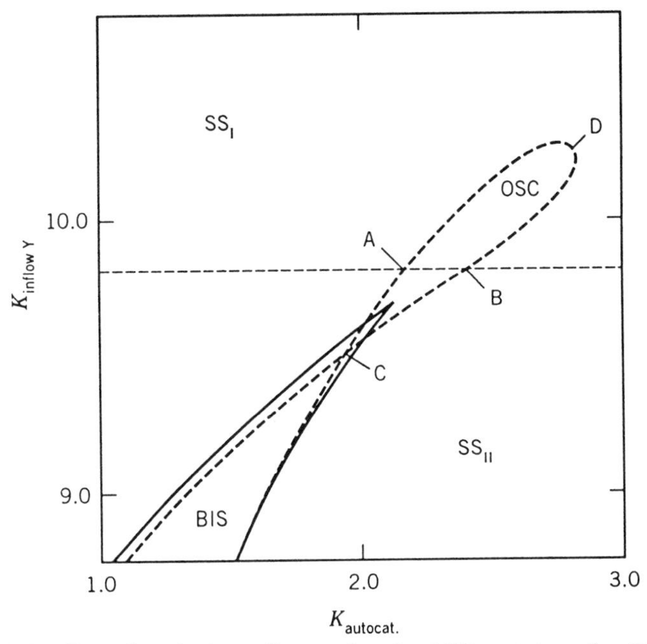

Figure 3.5. Cross-shaped phase diagram for the 1CX prototype (modified Franck oscillator). Solid lines indicate saddle–node bifurcations; dashed lines, Hopf bifurcations; OSC, the oscillatory region; BIS, the bistable region; A and B, bifurcation points discussed in the text; SS_I, the steady-state region with low [X] and high [Y]; and SS_{II}, the steady-state region with high [X] and low [Y]. (Reproduced from [19].)

(e.g., point D in Fig. 3.5), are inappropriate points at which to apply this test in order to assign roles of species, since the s/d behavior will change with the slope.

For a discussion of possible deductions concerning the reaction mechanism, assume that a species Q, monitored near the Hopf bifurcation, has a low concentration, and that increasing the inflow of Q is found to be destabilizing. From this information, we may assign Q as type X [Table V(b)] or type Y [Table V(a)]. To make a clear choice between these possibilities, additional experiments (e.g., concentration shift destabilization or phase relation measurements) are necessary to identify the species of type Z and to know whether the system can produce oscillations when run in batch. {The distinction between 1B and 1C is related to the distinction between systems that can produce oscillations when run in batch and those that show oscillations only when run in a CSTR, although there may be exceptions (see Eiswirth et al. [19]).} For a 1B oscillator, Y shows the same s/d behavior as Z and X shows the

TABLE V
Effect of Essential Species Influx on the Hopf Bifurcation in Oscillators of Category 1

(a) High [X] on Steady-State Side of Hopf Bifurcation

	X	Y	Z	W
1B	s	d	d	
1CW	s	d	s	s
1CX	s	d	s	

(b) Low [X] on Steady-State Side of Hopf Bifurcation

	X	Y	Z	W
1B	d	s	s	
1CW	d	s	d	d
1CX	d	s	d	

opposite behavior from Z. For a 1C oscillator, X (and W) shows the same s/d behavior as Z and Y shows the opposite behavior from Z.

From the observed concentration shift destabilization behavior for the chlorite–iodide reaction, the one destabilizing species I^- is at low concentration on the steady-state side of the supercritical Hopf bifurcation. This destabilization is consistent with I^- being an essential species of type X [as in Table V(b)], in which case the oscillator may be in Category 1B, or there exists at least one essential species of type Z that has not been used as a perturbant in the experiments. It is also possible that I^- is an essential species of type Y [suggesting a high concentration of the autocatalytic species X at the Hopf bifurcation, Table V(a)]. In this case, the oscillator is consistent with Categories 1CX or 1CW, but may be in Category 1B if there is an essential species of type Z in the reaction whose inflow has not been perturbed in the experiments. Without further information to identify the type Z species in this system, no definitive conclusions can be drawn.

Categorization of the roles of species is not (and probably cannot be) based on s/d behavior in the absence of an XPD, where a choice of either of two Hopf branches leads to clear interpretations of the experimental findings. However, comparison of experimental results with the predictions of a model may still be useful.

F. Qualitative Pulsed Species Response

In this experiment, a pulsed perturbation of one species at a time is applied to a system at a steady state near a supercritical Hopf bifurcation. The relaxation to the stable steady state is measured, and the behavior during this relaxation is characterized as oscillatory response, monotonic decay, or no response.

In experiments on the HRP system, Hung and Ross [25] measured the response of O_2, NADH, and compound III (CompdIII) to pulse perturbation with O_2, NADH, and H_2O_2. Upon perturbation, each of the monitored species shows a damped oscillatory response, as shown in Table VI(a). In similar experiments on the chlorite–iodide system, Stemwedel and Ross [29] observed the iodide potential in response to the addition of pulsed perturbations in I^-, ClO_2^-, IO_3^-, Cl^-, I_2, HIO_2, and H_2OI^+. The observed responses of the I^- evolution are shown in Table VI(b).

The results of these experiments give information toward the assignment of species as essential or nonessential, and the determination of the types of the nonessential species [6]. A nonessential species of type A shows a negligible response to perturbation by any species but itself, for which it exhibits monotonic decay; a perturbation by a type A species causes a damped oscillatory response in all essential species and in nonessential species of type B. A perturbation by a nonessential species of type B causes negligible response in all species except itself, for which it shows monotonic decay. A nonessential species of type C is in time-independent quasisteady state to which it returns nearly instantaneously after perturbation by any species. Finally, perturbation by essential species leads to damped oscillatory response of all essential species and of nonessential species of type B. This information is summarized in Fig. 3.6 and Table VII.

Solely from the observations of the HRP reaction presented above, O_2 and NADH, species whose concentrations show damped oscillatory responses to pulse perturbations in themselves, are identified as essential species. (Recall, however, that a comparison of relative amplitudes suggests that NADH is nonessential.) Since CompdIII is not added as a perturbation but responds with damped oscillations to perturbations in the essential species O_2 and NADH, it is unclear from this experiment whether it is an essential species or a nonessential species of type B. Similarly, because H_2O_2 is not monitored but causes damped oscillations in O_2 and NADH when it is added as a perturbation, it may be either an essential species or a nonessential species of type A.

Only a single species I^- is monitored to give the results presented

TABLE VI
Experimental Observations for Qualitative Studies on the Response of Species Due to Pulsed Perturbations of Species

(a) Horseradish Peroxidase System

		Response of Measured species		
		O_2	NADH	CompdIII
Perturbing species	O_2	damped oscillatory	damped oscillatory	damped oscillatory
	NADH	damped oscillatory	damped oscillatory	damped oscillatory
	H_2O_2	damped oscillatory	damped oscillatory	damped oscillatory

(b) Chlorite–Iodide System

		Response of Measured species
		I^-
Perturbing species	I^-	damped oscillatory
	ClO_2^-	damped oscillatory
	IO_3^-	no response
	Cl^-	small amplitude damped oscillatory
	I_2	no response
	HIO_2 [a]	damped oscillatory
	H_2OI^+ [a]	damped oscillatory

[a] Unstable species synthesized immediately prior to addition.

above for the chlorite–iodide system, but these results provide useful information. Since a pulse perturbation in I^- leads to a damped oscillatory response in I^-, I^- is an essential species. The species whose perturbations produce no response in I^-, that is, IO_3^- and I_2, are then nonessential species of type B, and the one producing small amplitude damped oscillatory decay, Cl^-, is most likely a nonessential species of type C. The species that are not monitored but whose perturbations cause damped oscillations in I^-, that is, ClO_2^-, HIO_2, and H_2OI^+, are either essential species or nonessential species of type A.

Determination of essential and nonessential species by this method relies on measurement of most or all species and requires the addition of pulsed perturbations of most or all species for complete determination. However, a complete determination may be deducible from partial information, and, as seen in the two examples discussed here, a partial determination may be useful.

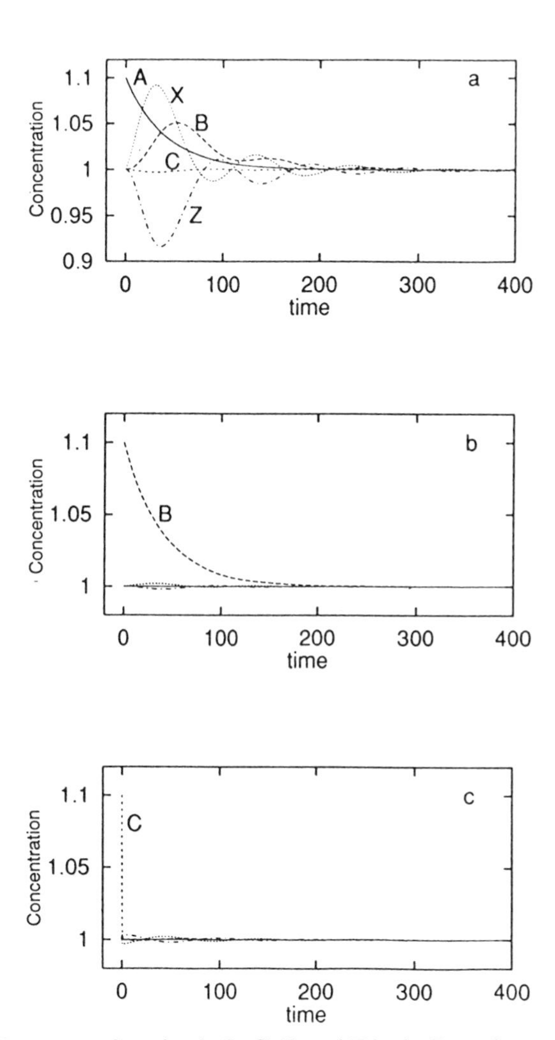

Figure 3.6. Responses of species A, B, C, X, and Z in the Brusselator model to pulsed perturbations in each species. Curves: (full line) A, (long dashed line) B, (dotted line) C, (short dashed line) X, (dashed–dotted line) Z. (*a*) Perturbation by A. (*b*) Perturbation by B. (*c*) Perturbation by C. (*d*) Perturbation by X. (*e*) Perturbation by Z. (Reproduced from [6].)

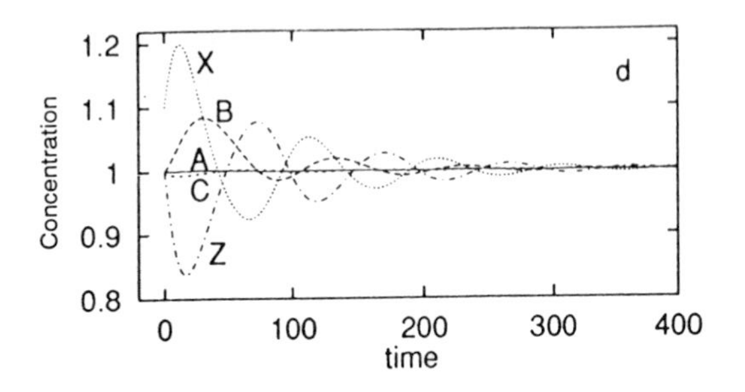

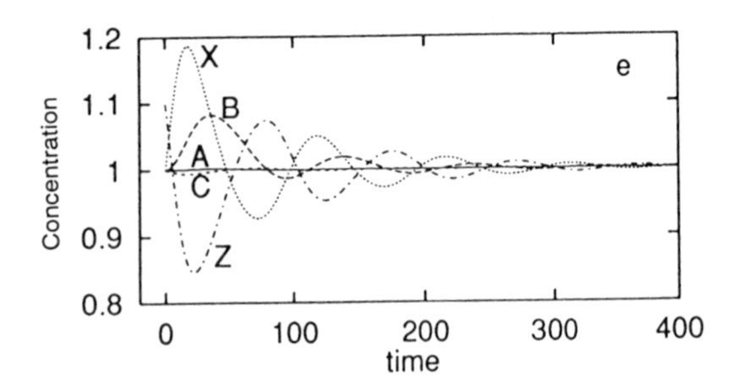

Figure 3.6 (*Continued*)

TABLE VII
Expected Responses to Pulsed Perturbations for Essential and Nonessential Species of Types A, B, and C

		Response of Measured species			
		Essential	A	B	C
Perturbing species	Essential	damped oscillatory	no response	damped oscillatory	no response
	A	damped oscillatory	monotonic decay	damped oscillatory	no response
	B	no response	no response	monotonic decay	no response
	C	small amplitude damped oscillatory	no response	small amplitude damped oscillatory	monotonic decay

G. Quantitative Pulsed Species Response

This experiment consists of the application of a pulse perturbation of each species one at a time to the system in a steady state near a supercritical Hopf bifurcation. The initial change in concentration of each measureable species upon perturbation is determined, and a distinction is made among three qualitatively different responses: an increase (+), a decrease (−), or no change (0). If possible, the initial slope of the temporal evolution of each species upon perturbation is measured.

From the experiments on the HRP system, discussed in Section III.F, Hung and Ross [25] determined by inspection the initial response of O_2, NADH, and CompdIII to pulse perturbation with O_2, NADH, and H_2O_2. The signs of the initial responses are shown in Table VIII. Two of the responses are ambiguous; it is unclear from the measurements whether the responses are decreases or no change. This difficulty arises because a reponse may appear due to a number of elementary reactions after some time lag, but it is only the *initial* response that determines the entries of the sign-symbolic Jacobian matrix.

The numerical Jacobian matrix (or portions of it) may be calculated from the measured initial slopes [6]. The kth column of the Jacobian is computed from the initial slopes of the temporal variation of the concentration of species i after a perturbation in species X_k, by the use of the relation $d\delta X_i/dt = (\partial F_i/\partial X_k)\,\delta X_k$. This finding can be compared to the Jacobian matrix predicted by a model reaction mechanism at the steady state.

The sign-symbolic Jacobian matrices yield information on the connectivity of the species in the reaction [4, 6], as summarized in Table IX. The diagonal elements give information about the self-production and self-consumption of each species. A diagonal entry of + indicates direct autocatalysis of a species; that of − suggests no autocatalysis or, if

TABLE VIII
Observed Initial Responses to Pulsed Perturbations in the HRP System

		Perturbing species		
		O_2	NADH	H_2O_2
Measured species	O_2	−	0	−/0
	NADH	0	−	−
	CompdIII	0	−/0	+

TABLE IX
Reaction Connectivity from JMEs

	$\mathbf{J}_{ij} = 0$	$\mathbf{J}_{ij} < 0$	$\mathbf{J}_{ij} > 0$
$i = j$	Self-production of ith species balances out self-consumption of ith species	No autocatalysis OR reactions consuming ith species dominate	Autocatalysis with respect to ith species
$i \neq j$	No elementary step with jth species as a reactant and ith species as a product	Reaction consuming ith species in which jth species takes part[a]	Elementary step with jth species as a reactant and ith species as a product

[a] The parameters $\mathbf{J}_{ij} < 0$ and $\mathbf{J}_{ji} < 0 \Rightarrow$ reaction consuming both ith species and jth species; check signs of $\mathbf{J}_{ii}$ and $\mathbf{J}_{jj}$ to determine whether reaction is autocatalytic in either of these species.

autocatalysis of the species takes place, that the reactions consuming that species are predominant; and 0 indicates that self-production and self-consumption of the species are balanced. The off-diagonal elements give information about how the various species interact with each other. An ith row, jth column entry of 0 indicates that there is no reaction involving the jth species that produces the ith species; one of +, that there is a reaction producing the ith species from a set of species including the jth species; and one of −, that there is a reaction consuming the ith species in which the jth species takes part, such that an increase in the concentration of the jth species will lead to a decrease in the concentration of the ith species. The most straightforward way to fulfill this last condition is an elementary step in which both the ith and the jth species are reactants (for which both $\mathbf{J}_{ij} < 0$ and $\mathbf{J}_{ji} < 0$), but such a condition could also arise from an elementary step in which the jth species appears as both a reactant and a product. For example, the elementary step

$$X_i + X_j \rightarrow X_j + X_k$$

has $\mathbf{J}_{ij} < 0$ and $\mathbf{J}_{ji} = 0$, whereas the elementary step

$$X_i + X_j \rightarrow 2X_j$$

has $\mathbf{J}_{ij} < 0$ and $\mathbf{J}_{ji} > 0$.

From the elementary mechanistic steps of prototypes shown in Fig. 2.4, we can deduce the signs of the JMEs for the essential species in the prototype of each category. The resulting sign-symbolic Jacobian matrices are given in Table X.

Determination of the complete Jacobian by this method relies on measurement of all species and the ability to add pulsed perturbations of all species, although submatrices may be determined if some species cannot be monitored or added. Furthermore, the method of initial slopes may require very fast measurement techniques.

H. Delay Experiments

Consider an oscillatory reaction run in a CSTR. One or more species in the system are measured as a function of time. A delayed feedback related to an earlier concentration value of one species is applied to the inflow of a species in the oscillatory system. The feedback takes the form $X_j^0(t) = f(X_i(t-\tau))$, where the input concentration $X_j^0(t)$ of the jth species at time t depends on $X_i(t-\tau)$, the instantaneous concentration of the ith species at time $t-\tau$. Once the delayed feedback is applied, the system may remain in the oscillatory regime or may settle on a stable steady state (or even on a chaotic attractor). The delays τ at which the system crosses a Hopf bifurcation and frequencies y of the oscillations at the bifurcation are determined. The feedback relative to a measured reference species is applied to the inflow of each of the species in turn. For each species used as a delayed feedback, Hopf bifurcations equal to the number of species in the system are located, if possible; for a system with n species, this would require locating n Hopf bifurcations with feedbacks applied to each of n species, for a total of n^2 sets of delay parameters. These experiments are performed by measuring only a single oscillating species in the system.

Chevalier et al. [30] applied a sinusoidal delayed feedback to the flow rate of sodium bromide (NaBr) in experiments on the minimal bromate oscillator (MBO). For a particular choice of parameters in the feedback equation, the system at a stable steady state begins to oscillate. As the delay time is increased, the amplitude of the oscillation increases. Analysis of the relation between amplitude and delay time indicates that a supercritical Hopf bifurcation has been located. Although Chevalier et al. [30] report the value of τ corresponding to the bifurcation point, the frequency y of the oscillations is not reported. Feedbacks were not applied to any other species in the reaction.

As described in detail by Chevalier et al. [6], experimentally determined τ and y values can be used to find traces of submatrices of the Jacobian; if n Hopf bifurcations are located for a feedback to each of the species in the system, this allows determination of the complete Jacobian

TABLE X

The Signs of the JMEs $J_{ij} = \partial \dot{X}_i / \partial X_j$ for the Prototype of Each Oscillatory Category. Entries of * Indicate a Sign That Is Determined by the Balance of Rate Constants and Instantaneous Species Concentrations

Category 1B

$X_i =$ \ $X_j =$	X	Y	Z
X	*	*	0
Y	−	−	+
Z	+	0	−

Category 1CX

$X_i =$ \ $X_j =$	X	Y	Z
X	*	−	+
Y	−	−	0
Z	−	0	−

Category 1CW

$X_i =$ \ $X_j =$	X	Y	Z	W
X	*	*	+	0
Y	−	−	0	−
Z	−	0	−	0
W	0	−	0	−

Category 2

$X_i =$ \ $X_j =$	X	Z
X	+	+
Z	−	−

from measurements of a single species in the reaction. The connectivity of the species may then be analyzed as discussed in Section III.G.

I. Quenching

The variation in time of concentrations of species in an oscillatory reaction can be stopped temporarily by application of a pulse perturbation of a species to the system. If the system is near a supercritical Hopf bifurcation, an interpretation of the addition may yield information about the Jacobian matrix at the bifurcation. In quenching experiments, the phase of oscillation at which the perturbation is added and the amount of perturbant added are varied for each perturbing species, one at a time.

Figure 3.7 shows the results of quench experiments performed by Hynne and Sørensen [31] on the BZ reaction. Additions of Br^-, Ce^{4+},

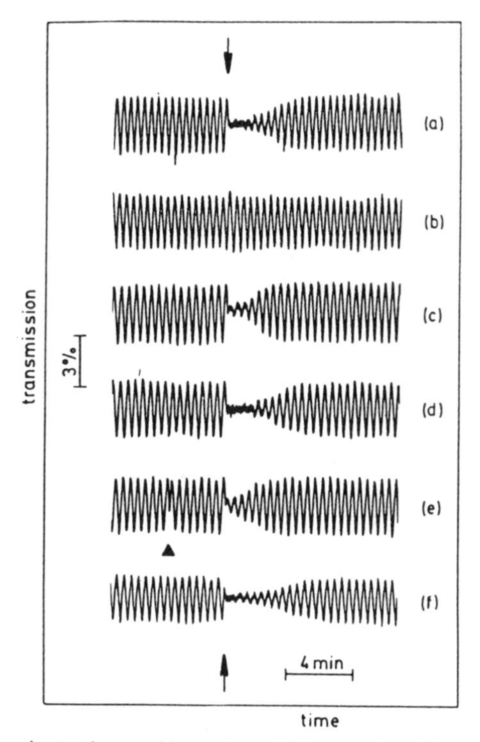

Figure 3.7. Experimental quenchings of oscillations in the BZ reaction by addition of pulse perturbations of species at particular concentrations and phases. Each trace is transmission of light due to Ce^{4+} versus time. (*a*) Addition of Br^- at $-110°$ (quenches oscillation). (*b*) Addition of same amount of Br^- at $+70°$ (does not quench oscillation at this phase). (*c*) Quenching by Ce^{4+}. (*d*) Quenching by $CHBr(COOH)_2$. (*e*) Quenching by HOBr. (*f*) Quenching by relative dilution by 0.01. (Reproduced from [31].)

$CHBr(COOH)_2$, and HOBr were found to quench the oscillations. The quench amplitude of $CHBr(COOH)_2$ is nearly 25 times that of Ce^{4+}, and the quench amplitude of HOBr is roughly 3 times that of Ce^{4+} [32].

The theory underlying quench studies is detailed in [33, 34] and relies on the properties of the Jacobian matrix near a Hopf bifurcation point, in particular the separation of slow and fast variables to divide concentration space into the center (unstable) manifold, which contains the limit cycle oscillations, and the stable manifold. The following information can be obtained from a complete set of experimental quench data (i.e. quench amplitudes and phases for each of the species capable of quenching the oscillations): (a) the eigenvectors corresponding to the pair of pure imaginary eigenvalues (at the supercritical Hopf bifurcation) of $\mathbf{J}^T$, the transpose of the Jacobian matrix (present at the bifurcation); (b) in the trivial case of systems with two essential species, the Jacobian matrix, calculated from the experimentally determined eigenvectors (in larger systems, knowledge of *all* the eigenvectors is necessary to recover the Jacobian, and as of yet this information cannot be obtained by quench experiments); and (c) reconstruction of the time series of oscillations of up to two species if, in addition to complete quench data, the oscillations of the other $n-2$ species in the system have been measured. An experimentally obtained Jacobian matrix allows an analysis of the connectivity of species as described in Section III.G and showed by example in Section IV. Reconstruction of the time series of oscillations of additional species may make possible a complete analysis of amplitude relations and phase relations, detailed in Sections III.B and C. As noted in [19] and summarized in Table XI, quench data is also useful in distinguishing essential from nonessential species: nonessential species of types B and C have much higher quench amplitudes than nonessential species of type A and essential species. The distinction between nones-

TABLE XI
Behavior of Quench Vectors for Nonessential and Essential Species as a Limiting Case Is Approached

Species	Quench Vector
A	finite
B	$\rightarrow\infty$
C	$\rightarrow\infty$
essential	finite

sential species due to quench amplitudes can be seen in terms of the position of the coordinate axis of the species relative to the stable and center eigenspaces (Fig. 2.4 in Section II.C).

In the quenching data discussed above for the BZ reaction, $CHBr(COOH)_2$ is almost certainly a nonessential species of type B or C, since its quench amplitude is so much larger than those of the other species. For the same reason, it is likely that HOBr may also be a nonessential species of type B or C. The species Br^- and Ce^{4+} are identified as essential species. This finding is consistent with the assignments made by Eiswirth et al. [19] of species in the SNB and FKN models of the BZ reaction.

Recently, Hynne et al. [35, 36] outlined a method by which to combine quenching data and representations of stationary states by extreme currents to optimize sets of rate constants for a particular reaction network. They applied this method to optimizations of models of the BZ reaction at a Hopf bifurcation. The set of EC are searched for Hopf bifurcations, which can be used to calculate quenching data for comparison with experimental data. Rate constants in a model can then be optimized to give a better fit between predictions and theory. They also present an alternate approach, which uses experimental quenching data (i.e., the experimentally determined left eigenvectors of the Jacobian) to analyze a network and determine whether there exists *any* set of real, nonnegative rate constants for that network that is consistent with the quenching data.

J. Phase Response Experiments

A pulsed perturbation of one species at a time is applied to an oscillatory system, often but not necessarily near a supercritical Hopf bifurcation. The phase of oscillation at which the perturbation is added and the amount of perturbant added are varied. It is determined whether each perturbation results in an advance or delay of the next oscillatory peak relative to the period of the unperturbed autonomous system. If we let t_n denote the time of the nth reference event (an easily followed feature of the oscillations, e.g., a sharp rise or fall in the concentration of a monitored species), the period of the unperturbed limit cycle is given by

$$T_0 = t_n - t_{n-1}$$

A pulse perturbation is applied at some time t_{stim}, between t_n and t_{n+1}, from which we defined a phase of stimulation as

$$\varphi_{\text{stim}} = \frac{t_{\text{stim}} - t_n}{T_0}$$

Another reference event of the same type as the one at t_n follows the perturbation at time t_{occ} (generally not equal to t_{n+1}). Perturbations for which $t_{occ} > t_{n+1}$ are phase delays, while those for which $t_{occ} < t_{n+1}$ are phase advances.

A plot of φ_{stim} along the abscissa and $\Delta\varphi$ along the ordinate is a phase response curve (PRC), where

$$\Delta\varphi = -\frac{(t_{occ} - t_{n+1})}{T_0}$$

as defined in Strasser et al. [8]. In a phase transition curve (PTC), φ_{stim} is plotted along the abscissa and φ' along the ordinate, where

$$\varphi' = \varphi_{stim} + \Delta\varphi$$

as described in [8].

Ruoff et al. [37] report phase response experiments in which the oscillatory BZ reaction is perturbed with $NaBrO_2$. Figure 3.8 shows the response of the system to perturbations added at different phases of the oscillation. Figure 3.9 gives a plot of the initial phase shift (the difference between the time of appearance of the oscillatory peak that follows the perturbation and the time at which the peak is expected in the absence of

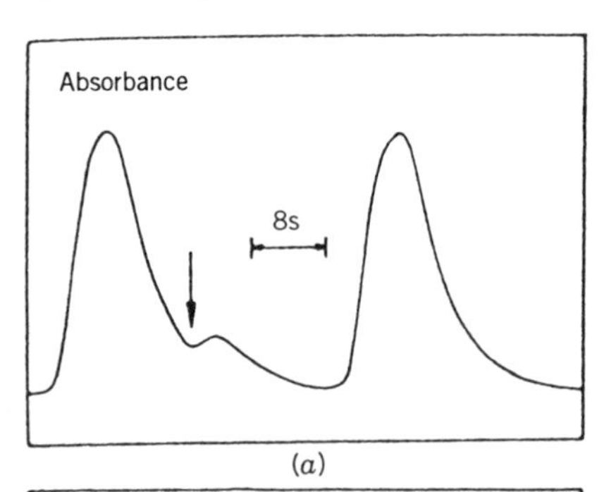

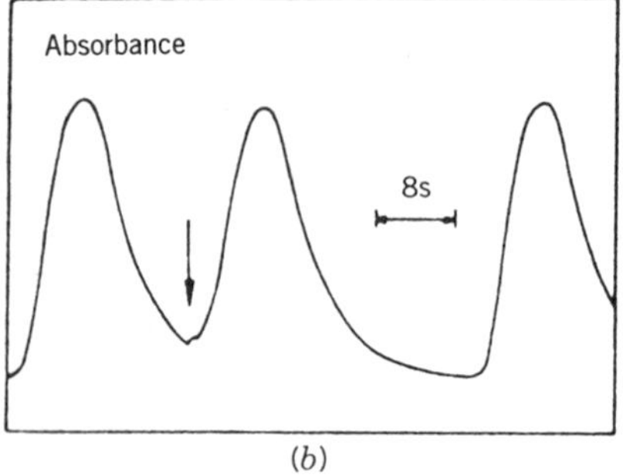

Figure 3.8. Phase response experiments on the BZ reaction. Plots show absorbance at 400 nm versus time. (*a*) Subcritical $NaBrO_2$ perturbation. Addition is made on decreasing side of oscillatory peak. (*b*) Supercritical $NaBrO_2$ perturbation. Addition is made on increasing side of oscillatory peak. (Reproduced from [37].)

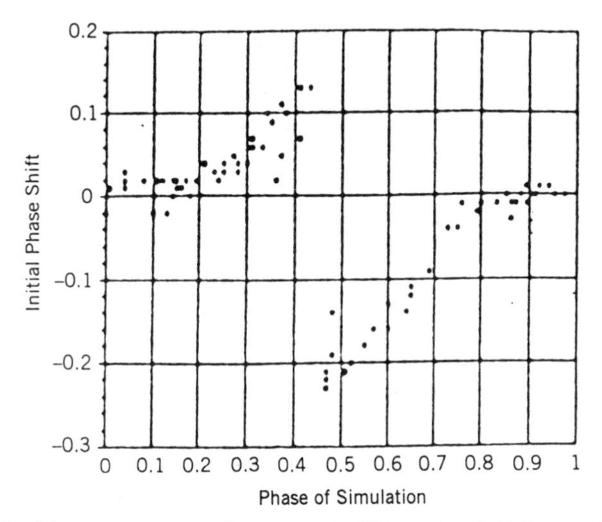

Figure 3.9. Phase response experiments on the BZ reaction. Initial phase shift is plotted against phase of stimulation for $1.2 \times 10^{-6}\,M$ $NaBrO_2$ perturbations. (Reproduced from [37].)

any perturbation) versus the phase of stimulation of the system. If the initial phase is divided by the period of the unperturbed oscillator and plotted against the phase of stimulation, a PRC results. Figure 3.10 shows the experimentally determined relation between perturbation size and the phase of stimulation at which a transition is seen between phase advances and phase delays in response to pulse perturbations. Ruoff et al. [37] compare these experimental results to the phase response behavior predicted by four different models for the BZ reaction. Two of the models examined, the original Oregonator and the Showalter, Noyes,

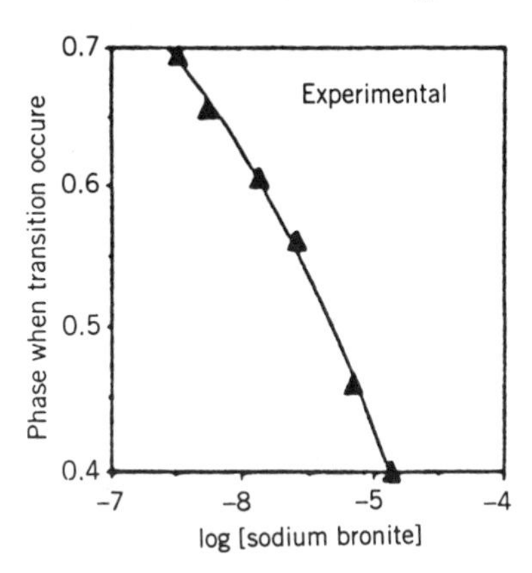

Figure 3.10. Phase response experiments on the BZ reaction. Experimental phases of transition from positive phase shifts to negative phase shifts as a function of perturbation strength in $NaBrO_2$. (Reproduced from [37].)

Bar-Eli model, did not account for experimentally observed phase delays. The other two models, a modified seven variable Oregonator and a 26 variable, 80 reaction model, not only agreed qualitatively with experiments, but also showed almost quantitative agreement between experimental and calculated phase response data. Phase response techniques utilizing two pulse perturbations (the first to initiate an excitation, the second to reset the phase of this cycle) have been applied to study excitable reaction systems [38], and periodic pulse stimulations have been used to study forced excitable systems [39].

This study provides a test of the behavior of the system with respect to individual species in the mechanism. Such behavior should be reproduced by a proposed mechanism. Whether different phase response behavior should be expected for different categories of oscillators or for perturbations in essential species of different roles has not been determined.

K. External Periodic Perturbation

A sinusoidal perturbation of the form $flow = flow_{av}[1 + \varepsilon \sin(\omega t)]$ is applied to an inflow of species to the oscillatory system. The amplitude ε and frequence ω of the perturbation are varied. The entrainment boundaries of the various bands, including the fundamental or 1:1 band (in which there is one response oscillatory peak for every perturbation oscillatory peak), are determined. Periodic (entrained), quasiperiodic, and chaotic responses are distinguished with an analysis tool such as next-phase maps. The widths of the fundamental entrainment bands for perturbations in each species are determined, as are the change in phase shift between the perturbation and the response of the system between one edge of the fundamental entrainment band and the other.

Skrumeda and Ross [27] calculated the changes in phase shift across entrainment bands for species in the PR model [40] for the low temperature (cool flame) oxidation of acetaldehyde. The results of these calculations are given in Table XII.

As discussed by Eiswirth et al. [19], entrainment data may be used to distinguish essential and nonessential species by means of relations set out by Loud [41]. If we define $\Delta\phi$ as the total change of the phase shift between the sinusoidal perturbation in the species in question and the response (taken as the distance between maxima) from one edge of the fundamental entrainment to the other, essential species have $\Delta\phi = \pi$ and nonessential species have $\Delta\phi < \pi$. In addition, nonessential species of type B have $\varepsilon \to \infty$. This information is summarized in Table XIII.

In the data calculated from the PR model, CH_3CHO has $\Delta\phi = 48.8°$ when perturbed by itself, and O_2 has $\Delta\phi = 54.0°$ when perturbed by itself. These changes in phase shift are significantly lower than 180°, so

TABLE XII
Change in Phase Shift Across the Fundamental Entrainment Band Calculated for the PR Model [40] of the Low-Temperature Combustion of Acetaldehyde[a]

		Perturbing species			
		CH_3CHO	O_2	CH_3CO_3H	Temperature
Measured species	CH_3CHO	48.8°	166.7°	NR	153.0°
	O_2	167.4°	54.0°	NR	NR
	CH_3CO_3H	NR	NR	154.2°	NR
	Temperature	146.2°	NR	NR	160.9°

[a] NR indicates data which is not reported.

TABLE XIII
Characteristics of the Fundamental Entrainment Band with Regard to Nonessential and Essential Species as a Limiting Case Is Approached

		Behavior	
		Change in Phase Shift ($\Delta\phi$)	Perturbation Amplitude (ε)
Species	A	$\rightarrow 0$	finite
	B	$\rightarrow 0$	$\rightarrow \infty$
	C	$\rightarrow 0$	finite
	essential	~180°	finite

both species are assigned as nonessential. Perturbation with CH_3CO_3H gives $\Delta\phi = 154.2°$ in CH_3CO_3H, and perturbation in temperature gives $\Delta\phi = 160.9°$ in temperature. These changes in phase shift are close enough to 180° that both species are assigned as essential. It should be noted that, even in simulations, none of the changes in phase shift is exactly 180 or 0°; the changes in phase shift of 146.2° and greater should probably theoretically be 180°, while those near 50° should be closer to 0°. Skrumeda gives four possible explanations for this discrepancy, some of which were also noted by Tsujimoto et al. [42]: (a) rapid change of the phase of response near the edges of the entrainment band, (b) critical slowing down that hampers attempts to locate the edge of the entrainment band, (c) an amplitude of oscillation not sufficiently small for

Loud's theorem [41] to hold, and (d) nonessential species that do not fall into the strict limiting case (i.e., which have some amount of "essential character").

In normal form analysis of autonomous and externally perturbed oscillatory reactions near a Hopf bifurcation there appear JMEs. Hence, from measurements on, say, the response of an oscillatory mechanism near a Hopf bifurcation to external periodic perturbation it may be possible to infer information on these matrix elements. While some theoretical work has been done in normal form methods, especially with regard to externally perturbed systems [43–45], necessary studies on the connection to JMEs remain to be done.

L. Bifurcation Analysis

Control parameters (a single control parameter for 1-D bifurcation analysis, two for 2-D) are varied and the dynamic behavior (steady state, oscillatory, quasiperiodic, or chaotic) of the autonomous system is observed. At steady states, pulsed perturbations of various species are added in varying amounts to test for excitability. Regions of distinct behavior are located (most often oscillatory, steady state, or excitable). Near transitions, hysteresis and increase or decrease in amplitude and/or period of oscillation are observed, and this information is used to identify bifurcation types. For a supercritical Hopf bifurcation, one observes that the period of oscillation remains constant, while the amplitude decreases as the square root of the distance from the bifurcation point to zero; no hysteresis is observed. For a subcritical Hopf bifurcation, the period of oscillation remains constant, while the amplitude goes suddenly to zero; hysteresis is observed. For a saddle–node infinite period bifurcation, the amplitude of oscillation remains constant, while the period increases exponentially to infinity: no hysteresis is observed. Finally, for a saddle-loop infinite period bifurcation, the amplitude of oscillation remains constant, the period increases exponentially to infinity, and hysteresis is observed.

Figure 3.11 shows an experimentally generated bifurcation diagram for the chlorite–iodide reaction [29]. The bifurcation features are plotted on axes of the input concentration ratio $[ClO_2^-]_0/[I^-]_0$ versus reciprocal residence time k_0, and are compared to those predicted by continuation techniques [8] from the model proposed by Citri and Epstein [24]. Figure 3.12 shows a trace of the iodide potential from which the supercritical Hopf bifurcation boundary is located. The amplitude decreases and goes to zero, leaving the system on a steady state. No hysteresis is observed on returning from the steady state to the oscillatory region. This transition to a steady state, via oscillations of finite constant period with amplitude

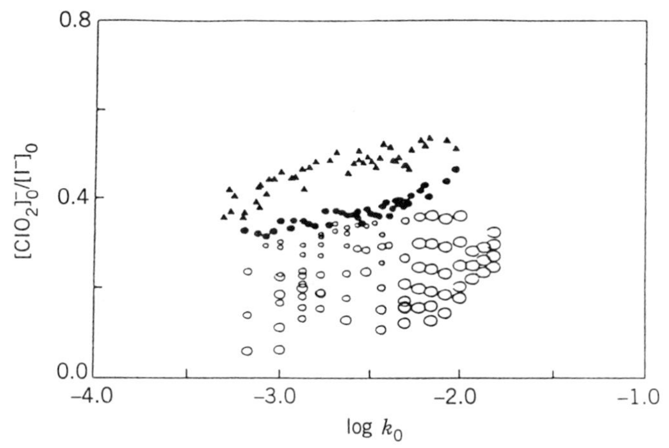

Figure 3.11. The experimentally determined bifurcation structure of the chlorite–iodide reaction. The ratio of input concentrations, $[ClO_2^-]_0/[I^-]_0$, is plotted versus the logarithm of the reciprocal residence time, $\log k_0$. Saddle–node infinite period bifurcations are indicated by filled circles. Supercritical Hopf bifurcations are indicated by filled triangles. In the region between the supercritical Hopf and saddle–node infinite period bifurcations, stable oscillations were observed. Sites of excitations are indicated by open circles, with the smallest corresponding to perturbations of $2 \times 10^{-3}\ M$ in $NaClO_2$; the next smallest, $6 \times 10^{-3}\ M$ in $NaClO_2$; the next smallest, $1.25 \times 10^{-3}\ M$ in $AgNO_3$; and the largest, $2 \times 10^{-2}\ M$ in $AgNO_3$. (Reproduced from [29].)

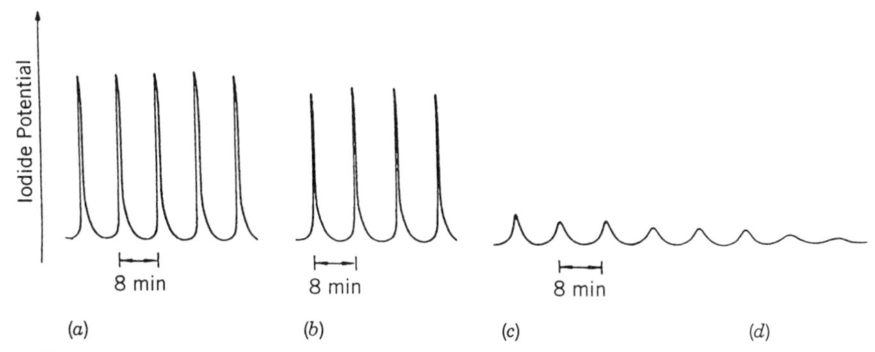

Figure 3.12. Experimental values of iodide electrode potential versus time for high $[ClO_2^-]_0/[I^-]_0$, where $k_0 = 8.333 \times 10^{-4}\ s^{-1}$. (*a*) $[ClO_2^-]_0/[I^-]_0 = 0.3791$; (*b*) $[ClO_2^-]_0/[I^-]_0 = 0.4129$ (14 min after parameter change); (*c*) $[ClO_2^-]_0/[I^-]_0 = 0.4129$ (109 min after parameter change); and (*d*) $[ClO_2^-]_0/[I^-]_0 = 0.4500$. As the system moves away from a saddle–node infinite period bifurcation, the period decreases somewhat. Then, as the system approaches a supercritical Hopf bifurcation, the period stays constant and the amplitude decreases, going to zero when the steady state is reached. To the right of (*d*) the system remains in the steady state.

decreasing to zero, is indicative of a supercritical Hopf bifurcation. For a Hopf bifurcation, the amplitude of oscillation is predicted [46] to be proportional to $(\mu - \mu_c)^{1/2}$, where $\mu - \mu_c$ is the distance between the control parameter μ and the critical value μ_c at which the bifurcation occurs; this explains the decrease in amplitude as the system approaches the bifurcation. Similar analyses are used to locate saddle–node infinite period bifurcation boundaries and to determine the extent of the excitable region, as detailed in [8].

For 2-D bifurcation analyses, a phase plane may exist, which contains an XPD, in which bifurcations divide parameter space into two separate stable steady states, a region of bistability, and a region of stable oscillations. An experimentally determined XPD [16] for the chlorite–iodide reaction is presented in Fig. 3.13. This bifurcation diagram, plotted in the $[ClO_2^-]_0$–$[I^-]_0$ plane, shows two distinct steady states, SS I and SS II, a region of bistability between the two steady states, and an oscillatory region.

If an XPD is located, it may be of use in assigning roles of species and the category of oscillator [19]. An XPD with axes of inflow concentrations of essential species type X and Y can produce a cusp; essential species

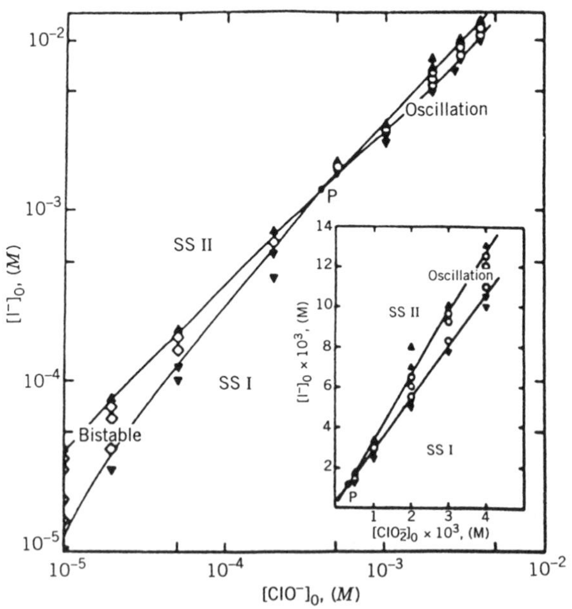

Figure 3.13. Experimental [51] XPD for the chlorite–iodide reaction in the $[ClO_2^-]_0$–$[I^-]_0$ plane with pH 2 and $k_0 = 1.1 \times 10^{-3}\ s^{-1}$. Filled triangles indicate the high iodide steady state, filled inverted triangles indicate the low iodide steady state, open diamonds indicate bistability, and open circles indicate oscillation. (Reproduced from [16].)

type Z cannot produce a cusp. If possible, an XPD is found on axes of the inflow concentration of essential species type Y and of the rate of the autocatalytic reaction. In such an XPD, Category 1B oscillators exhibit oscillatory behavior at lower parameter values and bistability at high parameter values, while Category 1C oscillators exhibit bistability at low parameter values and oscillatory behavior at high parameter values. (The directions of cusp openings for various choices of axes are given for Category 1B, 1CX, and 1CW in Table II, [19]. This information does not allow one to distinguish 1CX and 1CW, and the axes that are experimentally accessible may not correspond to those used in this approach to categorization.) In general, while Category 2 oscillators exhibit regions of bistability and of oscillatory behavior, their bifurcation structures are more complex than XPDs, resembling, for example, the unfolding of a Takens–Bodganov–Cusp point. In general, the existence of an XPD for a system is considered indicative that the oscillator is in Category 1, although we are aware of Category 2 oscillators (e.g., the reversible Sel'kov model) in which XPDs may occur.

Ringland [17] uses a 2-D bifurcation diagram generated from the SNB model for the BZ reaction to locate features for experimental verification. In particular, he locates a "crossing" feature or connection between upper and lower branch Hopf bifurcations, which has either of two possible characters depending on the location of the upper branch double zero point relative to the lower branch Hopf curve. In the experimentally determined crossing, the upper branch steady state disappears at a saddle–node bifurcation rather than first losing stability at a supercritical Hopf bifurcation. Only a limited region of the parameter space of the SNB model is in qualitative agreement with this finding, and adjustment of rate constants and stoichiometric parameters within the model produces quantitative agreement no better than an order of magnitude off. A different set of rate constants for the model produces neither quantitative nor qualitative agreement with the experimentally observed crossing feature.

Recently, Clarke and Jiang [18] reexamined the SNB model, generating bifurcation curves from currents in the reaction mechanism. By using different combinations of extreme currents, it is possible to determine the effect of deleting or including particular reactions on the shape, location, and nature of the bifurcation curves. Clarke and Jiang report that the SNB model cannot be brought into agreement with Ringland's experimental results without including Br^--consuming reactions in the network; otherwise, changes in parameters that improve the agreement between experimental and theoretical location for the region of bistability worsen the agreement on the location of the high-flow Hopf bifurcation and vice versa.

M. Stabilization of the Unstable Branch of Steady States

There are chemical reactions with multiple stationary states, of which some are stable and some are unstable. Oscillatory chemical reactions may have an unstable stationary state within the attractor of a limit cycle. There are experimental techniques for stabilizing unstable stationary states without affecting the location of the unstable stationary state in constraint space.

In this experiment, a feedback of the form $f_{X_i} = ampl + gain[X_i]$ is applied to the flow rate f_{X_i} in the autonomous oscillating system. This feedback introduces new variables, and thereby may stabilize the unstable attractors that exist in the system without feedback. Parameters are varied to trace out a stabilized unstable branch. Experimental behavior that is seen only when feedback is applied is assumed to be due to the unstable branch; such behavior should cease if feedback is shut off. The behavior under feedback conditions is used to deduce the type of unstable branch being stabilized (e.g., node, focus, or limit cycle).

Hjelmfelt and Ross [47] report experiments in which the use of feedback to I^- inflow rate is used to stabilize unstable steady states in the oscillatory and excitable regimes of the chlorite–iodide system. Feedback applied in the oscillatory regime places the system on a new steady state (i.e., the stabilized unstable steady state). When the feedback is switched off, the oscillations return after a slow change in $[I^-]$, indicative that the system is initially at or near an unstable steady state. When the feedback is switched back on, the system returns once again to the fixed point that has been stabilized. An illustration of this phenomenon is given in Figure 3.14.

Such an investigation allows comparison of unstable branches between model and experiment, and may lead to a more precise assignment of bifurcation types than would mapping of experimental bifurcation structure alone. Furthermore, an examination of the unstable branches may allow for more reliable determination of the boundaries of excitable regions than pulsed additions, for which fluctuations in the system may produce an "excitatory" response at a point in phase space outside the mathematically defined excitable region.

IV. EXAMPLE OF DEDUCTION OF REACTION MECHANISM FROM EXPERIMENTS

In this section we turn to a computational test of some of the operational procedures designed to lead to deduction from experiments to a formulation of a reaction mechanism for oscillatory reactions. We do so by

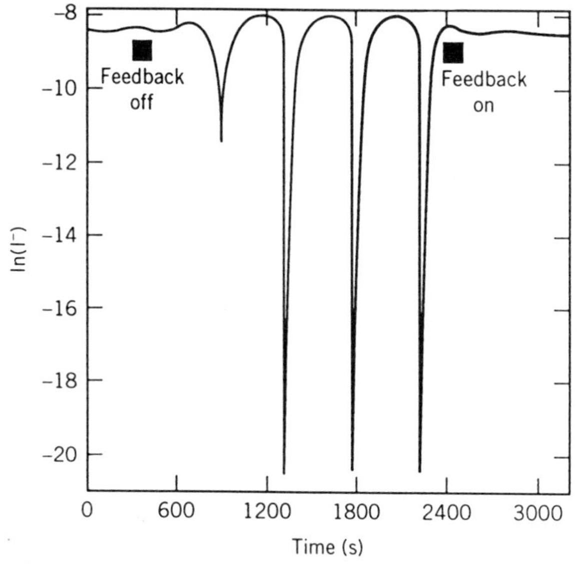

Figure 3.14. Experimental time series for the chlorite–iodide system. The autonomous system is in an oscillatory state, with an unstable stationary state within the limit cycle. In the system with delayed feedback to the inflow rate of iodide, this unstable stationary state becomes a stable stationary state. The feedback is initially on (stabilized unstable steady state). The black square on the left indicates when the feedback is shut off; limit cycle oscillations resume. The black square on the right indicates when the feedback is turned on; the oscillations disappear and the stabilized steady state reappears. (Reproduced from [47].)

choosing a model mechanism for the HRP reaction, and then carry out calculations on this model to simulate results of suggested experiments. From the results of these calculations, we then attempt to construct a reaction mechanism, and compare that with the model mechanism chosen for the demonstration of the method.

The HRP reaction is the oxidation of the enzymic cofactor NADH and has the overall stoichiometry

$$2\text{NADH} + \text{O}_2 + 2\text{H}^+ \xrightarrow{\text{HRP}} 2\text{NAD}^+ + 2\text{H}_2\text{O}$$

The reaction is oscillatory in certain ranges of influx rates of O_2 and NADH in an open reactor (a CSTR) and also shows other interesting dynamic behavior, such as quasiperiodicity and chaos. The reaction mechanism is complex, as illustrated in Fig. 4.1, which shows, in the language of stoichiometric network analysis, the so-called Model C mechanism [48].

For the purpose of illustration, we choose a simpler version of the

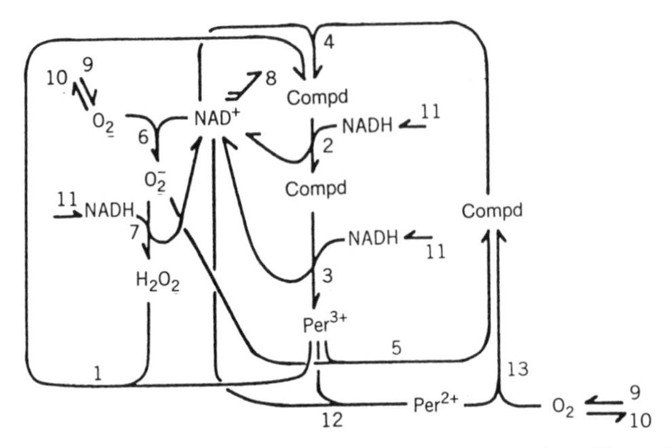

Figure 4.1. Network diagram for Model C of the HRP reaction. (Reproduced from [48].)

reaction mechanism for this reaction, the so-called DOP model [26], for which the mechanism is

$$A + B + X \xrightarrow{k_1} 2X \tag{4.1}$$

$$2X \xrightarrow{k_2} 2Y \tag{4.2}$$

$$A + B + Y \xrightarrow{k_3} 2X \tag{4.3}$$

$$X \xrightarrow{k_4} P \tag{4.4}$$

$$Y \xrightarrow{k_5} Q \tag{4.5}$$

$$X_0 \xrightarrow{k_6'} X \tag{4.6}$$

$$A_0 \xrightarrow{k_7'} A \tag{4.7}$$

$$A \xrightarrow{k_{-7}} A_0 \tag{4.8}$$

$$B_0 \xrightarrow{k_8'} B \tag{4.9}$$

where the symbol $A = O_2$, $B = \text{NADH}$, and X and Y represent unspecified intermediates, possibly the enzyme HRP in the form of the so-called CompdIII or the radicals NAD· or $O_2\cdot$, and P and Q are products of the reaction. The deterministic kinetic equations for this mechanism are

$$\dot{A} = -k_1ABX - k_3ABY + k_7'A_0 - k_{-7}A \tag{4.10}$$

$$\dot{B} = -k_1ABX - k_3ABY + k_8'B_0 \tag{4.11}$$

$$\dot{X} = k_1ABX - 2k_2X^2 + 2k_3ABY - k_4X + k_6'X_0 \tag{4.12}$$

$$\dot{Y} = -k_3ABY + 2k_2X^2 - k_5Y \tag{4.13}$$

First we consider the idealized situation, in which we are able to measure all the species in the system and to add each species in turn as a perturbant to the system. We assume that X corresponds to NAD· and Y corresponds to the CompdIII form of the enzyme, as suggested by Aguda and Larter [49]. Using these assignments, we present the results of simulated experiments on the DOP model, many of which were first given in [6]. We have not performed tests, such as phase-response studies, which do not classify the roles of the species or the category of the oscillator. We also have not applied external periodic perturbations, which duplicates information about essential and nonessential species available from other, more straightforward tests. As described in [6], the application of delayed feedback to the DOP yields only partial information about the JMEs, since the model seems insensitive to feedbacks applied to at least one variable (B); equivalent information is available from pulsed perturbations of the steady state.

A 2-D bifurcation diagram for the system is shown in Fig. 4.2. The abscissa is the input concentration of O_2 and the ordinate is the input concentration of NADH. The solid line is a supercritical Hopf bifurcation curve, while the dashed line indicates a subcritical Hopf bifurcation. The region of stable oscillations lies below the curve, while the stable steady state lies above the curve. An XPD has not been found in this model in searches that included ranges of parameters. As has been noted in [6], the behavior of the variables in the system changes dramatically for different input concentrations of O_2. Most of the tests described above that lead to categorization are generally performed at a supercritical Hopf bifurcation. Tests such as steady-state perturbations (see Sections III.D–G) are valid at subcritical Hopf bifurcations as well, providing the perturbation strengths are sufficiently small that the system relaxes to the steady state after a transient rather than being pushed into a limit cycle. However, tests that require stable sinusoidal oscillations (e.g., quenchings, Section III.I) cannot be made at subcritical Hopf bifurcations, at which the stable oscillations are generally not sinusoidal.

We made calculations with the DOP model near the supercritical Hopf

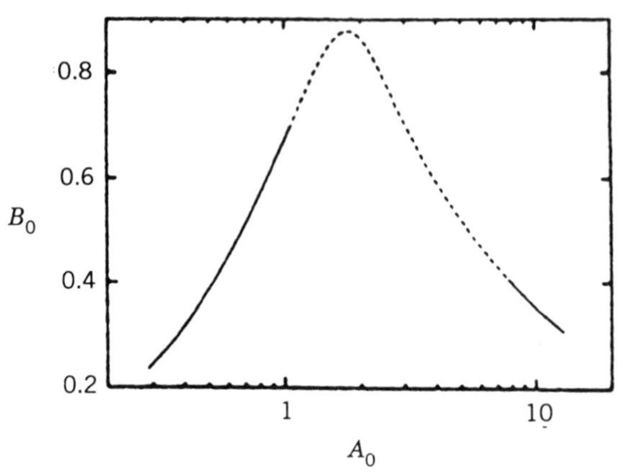

Figure 4.2. Bifurcation diagram generated from the DOP model in the A_0–B_0 plane of parameter space. The region of stable oscillations lies below the curve, while the stable steady state lies above the curve. The solid line indicates a supercritical Hopf bifurcation, the dashed line, a subcritical Hopf bifurcation. The fixed parameter values used to generate the diagram are $k_1 = 0.135$, $k_2 = 1250$, $k_3 = 0.046875$, $k_4 = 20$, $k_5 = 1.104$, $k_6 = 0.001$, $k_7' = 1$, $k_{-7} = 0.1175$, and $k_8' = 1$. (Reproduced from [6].)

branch using two different sets of input concentration values, $A_0 = 0.6$ and $B_0 = 0.5$ (low O_2), and $A_0 = 9.0$ and $B_0 = 0.4$ (high O_2). The oscillations of the autonomous system are calculated for both sets of inflow conditions. The relative amplitudes are given in Table XIV, along with the steady-state concentrations near the Hopf bifurcations. The

TABLE XIV
Relative Amplitudes of Oscillations and Steady-State Concentrations in the Autonomous DOP Model for Low O_2 ($A_0 = 0.6$, $B_0 = 0.5$) and High O_2 ($A_0 = 9.0$, $B_0 = 0.4$) Values of Influx

Concentration	Species	Relative Amplitude	SS Concentration (M)
Low O_2	O_2	1.3	0.851
	NADH	6.5×10^{-2}	57.1
	NAD·	1.6	1.54×10^{-2}
	CompdIII	3.4	0.175
High O_2	O_2	2.3×10^{-2}	73.2
	NADH	2.8	0.724
	NAD·	3.1	1.32×10^{-2}
	CompdIII	6.4	0.123

matrices of phase shifts and the sign-symbolic phase shift matrices are given in Table XV.

We applied pulse perturbations to each of the four species in the simulation and determined the response of all four species. Assuming sufficiently fast measurement of the initial slopes of the responses, we can calculate the Jacobian matrices shown, with the sign-symbolic Jacobian, in Table XVI.

On the steady-state side of the Hopf bifurcation, constant perturbing inflows are applied to each of the species in turn. The steady-state concentrations of the perturbed system are calculated, and this yields the concentration shift regulation matrices shown in Table XVII.

Application of slightly larger perturbing inflows to each of the species

TABLE XV
Phase Relations of Oscillations in the Autonomous DOP Model for Low O_2 and High O_2 Values of Influx

(a) Numerical Phase Relations (in degrees):

		O_2	NADH	NAD·	CompdIII
Low O_2	O_2	0	−11	−104	−129
	NADH	11	0	−93	−118
	NAD·	104	93	0	−25
	CompdIII	129	118	25	0
High O_2	O_2	0	−8	−103	−133
	NADH	8	0	−95	−126
	NAD·	103	95	0	−31
	CompdIII	133	126	31	0

(b) Sign-Symbolic Phase Shift Matrix for Both Low O_2 and High O_2 Values of Influx:

	O_2	NADH	NAD·	CompdIII
O_2	I	I	−	−
NADH	I	I	−	−
NAD·	+	+	I	I
CompdIII	+	+	I	I

TABLE XVI
Jacobian Matrices for the DOP Model for Low O_2 and High O_2 Values of Influx

(a) Numerical Values of JMEs:

		O_2	NADH	NAD·	CompdIII
Low O_2	O_2	−0.44	-1.6×10^{-2}	−6.6	−2.3
	NADH	−0.32	-1.6×10^{-2}	−6.6	−2.3
	NAD·	0.58	2.9×10^{-2}	−90	4.6
	CompdIII	−0.26	-1.3×10^{-2}	77	−3.4
High O_2	O_2	−0.12	−0.52	−7.5	−2.6
	NADH	-4.3×10^{-3}	−0.52	−7.5	−2.6
	NAD·	7.5×10^{-3}	0.91	−74	5.2
	CompdIII	-3.2×10^{-3}	−0.39	61	−3.7

(b) Sign-Symbolic Jacobian Matrix for Both Low O_2 and High O_2 Values of Influx:

	O_2	NADH	NAD·	CompdIII
O_2	−	−	−	−
NADH	−	−	−	−
NAD·	+	+	−	+
CompdIII	−	−	+	−

in turn give concentration shift destabilization data. These inflows are applied to the steady-state side of the bifurcation and to the oscillatory side of the bifurcation, increasing the perturbation strength until either the perturbation on the steady-state side shifts the system to an oscillatory regime or the perturbation on the oscillatory side shifts the system to a steady state. The resulting concentration shift destabilization data is given in Table XVIII.

Simulations of quench experiments were unsuccessful for low O_2 inflow parameters. Here, the perturbation was followed by a complex transient resembling quasiperiodicity before settling to the limit cycle. Since quasiperiodicity has been seen in the DOP model near the left Hopf branch, this branch is probably unsuitable for quench experiments. Quenching data located near the high O_2 Hopf branch are given in Table XIX.

TABLE XVII
Concentration Shift Regulation Matrices for the DOP Model for Low O_2 and High O_2 Values of Influx

(a) Numerical Values of Concentration Shift Matrix Elements:

		O_2	NADH	NAD·	CompdIII
Low O_2	O_2	8.5	−8.5	0	0
	NADH	−170	150	−100	−120
	NAD·	0	2.0×10^{-2}	1.3×10^{-2}	3.8×10^{-3}
	CompdIII	0	0.54	0.67	0.84
High O_2	O_2	8.5	−8.5	0	0
	NADH	-7.0×10^{-2}	−0.73	−3.3	−4.1
	NAD·	0	2.4×10^{-2}	$1.6 \times 10^{-}2$	6.2×10^{-3}
	CompdIII	0	0.47	0.61	0.79

(b) Sign-Symbolic Concentration Shift Matrix:

		O_2	NADH	NAD·	CompdIII
Low O_2	O_2	+	−	0	0
	NADH	−	+	−	−
	NAD·	0	+	+	+
	CompdIII	0	+	+	+
High O_2	O_2	+	−	0	0
	NADH	−	−	−	−
	NAD·	0	+	+	+
	CompdIII	0	+	+	+

First, we turn to the results for the high O_2 values of influx. The relative amplitude of the oscillations in O_2 (Table XIV) is two orders of magnitude smaller than those of the other species, while the steady-state concentration of O_2 is two orders of magnitude larger than those of the other species. From this, we identify O_2 as a nonessential species of type A or C. This identification is supported by the small entries in the first column of the numerical Jacobian matrix (Table XVI) for high O_2 values

TABLE XVIII
Concentration Shift Destabilization Assignments for Low O_2 and High O_2 Values of Influx

	Low O_2	High O_2
O_2	d	s
NADH	s	s
NAD·	s	s
CompdIII	s	s

TABLE XIX
Quenching Data for High O_2 Values of Influx (No Successful Quenching Was Found at Low O_2 Values of Influx)

	Quench Amplitude	Quench Phase
O_2	19	110°
NADH	0.16	110°
NAD·	0.07	160°
CompdIII	0.55	160°

of influx, corresponding to the change in the rate of species production with respect to a change in $[O_2]$. The assignment of O_2 is further strengthened by the quenching data given in Table XIX, which shows a much higher quench amplitude for O_2 than for the other species. Thus, since it has a small relative amplitude and a large quench vector, we conclude that O_2 is a nonessential species of type C.

We neglect the nonessential species in our attempts to determine the roles of essential species and the category of the oscillator. From the phase relations of the autonomous system (Table XV), we see that NAD· and CompdIII oscillate nearly in-phase with each other but lag behind the oscillations in NADH. No antiphase species are seen, which excludes Category 1C; the system does not oscillate when modeled as a batch system (i.e., no inflows or outflows), which probably excludes Category 1B. Thus, this data suggests the oscillator is in Category 2, with NAD· and CompdIII as the essential species of type X and NADH as the essential species of type Z. This assignment is supported by the calculated concentration shift matrix (Table XVII), in which the diagonal entry for NADH is negative (i.e., NADH exerts inverse self-regulation). Moreover, as for the Category 2 prototype, the concentration shift matrix for the DOP at high O_2 parameters indicates that the type X essential species (NAD· and CompdIII) show an increase in steady-state concentration in response to inflows of type Z species, whereas the type Z essential species (NADH) shows a decrease in steady-state concentration in response to inflows of type X species.

Finally, we use the sign-symbolic JMEs to construct elementary mechanistic steps consistent with the calculated data presented here. We consider the relations among all four species, essential and nonessential, since understanding how the nonessential species interact with essential species in the elementary steps gives a more complete picture. First, we consider the 2×2 submatrix of NAD· and ComptIII:

	NAD·	CompdIII
NAD·	−	+
CompdIII	+	−

The off-diagonal elements indicate that increasing the concentration of NAD· increases the concentration of CompdIII, and that increasing the concentration of CompdIII increases the concentration of NAD·. This finding is consistent with the pair of elementary steps:

$$\text{NAD}\cdot + \_\_\_ \rightarrow \text{Compd} + \_\_\_ \tag{4.14}$$

$$\text{CompdIII} + \_\_\_ \rightarrow \text{NAD}\cdot + \_\_\_ \tag{4.15}$$

Equation (4.14) is consistent with Eq. (4.2) of the DOP model presented above; Eq. (4.15) will be discussed shortly. Next, we consider the 3×3 submatrix of O_2, NADH, and NAD·:

	O_2	NADH	NAD·
O_2	−	−	−
NADH	−	−	−
NAD·	+	+	−

The entries in the first and second columns indicate that increasing the concentration of O_2 or of NADH decreases the concentrations of O_2 and NADH but increases the concentration of NAD·. In contrast, increasing the concentration of NAD· decreases the concentrations of O_2, NADH, and NAD·. This is consistent with the elementary step:

$$O_2 + \text{NADH} + \text{NAD}\cdot \rightarrow n\text{NAD}\cdot \tag{4.16}$$

provided we also include a reaction that consumes NAD·, for example, Eq. (4.14). Equation (4.16) is the same form as seen in Eq. (4.1) in the model. Next, we turn to the 3×3 matrix of O_2, NADH, and CompdIII:

	O_2	NADH	CompdIII
O_2	−	−	−
NADH	−	−	−
CompdIII	−	−	−

Here, increasing the concentrations of O_2, NADH, and CompdIII leads to a decrease in the concentrations of these three species, which is consistent with the elementary step:

$$O_2 + \text{NADH} + \text{CompdIII} \rightarrow \_\_\_\_ \tag{4.17}$$

Returning to the original matrix, we look for positive elements (products) in the columns corresponding to O_2, NADH, and CompdIII. These are all found in the row corresponding to NAD·:

	O_2	NADH	CompdIII
NAD·	+	+	+

These data allow us to elaborate on Eq. (4.17) to propose the elementary step:

$$O_2 + \text{NADH} + \text{CompdIII} \rightarrow p\text{NAD}\cdot \tag{4.18}$$

Equation (4.18) effectively combines the information from Eqs. (4.15) and (4.17). This elementary step has the same form as Eq. (4.3) in the model. Finally, because of the experimental arrangement, we can posit inflows of O_2 and NADH [corresponding to Eqs. (4.7) and (4.9) in the model] and outflows of NAD· and CompdIII [corresponding to Eqs. (4.4) and (4.5) in the model]. Thus, from the signs of the JMEs, we can propose a full set of elementary mechanistic steps consistent with the evidence from simulated experiments.

Next, we consider the inferences that may be drawn from these results for low values of O_2 influx. The relative amplitude of the oscillations in NADH (Table XIV) is two orders of magnitude smaller than those of the other species, while the steady-state concentration of NADH is two orders of magnitude larger than those of the other species. From these facts, we identify NADH as a nonessential species. This identification is supported by the small entries in the second column of the numerical Jacobian matrix (Table XVI) for high O_2 values of influx, corresponding to the change in the rate of species production with respect to a change in [NADH]. The switch of NADH from an essential species at high O_2 influx to a nonessential species at low O_2 influx is merely due to the issue of which species is present in excess: O_2 at high O_2 influx, NADH at low O_2 influx.

Since NADH is nonessential, we may neglect it in our attempts to determine the roles of essential species and the category of the oscillator. From the phase relations of the autonomous system (Table XV), we see that NAD· and CompdIII oscillate nearly in-phase with each other but lag behind the oscillations in O_2. No antiphase species are seen, which excludes Category 1C; the system does not oscillate when modeled as a batch system, which probably excludes Category 1B. Thus, this data suggests the oscillator is in Category 2, with NAD· and CompdIII as the essential species of type X and O_2 as the essential species of type Z.

However, the assignment of this system from phase relations is not supported by the calculated concentration shift matrix (Table XVII), from which *none* of the species appears to exert inverse self-regulation. This would suggest the absence of a type Z species altogether (which would not be consistent with any category of oscillator proposed thus far). Examination of the shift matrix reveals an unusual property of the DOP model as formulated above: there are zero elements in the first row (third and fourth column) and the first column (third and fourth row).

These entries indicate that, for our assignment of chemical species to the variables, the steady-state value of O_2 does not depend on the influx values of NAD· and CompdIII, nor do the steady-state values of NAD· and CompdIII depend on the influx value of O_2. This insensitivity is not a cause for concern at high O_2 values of influx, where O_2 is nonessential. However, such a degenerate situation does not yield the expected inverse self-regulation for the essential species of type Z. Adding an outflow of NADH to the model (arguably a more realistic formulation) removes this degeneracy and yields the expected inverse self-regulation of O_2 for the low O_2 values of influx, and of NADH for the high O_2 values of influx. Thus, with the addition of an outflow of NADH, the calculated data at low O_2 influx lead to the formulation of the same reaction mechanism as we obtained from the data at high O_2 influx.

In most cases, however, it is not possible to make measurements of all the species in a reaction mechanism. Moreover, some species in a mechanism may be fairly unstable compounds, which are not easily utilized in perturbing a system. Thus, we consider what may be deduced from a more limited set of measurements. In the case of the HRP reaction the concentrations of O_2, NADH, and the CompdIII form of the enzyme can be measured simultaneously, and it is possible to add O_2 and NADH as perturbants. Thus, we next consider the information available from simulated experiments taking these limitations into account.

In calculated autonomous oscillations of the system at high values of O_2 influx, O_2 has a smaller relative amplitude than NADH or CompdIII. The oscillations In O_2 and NADH are in-phase with each other, while the oscillatory peaks of CompdIII lag behind those of the other two species. In simulations of concentration shift regulation experiments, an additional inflow of O_2 shifts the steady-state concentrations of itself and NADH to lower values and shifts that of CompdIII to a higher value; an additional inflow of NADH shifts the steady-state concentration of itself to a higher level, shifts the steady-state concentration of O_2 to a lower level, and does not affect the steady-state concentration of CompdIII. Larger perturbing inflows in O_2 and NADH stabilize the system (i.e., push the system from the oscillatory regime to a steady state). Calculations are made of the initial slopes of the concentration traces of O_2, NADH, and CompdIII after pulse perturbations in O_2 and NADH, all of which turn out to be negative. Simulations of quenching experiments are performed, adding pulse perturbations of O_2 and NADH. These calculated data are summarized in Table XX.

From the relative amplitudes of the autonomous oscillations, NADH and CompdIII are likely essential species, while O_2 is nonessential. The large quenching amplitude of O_2 indicates that this species is nonessential

TABLE XX
Summary of Simulated Experimental Data for the HRP System, Calculated from the DOP Model for High O_2 Values of Influx

(a) Relative Ampitudes of Oscillations and Steady-State Concentrations in the Autonomous System:

Species	Relational Amplitude	SS Concentration
O_2	2.3×10^{-2}	73.2 *M*
NADH	2.8	0.724 *M*
CompdIII	6.4	0.123 *M*

(b) Sign-Symbolic Phase Shift Matrix:

	O_2	NADH	CompdIII
O_2	I	I	–
NADH	I	I	–
CompdIII	+	+	I

(c) Sign-Symbolic Jacobian Matrix:

	O_2	NADH
O_2	–	–
NADH	–	–
CompdIII	–	–

(d) Sign-Symbolic Concentration Shift Matrix:

	O_2	NADH
O_2	+	–
NADH	–	–
CompdIII	0	+

(e) Concentration Shift Destabilization Data:

	Response
O_2	s
NADH	s

(f) Quenching Data:

	Quench Amplitude	Quench Phase
O_2	19	110°
NADH	0.16	110°

of type C. The concentration shift regulation data indicate that NADH shows inverse self-regulation, so it must be an essential species of type Z. If we apply the knowledge that the HRP system will not oscillate under batch conditions to exclude the possibility that this system is in Category 1B, we may use the information that CompdIII delays NADH and that an additional inflow of NADH raises the steady-state concentration of CompdIII to deduce that CompdIII is an essential species of type X. Using this information to consider the phase shift data, we note that the lag in CompdIII oscillations with respect to NADH oscillations is consistent with our assignment of the former as X and the latter as Z in any of the oscillatory categories except 1B (which we have already excluded for experimental reasons). The fact that a large enough perturbing inflow of NADH pushes the system from oscillations to a steady state supports our elimination of the Category 1B, in which the inversely regulated species would be destabilizing. (Unfortunately, concentration shift destabilization data are not useful in differentiating 1C and 2 oscillators. Moreover, the absence of a cusp in the bifurcation diagram for the system make the use of such data for the classification of the roles of species ambiguous.) Without knowledge of additional species concentrations or the effects of these species as perturbants, we may not be warranted in distinguishing between Categories 1C and 2. However, from these simulated measurements we have been able to distinguish O_2 as a nonessential species of type C and to deduce the roles of NADH and CompdIII.

ACKNOWLEDGMENTS

We thank Lisa L. Skrumeda, Allen Hjelmfelt, and Yu-Fen Hung for sharing their experimental findings with us, and Yu-Fen Hung for performing additional calculations on the DOP model. This work was supported in part by the National Institutes of Health and the Air Force Office of Scientific Research.

REFERENCES AND NOTES

1. J. O. Edwards, E. F. Greene, and J. Ross, *J. Chem. Educ.* **45**, 381 (1968).
2. R. S. Berry, S. A. Rice, and J. Ross, *Physical Chemistry*, Wiley, New York, 1980.
3. This lack of uniqueness is not confined to reaction mechanisms; substantiation by experiments of any proposed model, hypothesis, or theory constitutes a sufficient, but not necessary, condition of validation.
4. J. J. Tyson, *J. Chem. Phys.*, **62**, 1010 (1975).
5. Y. Luo and I. R. Epstein, *Adv. Chem. Phys.* **79**, 269 (1990).

6. T. Chevalier, I. Schreiber, and J. Ross, *J. Phys. Chem.*, **97**, 6776 (1993).
7. Numerical continuation methods have been described and utilized by R. J. Olsen and I. R. Epstein [*J. Phys. Chem.*, **89**, 3083 (1991)] and by J. Ringland in [17]. An available program for performing numerical continuation is AUTO: Software for Continuation and Bifurcation Problems in Ordinary Differential Equations by E. J. Doedel.
8. P. Strasser, J. D. Stemwedel, and J. Ross, *J. Phys. Chem.*, **97**, 2851 (1993).
9. J. Guckenheimer, *Physica D*, **20**, 1 (1986).
10. J. Maselko, *Chem. Phys.*, **67**, 17 (1982).
11. J. Maselko, *Chem. Phys.*, **78**, 381 (1983).
12. M. Sheintuch and D. Luss, *Chem. Eng. Sci.*, **42**, 41 (1987).
13. M. Sheintuch and D. Luss, *Chem. Eng. Sci.*, **42**, 233 (1987).
14. M. Sheintuch and D. Luss, *J. Phys. Chem.*, **93**, 5727 (1989).
15. J. Boissonade and P. DeKepper, *J. Phys. Chem.*, **84**, 501 (1982).
16. P. DeKepper, J. Boissonade, and I. R. Epstein, *J. Phys. Chem.*, **94**, 6525 (1990).
17. J. Ringland, *J. Chem. Phys.*, **95**, 555 (1991).
18. B. L. Clarke and W. Jiang, *J. Chem. Phys.*, **99**, 4464 (1993).
19. M. Eiswirth, A. Freund, and J. Ross, *Adv. Chem. Phys.*, **80**, 127 (1991).
20. B. L. Clarke, *Adv. Chem. Phys.*, **43**, 1 (1980).
21. B. L. Clarke, *Cell Biophys.*, **12**, 237 (1988).
22. B. L. Clarke, *Adv. Chem. Phys.*, **43**, 1 (1990).
23. I. R. Epstein and K. Kustin, *J. Phys. Chem.*, **89**, 2275 (1985).
24. O. Citri and I. R Epstein, *J. Phys. Chem.*, **91**, 6034 (1987).
25. Y.-F. Hung and J. Ross, in preparation.
26. H. Degn, L. F. Olsen, and J. W. Perram, *Ann. N.Y. Acad. Sci.*, **316**, 623 (1979).
27. L. L. Skrumeda and J. Ross, in preparation.
28. P. DeKepper and J. Boissonade, in *Oscillations and Traveling Waves in Chemical Systems*, R. J. Field and M. Burger, Eds., Wiley-Interscience, New York, 1983, pp. 223–256.
29. J. D. Stemwedel and J. Ross, *J. Phys. Chem.*, **97**, 2863 (1993).
30. T. Chevalier, A. Freund, and J. Ross, *J. Chem. Phys.*, **95**, 308 (1991).
31. F. Hynne and P. G. Sørensen, *J. Phys. Chem.*, **91**, 6573 (1987).
32. P. G. Sørensen and F. Hynne, *J. Phys. Chem.*, **93**, 5467 (1989).
33. F. Hynne, P. G. Sørensen, and K. Nielsen, *J. Chem. Phys.*, **92**, 1747 (1990).
34. P. G. Sørensen, F. Hynne, and K. Nielsen, *J. Chem. Phys.*, **92**, 4778 (1990).
35. F. Hynne, P. G. Sørensen, and T. Møller, *J. Chem. Phys.*, **98**, 211 (1992).
36. F. Hynne, P. G. Sørensen, and T. Møller, *J. Chem. Phys.*, **98**, 219 (1992).
37. P. Ruoff, H.-D. Forsterling, L. Gyorgyi, and R. M. Noyes, *J. Phys. Chem.*, **95**, 9314 (1991).
38. J. Finkeova, M. Dolnik, B. Hrudka, and M. Marek, *J. Phys. Chem.*, **94**, 4110 (1990).
39. M. Dolnik and M. Marek, *J. Phys. Chem.*, **95**, 7267 (1991).
40. S. A. Pugh, B. DeKock, and J. Ross, *J. Chem. Phys.*, **85**, 879 (1986).
41. W. S. Loud, *Q. J. Appl. Math.*, **22**, 222 (1967).

42. K. K. Tsujimoto, A. Hjelmfelt, and J Ross, *J. Chem. Phys.*, **95**, 3213 (1991).
43. G. Tsarouhas and J. Ross, *J. Chem. Phys.*, **87**, 6538 (1987).
44. G. E. Tsarouhas and J. Ross, *J. Chem. Phys.*, **89**, 5715 (1988).
45. G. Tsarouhas and J. Ross, *J. Phys. Chem.*, **93**, 3657 (1989).
46. J. Guckenheimer and P. Holmes, *Nonlinear Oscillations, Dynamical Systems, and Bifurcations of Vector Fields*, Springer-Verlag, New York, 1983.
47. A. Hjelmfelt and J. Ross, in preparation.
48. R. Larter and B. D. Aguda, *J. Am. Chem. Soc.*, **113**, 7913 (1991).
49. B. D. Aguda and R. Larter, *J. Am. Chem. Soc.*, **112**, 2167 (1990).
50. J. D. Stemwedel and J. Ross, in preparation.
51. C. E. Dateo, M. Orban, P. DeKepper, and I. R. Epstein, *J. Am. Chem. Soc.*, **104**, 504 (1982).

AUTHOR INDEX

Numbers in parentheses are reference numbers and indicate that the author's work is referred to although his name is not mentioned in the text. Numbers in *italic* show the pages on which the complete references are listed.

SUBJECT INDEX